Migration and mobility in Britain since the eighteenth century

Migration and mobility in Britain since the eighteenth century

Colin G. Pooley and Jean Turnbull
Lancaster University

UCL PRESS
Taylor & Francis Group

First published in 1998 by UCL Press

UCL Press Limited
1 Gunpowder Square
London EC4A 3DE
UK

and

1900 Frost Road, Suite 101
Bristol
Pennsylvania 19007–1598
USA

The name of University College London (UCL) is a registered
trade mark used by UCL Press with the consent of the owner.

British Library Cataloguing-in-Publication Data
A catalogue record for this book is available from the British Library.

Library of Congress Cataloging-in-Publication Data are available

ISBNs: 1–85728–867–X HB
      1–85728–868–8 PB

Printed and bound by T.J. International Ltd, Padstow

Contents

Preface

The research on which this book is based originated from chance discussions at conferences attended by academics, family historians and genealogists. Whilst mulling over the relative lack of detailed information about individual migration patterns in the past, and complaining about the difficulties which frequent residential moves pose for family historians tracing their ancestors, it became obvious that there was scope for collaboration between the various groups. Family historians and genealogists who trace their ancestors invest a vast amount of effort in reconstructing individual life histories, including details of all residential moves. Although collected mainly out of personal interest, we realized that such data could collectively provide a rich source for historical analysis of migration. The research on which this book is based uses a large volume of information collected by family historians and systematically analyses these data to try to address some of the key issues surrounding population migration in the past. In particular, these data allow research to go beyond the interpretation of static census and similar sources to focus on the life-time experiences of migration for a large and representative population.

We also undertook this research in the belief that individual migration experiences were important, and are therefore worth a considerable investment of effort to study. Any understanding of social, economic and cultural change in the past must take account of the process of migration. Population movement had a fundamental impact on individuals, families, places, and the wider societal structures within which such change took place. The migration process can only be understood in relation to such changes; but the migration experience itself provides a window on broader social, economic and cultural changes which were occurring within British society from the eighteenth to the twentieth centuries. Thus, the analysis presented in this volume is, wherever appropriate, set within a context which relates the causes and effects of migration to other aspects of society, economy and environment.

The research itself has been undertaken over a period of some four years. We began with a pilot survey using information gathered from family historians based in North-West England, and then refined our research methodology for the main phase of data collection from December 1993 to September 1995. During this period we collected data from respondents in all parts of Britain and overseas. The success of the research is mainly due to the enthusiastic way in which family historians and genealogists responded to our requests for information, and the excellent detail that these responses contained. The full research methodology and the accuracy of the data are evaluated in the body of the book, but this research has certainly demonstrated that there is great potential for collaboration between academic researchers and family historians. There is also scope to extend the longitudinal analysis of these data beyond the areas covered in this volume, and the material on which analysis is based is available through the ESRC Data Archive.

Any large research project such as this incurs many debts of gratitude. We are particularly indebted to the many family historians and genealogists who completed data forms, and to the family history and genealogy societies which willingly publicized the project. Without this support the research would simply not have been possible. We would also like to thank the funding bodies that supported the research: The Nuffield Foundation for two small grants to allow collection and analysis of the pilot data, and the Economic and Social Research Council (ESRC) which supported the main project. During the life of the project a large number of individuals provided us with help and advice. Casual research assistance with coding, data entry and data checking was at various times provided by Amanda Topps, Geraldine Byrne, Fiona White, Janet Darrall, Margaret Stelfox, Jan Hopkins, Mike Bullock, Alan Gunton, Lucy Berger and Anne Heaven. Assistance with computer mapping, and the establishment of the ARC/INGRES link to enable us to plot and analyze spatial data within a Geographical Information System, was provided by staff of the North West Regional Research Laboratory at Lancaster University, and we are especially indebted to Barry Rowlingson and Adrian Maddocks. Further cartographic assistance was provided by Nicola Higgitt and Claire Jeffery in the Department of Geography Cartographic Unit, and computing assistance was given by staff in Lancaster University's Information Systems Service, especially Lynne Irvine.

During the life of the project we have gained immeasurably from a large number of informal discussions with colleagues in the Geography and History Departments at Lancaster, and from comments and questions at seminar and conference presentations to academics and family historians in various locations. We are especially grateful for the advice given by Kevin Schurer and colleagues at the Cambridge Group for the Study of Population and Social Structure in the early stages of the project; to Ken Prandy for information on computer-assisted occupational coding; to Richard Dennis for his constructive comments on an early draft of the book manuscript; and to Steven Gerrard and other staff at UCL Press. Last, but by no means least, we should also thank our respective families for putting up with the times when the project spilled over into evenings and weekends. Any errors and misinterpretations in the book are, of course, entirely our own responsibility.

We hope that this book will begin to fill some gaps in our knowledge of migration in the past, and will stimulate further research on the important topic of population mobility. However, perhaps the greatest benefit we have gained from the project is the establishment of a wide range of contacts with family historians who are interested in placing the information they have about their ancestors in a wider historical context. We hope to be able to maintain these contacts, and to continue research which benefits both academics and family historians.

Lancaster University Colin G. Pooley
September 1997 Jean Turnbull

List of figures

List of tables

CHAPTER 1

Introduction: why study migration?

The significance of migration

Almost everyone moves home at some time in their lives. Together with experiences such as the birth of a child, the death of a relative, marriage, starting a new school and finding fresh employment, moving to a new house is a relatively infrequent but everyday occurrence which, whilst commonplace, can also have wide-ranging ramifications for all concerned. Migration not only affects individuals, and in some cases fundamentally alters people's lives, but large scale population movement also has implications for wider society, changing population distributions and placing demands on housing, labour markets and services. Such consequences stem most obviously from mass flows of refugees or migrants fleeing events such as famine or war, but even short distance movement within national boundaries can have significant and wide ranging impacts (Black and Robinson, 1993). Moving home is also a complex process, undertaken for a large number of different reasons, and causing many potential difficulties or disruptions which have varied effects on members of a mobile household.

This was no less true in the past than it is in the present; but because much migration left little in the way of written records, and because the reasons for and impacts of residential movement are so varied and personal, we know relatively little in detail about migration in the past. However, where oral or written testimonies exist it is possible to gain some inkling of the significance of moving home.

Much migration was undertaken for work, and for some this seemed a relatively unproblematic event. In the early-twentieth century, Trafford Park in Manchester was being developed as a massive industrial estate with associated housing for workers. Labour migrants were attracted to new opportunities in Manchester and Salford from a relatively wide area:

> My parents – they were in Liverpool, on the docks. Me father was a ships' carpenter and there was no work whatsoever, and he got the

information that there was work on the docks at Salford and he got a return ticket, half a crown, and he came and he got took on the same morning. In the evening he sent a telegram to me mother: 'Come, I've got a job and somewhere to live'. And thats the way they started, they came there with nothing – only bits of essentials in a bag. (Russell and Walker, 1979, p. 6)

Knowledge about new work opportunities could spread in many ways and, in some instances, employers deliberately recruited workers to fill a labour shortage. Thus in the late-eighteenth century, rural textile mills in North Lancashire attracted workers with domestic textile skills from the surrounding region:

Mr. Edmondson the managing master came over to dent to engage hands, several more families hired, & among the rest my father and family – we were 7 childrer & they liked large families the Best for the Childrers sake – my father engaged for him & me to work in the machanic shop, at our trade &c – So my father Sold the greater part of his goods, &some he left unsold, & came to this dolphinholme, with some others in July 1791. (Shaw life history, see Appendix 1)

Even when work was readily available, and movement was undertaken as a family group in conjunction with others from the locality, migration could cause some trepidation and concern:

this leaving our own Countery [*the move was only around 50 km*] was a great cross to my mother, for she was greatly atached to her Native town, & had she known what would follow, I am sure that she never would have left her relations & countery on any account. (Shaw life history)

However, for some migrants emigration to the other side of the world was undertaken with enthusiasm and encouragement to others:

i am sure that if laboren people ad the least idea of the colney they would not work hard in Ingland to starve whear in the colnies thear is plenty for whe ave plent of wheat and mutton and biefe and evrey thinks that whe can whish for the onley thing is crossing the salt warter. (Richards, 1991, p. 25)

Even with short distance migration, undertaken with optimism by a 16-year-old migrant moving some six km in London, moving away from home could cause understandable unhappiness and concern amongst relatives left behind:

Upon the appointed day May 10th 1858 my poor mother took me over and duly delivered me to my new masters, returning home alone. I cannot say that parting from those at home caused me any great sorrow. I felt most for my mother – and she felt my going away acutely. As for me, I was full of anticipation. (Jaques life history)

Such relatively rare testimonies provide evidence of the human impacts of migration. The research presented in this book combines a wide range of sources in an attempt to paint a broader picture of the ways in which moving home has affected people, places and wider society in Britain since the mid-eighteenth century.

One measure of the significance of migration is its volume. Although the British population censuses do not provide evidence for the total number of moves undertaken in an individual's life-time, birthplace statistics do indicate the extent to which people had moved outside their county of birth at a particular point in time (Baines, 1972, 1978). This is a poor surrogate for migration, combining long distance moves from one part of the country to another with short distance moves across a county boundary, ignoring movement within counties, and giving no inkling of life-time residential histories. However, even such flawed evidence indicates the extent to which people were mobile in the past. Thus in 1851, the more industrial regions such as Central Scotland, South Wales, North-West England and the West Midlands had more than a quarter of their population born outside their county of residence. London attracted the largest number of migrants from all over Britain (38.3 per cent of its population had been born outside London in 1851), but cities such as Glasgow (55.9 per cent), Liverpool (57.5 per cent), Manchester (54.6 per cent) and Birmingham (40.9 per cent) contained proportionately more migrants. By 1911, South-East England was attracting the largest numbers of migrants, with over 40 per cent of the population of many southern counties born outside their county of enumeration, but even in the older industrial regions of Central Scotland, South Wales and North-West England over one third of the population had been born in a county which was different from the one in which they lived on census night. This limited census evidence suggests that migration was a common experience for a large proportion of the population (Langton and Morris, 1986, pp. 10–29).

Although experienced by a smaller number of people, emigration overseas, especially to the British colonies in the nineteenth and early-twentieth centuries, was also a relatively common occurrence. Dudley Baines (1985) has calculated the volume of emigration from English and Welsh counties in the nineteenth century, and argues that although proportionately the greatest volume of emigrants left from impoverished rural areas, a quantitatively larger volume of emigrants originated in urban areas, although many of these may have been previous rural to urban migrants. Emigration had a particular impact on the demography of Scotland. The Highlands experienced an absolute loss of some 44,000 people from 1851 to 1961, and throughout the nineteenth century Britain as a whole experienced a net loss of population through emigration. In total, possibly some 11,662,300 people emigrated from England, Wales and Scotland between 1825 and 1930 (Baines, 1985, pp. 299–306; Harper, 1988; Erickson, 1994).

Immigration to Britain from overseas has an impact both on the individual who moves and on the communities which absorb newcomers who may bring with them different customs and beliefs. In the nineteenth century, the largest volume of immigration came from Ireland (although in some respects this should be viewed as internal migration). Many British cities had substantial Irish communities, and the Irish population generated considerable hostility and experienced prolonged discrimination (Swift and Gilley, 1985; 1989; Davis, 1991). Immigrants from elsewhere in the world were relatively few in number, and mainly concentrated in port cities such as London, Bristol, Liverpool, Newcastle and Glasgow (Holmes, 1988). In the twentieth century, and especially after the Second World War, Britain experienced substantial waves of labour migrants from its former colonies as immigrants from Asia and the Caribbean were encouraged to come to Britain to fill the post-war labour shortage (Walvin, 1984; Panayi, 1994). By the 1960s, London and many provincial cities had substantial communities of immigrants who had often brought with them their own customs, religion and culture. In the late-twentieth century, Britain is a truly multi-cultural society, although many immigrants and their children continue to experience similar problems to those faced by Irish migrants in the nineteenth century (Robinson, 1986; Peach, 1996).

Whilst the extent of population movement is undeniable, if sometimes difficult to pin down precisely in the past, the implications are more difficult to formulate. It can be suggested that all residential mobility has significance at three distinct levels: impacts on the individual and his/her family, impacts on places which lost and gained migrants, and impacts on wider social, economic and political structures. Figure 1.1 summarizes some of the connections between these different layers, and suggests that migration was central to the process of social, economic and cultural change in the past. At the societal level, population movement transferred labour to new sites of industry, restructured labour markets and affected the viability of new industrial ventures. In places where migrants concentrated, the social and cultural structure of society was changed as people with different outlooks and beliefs mixed on the streets and in workplaces. In some instances, large scale migration (such as Irish famine migration into mid-nineteenth century Britain) stimulated governments to develop specific migration policies or to use existing economic and social policy to target particular migrant groups. All places were affected by migration, and all locations both gained and lost

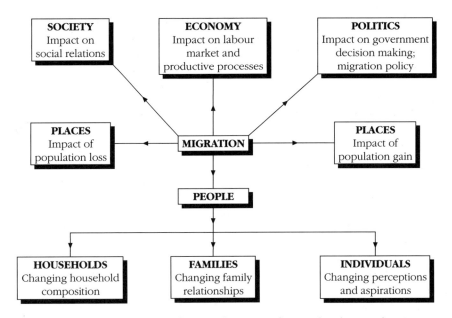

Figure 1.1 Diagram of the significance of migration for people, places and society

migrants over time, but where there was an imbalance between in- and out-movement, this could have a significant effect upon demographic and community structures. Some communities were left with an ageing population and diminished labour resources, whilst others suffered overcrowding, labour surpluses and housing shortages. Impacts at the individual level were the most varied and hard to define, but it can be suggested that the process of migration broadened horizons, changed attitudes and altered the nature of relationships between family members as kin became more widely scattered around the country. However, the ways in which individuals reacted to such changes varied from person to person and cannot easily be generalized.

It can be suggested that the process of population migration is of particular relevance to cultural change. If culture is about the sharing of meanings, and the ways in which different individuals and groups make sense of the world in which they live, then the process of migration – which brings different groups of the population together – is central to the process of sharing and mixing which creates cultural change. The links between migration and cultural change are most marked where immigration has brought groups with distinctive cultural characteristics into the country. Thus, Irish migrants into nineteenth-century cities had difficulty adjusting to an English urban way of life, and experienced significant hostility to their culture. This was clearly expressed by a Liverpool Medical Officer of Health commenting on Irish funeral customs:

5

The three houses were crammed with men, women and children, while drunken women squatted thickly on the flags of the court before the open door of the crowded room where the corpse was laid. There had been, in the presence of death, one of those shameful carousals which, to the disgrace of the enlightened progress and advanced civilization of the nineteenth century, still lingers as dregs of ancient manners amongst the funeral customs of the Irish peasantry. (Trench, 1886, p. 23)

New Commonwealth migrants to Britain since the 1950s have had similar experiences and, despite an initial desire to assimilate to British culture, have met with sufficient hostility to encourage second and third generation immigrants to look more towards their Afro-Caribbean and Asian origins for cultural support than to British society. A Punjab-born man who came to Bradford in 1956 at the age of two recalls his schooldays:

And I can remember vividly at Belle Vue one time when the Asian boys had to band together and march down the road as a phalanx column . . . to ensure their own safety. And I can also remember . . . things about skinheads coming to Bradford and beating up Asians and black people. And I can remember, you know, things about repatriation, and . . . leaflets going round school which were from something called the, was it 'The Yorkshire Campaign against Immigrants?' . . . which said things like, 'If you sit against, sit next to an Asian in a class you are bound to catch smallpox' and things like this. So yes, I was becoming more and more aware of myself, what I was, and, and these things that were happening . . . (Perks, 1987, p. 70)

In these instances the process of migration has fundamentally influenced both the lives of individual migrants and the nature of the society in which migrant communities have formed.

In the late-twentieth century we are increasingly aware of the processes which are creating a global society and economy. Easy communication, instant world news coverage and global marketing by multinational companies has meant that many people experience much the same influences wherever they are in the world. Population migration is part of this globalizing tendency, as at least some people move easily over long distances and, in Europe, economic and political integration has lessened barriers to movement (King, 1993; Hall and White, 1995; Clark, 1996). However, despite such trends, migration remains traumatic and difficult for some. There are increased barriers in Europe to refugees and those seeking political asylum, and the extent to which people are attached to particular places in which they have lived much of their lives remains significant. Thus refugees from zones of conflict such as the former Yugoslavia have experienced massive problems of dislocation and cultural adjustment (Glenny, 1993). For most people in late-twentieth century Britain the most important migration experience remains a short distance move within a particular locality, and attachment to place is likely to be an important part of cultural identity (Stillwell, Rees and Boden, 1992). Although the ways in which such processes operated may have changed over time, conflicts between globalizing and localizing influences – as migrants were stimulated to move by increasing opportunities and easier transport, but to some extent restrained by attachments to people and places within one locality – were as important in the past as in the present.

What is migration?

Although almost everybody experiences migration, and most people think that they know what it is, the definition of migration is not straightforward. Whereas other demographic events – notably births and deaths – are finite, clearly defined, and therefore relatively easy to measure, migration is 'a physical and social transition and not just an unequivocal biological event' (Zelinsky, 1971, p. 223). Apart from the fact that migration necessarily involves physical movement through space, there is little other agreement about what precisely constitutes migration.

The terms migration and mobility are often used interchangeably, but mobility is, if anything, an even more loosely defined term. It is possible to suggest that, during a life-time, any individual will undertake a variety of types of mobility. These could range from very short distance daily moves within a locality for social or educational purposes, through journeys to work which may be over considerable distances, short distance residential moves in which a person changes house but remains part of the same community, longer distance residential moves to another part of the country, to emigration to a new home on the other side of the world. Although such a classification also implies a division between temporary and permanent movement, clear-cut definitions are illusory. Whilst we are alive, few moves are truly permanent, and much residential change may involve population circulation, as people migrate between two locations perhaps related to seasonal employment opportunities or as students attending college during termtime but living elsewhere during vacations. Some people live in one location during the working week and elsewhere at weekends, whilst temporary visits to relatives can become prolonged if, for instance, there is need to care for a sick parent. The space–time categorization of different aspects of mobility thus becomes rather fuzzy at the edges, and it is not possible to draw clear definitional lines in terms of distance moved or degree of permanence.

In studies of contemporary and past migration many different definitions have been used. Data constraints sometimes require migration to be

7

defined simply as a move which crosses an administrative boundary, and which therefore is recorded as a statistic (for example Friedlander and Roshier, 1966). This is clearly unsatisfactory, especially where administrative units are large and/or variable in size. Other studies define migration as residential moves which also require a change in community affiliation and employment, thus excluding short distance intra-urban moves. Such definitions focus attention on the links between migration, community adjustment and wider society, but ignore the personal and local impacts of frequent short distance mobility (Bogue, 1959, p. 489). A broader definition of migration is one which includes all 'permanent or semi-permanent change of residence' with no restriction on distance moved (Lee, 1966, p. 49). Although, as noted above, the term permanent itself raises definitional problems, this seems a more appropriate definition.

In this book, migration is defined in its widest sense to include all changes of normal residence, irrespective of the distance moved or the duration of stay at an address. Thus movement to go on holiday or to visit friends for a few days or even weeks is excluded, but any move of personal possessions and household effects to a new location, even if this is only for a few days, is counted as a residential move. Non-residential mobility is excluded from the study, although some data on the relationship between home and workplace has been collected and is briefly analyzed in Chapter 5. Further information on how data were collected and defined is presented in Chapter 2. At all stages the term migration is used to refer to a change of normal residence within Britain, emigration refers to a residential move from Britain to another country, and immigration to a move from another country to take up residence in Britain.

Studies of developing societies have drawn attention to the importance of population circulation, as individuals move their normal residence between two or more locations in tune with the rhythm of employment opportunities, family responsibilities or other factors (Skeldon, 1990; Chant, 1992). It can be suggested that nineteenth-century Britain had some similarities with developing countries today and that, intuitively, population circulation may have been an important element in British migration in the past. The data collection methods adopted in this study were designed, where possible, to capture the process of population circulation even though this further blurs any distinction between permanent and temporary residence.

Changes in the pattern and process of migration over time and space are central to the analysis presented in this book, and the changing nature of migration in Britain from the 1750s can also be compared to changes which have been more recently experienced in developing countries. Zelinsky has focused attention on the ways in which levels of population mobility in different countries can be related to, and interact with, the process of modernization and the demographic transition from high mortality and fertility to low mortality and fertility. Although his meta-theoretical framework of a mobility transition draws evidence from a wide range of developing societies, Zelinsky argues that:

> there are definite, patterned regularities in the growth of personal mobility through space and time during recent history, and these regularities comprise an essential component of the modernization process (Zelinsky, 1971, pp. 221–2)

Such concepts are just as applicable to British society, which experienced the demographic transition in the nineteenth century, as they are to developing countries in the twentieth century.

Zelinsky proposed a five-stage model of mobility change which parallels and interacts with the demographic transition model. He argues that the nature and extent of personal mobility changed fundamentally – and just as significantly as changes in vital rates – as countries moved from what he termed the 'pre-modern transitional society', through 'early' and 'late' transitional societies to a fourth 'advanced' society and a future 'super-advanced' society. The key points which emerge from his analysis are, first, that there is a transition from relatively low levels of physical and social mobility in phase one of the mobility transition to higher levels of such movement as the process of modernization occurs. Second, that as mortality declines but rates of fertility remain high in the early stages of industrial and urban growth, rates of mobility increase and, thereafter, changes in mobility parallel changes in vital rates, with the two processes interacting during the modernization of society. Third, that there were significant changes in many aspects of the mobility process, including changes in the function, frequency, duration, periodicity and routing of migration; in the categories of migrants who moved; and in the types of origins and destinations. Fourth, that in phase two of the mobility transition rural to urban migration, together with what is termed 'frontierward migration' was especially important, but that in phases three and four this was replaced by inter-urban and intra-urban moves. Fifth, that there were important links between spatial and social mobility and, sixth, Zelinsky stressed the importance of increased levels of population circulation as well as residential migration.

Zelinsky's ideas have been subject to a number of criticisms, many of which mirror critiques of the demographic transition itself (Woods, 1993; Cadwallader, 1993). Five main points emerge from the literature. First, it can be suggested that definitional problems in measuring migration are so great that comparisons and generalizations between societies are meaningless. Second, it has been argued that stage models of development (as in the demographic transition and mobility transition) are by definition flawed and that, in particular, models such as the demographic transition, largely derived from empirical evidence in Western Europe, are not applicable to Third World countries. Third, much debate has centred around the inadequacy of modernization theory, to which Zelinsky's arguments are crucially linked. Amongst other criticisms, it can be suggested that modernization does not proceed in smooth stages but can also include catastrophic and unpredictable events. Mobility and demographic transition models do not take these into account. Fourth, it has been suggested that Zelinsky placed too much emphasis on the notion of the spatial diffusion of mobility patterns as modernization spread from one area to another. Fifth, more recent studies have seriously challenged Zelinsky's assumption of low levels of mobility in pre-industrial societies. In particular, evidence from Europe suggests that migration was frequent, and often over quite long distances, even before phase two of the demographic and mobility transitions (Lucassen, 1987; Moch, 1992; Hoerder and Moch, 1996).

However, despite these criticisms, it can be suggested that the notion of a mobility transition, operating in parallel with a demographic transition, provides a common-sense view of the links between population mobility, demographic change, urbanization and industrialization which may be more relevant to the changes that occurred in Britain from the mid-eighteenth century than it is to developing countries in the twentieth century. Data presented in this volume will allow the pattern and process of migration in Britain from the mid-eighteenth century to be compared directly with Zelinsky's hypothesis of the mobility transition, and will enable this important contribution to migration theory to be tested against relevant empirical evidence (Chapter 11). Moreover, by focusing on individual migration decisions and experiences it is possible to go beyond the possibly over-simplified concept of 'modernization' to explore the links between migration and 'modernity' as expressed through the extent to which everyday experiences took on different meanings for successive generations (Featherstone, Lash and Robertson, 1995).

One reason why ideas such as those of Zelinsky have not been comprehensively tested previously, is because of the inadequacy of much migration information. One aspect of migration which the hypothesis of the mobility transition does stress, is the way in which different aspects of the migration process are linked through time. Thus, an individual born in the 1790s may have grown up during phase one of the transition model, but may have lived through phases two and three, and thus any changes in mobility behaviour were, for that individual, real events which related to decisions that he or she made during their life-time. In order to test adequately any hypothesis about changes in migration behaviour over space and time, it is necessary to have longitudinal data which relate to the life-time residential histories of individuals. In the absence of British population registers such information is rare. Data collected for the project which forms the basis of this volume are the most comprehensive set of historical longitudinal migration records assembled for Britain.

What we know about migration and what we would like to know

The amount that we currently know about migration in the past is directly related to the range of sources that have been used to study migration, and the limitations that those sources impose. Unlike the situation in some European countries which have kept population registers (United Nations, 1959), in Britain there is no single source which provides complete (or even partial) information on all the residential moves undertaken by an individual. Sources available for the study of migration will be examined in detail in Chapter 2. This chapter focuses on the current state of knowledge about migration in the past, but recognizes that such knowledge is limited by the availability and use of sources.

Although sources are limiting, there is no lack of migration studies. These have been produced from a wide range of perspectives and, although geographers have provided more than most, there are significant contributions to the literature by social and economic historians, sociologists and local historians amongst others. Population migration frequently attracted attention from contemporary analysts in the nineteenth and early-twentieth centuries. In the late 1850s two statisticians, J.T. Danson and T.W. Welton (1857–60) produced four papers on population change in Lancashire and Cheshire using the census returns for 1801–51. They emphasized the short distance over which most migrants travelled, even to an industrial county such as Lancashire which had considerable attractive power for labour migrants. They concluded:

> ... the county of Lancaster has retained among its permanent inhabitants a larger proportion of those born within its bounds than any other county; and has also drawn very largely from the adjacent counties, and from Ireland and Scotland. It is also apparent that from those parts of England lying at a distance it has received comparatively small contributions to its population. (Danson and Welton, 1859, p. 41)

Some thirty years later E.G. Ravenstein made similar observations about the short-distance nature of most population migration in the nineteenth century, but also developed a series of now well-known 'laws' (more accurately hypotheses) of migration which, from empirical census evidence, additionally suggested that migration volume increased with the development of industry and commerce and that most moves were from agricultural to industrial areas (or from rural to urban areas); that long distance migration was focused on the major industrial and commercial centres and that longer distance moves proceeded by steps up the urban hierarchy; that each migration flow had a counter-flow in the opposite direction; that most migrants were adults and that families rarely migrated over long distances; that females were more migratory than males except for international moves; that most migrants had rural origins and that the principal motives for migration were economic (Ravenstein, 1885/1889; Grigg, 1977). These observations have been only partially tested in later work, due often to the limitations of sources and methods used. Thomas Welton (1911) produced further statistical analyses of population change in nineteenth-century Britain, emphasizing the extent of rural to urban movement and the significance of migration for new residential and industrial towns. However, he also demonstrated that although rural to urban movement was important, most towns grew more by natural increase than by in-migration. These themes were further developed in the twentieth century by A.K. Cairncross (1949) and R. Lawton (1968a).

In 1926 A. Redford published a pioneering study of labour migration in Britain in the first half of the nineteenth century, combining census data with evidence from a wide range of other sources. This work went largely unchallenged for almost half a century and in his preface to the third edition (revised 1975) W.H. Chaloner wrote that 'Not a great deal of further work has been published on internal migration since 1926'. Though perhaps rather too dismissive of research published in the 1950s and 1960s, this does emphasize the influence of Redford's contribution. Redford's work did not substantially challenge that of Ravenstein, though he explored in much more detail the process of labour migration and its relation to factors such as economic change and the operation of the Poor Law. In conclusion he wrote:

> All the rising centres of industry and commerce were attracting workers by a process of short-distance migration from the surrounding country; where the attractive force of a large town was exerted over a wide area the inward movement took place usually by stages. The majority of the migrants to the town came from the immediately surrounding counties, their places in turn being taken by migrants from places further away. (Redford, 1926, p. 183)

In the first six decades of the twentieth century improved computational techniques, together with the availability of individual census birthplace data from unpublished census enumerators' books from the 1950s, has encouraged a wider range of studies based both on published (Thomas, 1930; Darby, 1945; Smith, 1951; Lawton, 1958) and unpublished census data (Lawton, 1955, 1959; Shephard, 1961). However, these have not substantially challenged the conclusions of earlier researchers. More sophisticated statistical analyses of published census data attempted to calculate intercensal survivorship rates to enable net flows to be computed (Hill, 1925; Shannon, 1935, Baines, 1972, 1978; Pooley, 1978). The most detailed study of this type was that undertaken by Friedlander and Roshier (1966). They concentrated on flows between non-adjacent counties for the period 1851–1951, again stressing the importance of short distance movement for all areas apart from London and, for short periods, rapidly expanding industrial districts such as South Wales. They also demonstrated the extent to which regional migration flows changed from the nineteenth to the twentieth centuries as the northern industrial districts declined in importance and inter-regional flows to southern Britain increased.

Use of census enumerators' books has enabled more detailed analyses of movement into specific towns and regions, and of the characteristics of migrants (Pooley, 1983; Brayshay and Pointon, 1984; Doherty, 1986), although the limitations of using birthplace data are severe. The extent to which people moved frequently has been demonstrated by studies that have plotted the successive birthplaces of children born to migrant families, and analyses of the occupations of migrants have suggested that longer distance movement was mostly undertaken by those in higher socio-economic groups. Marketable skills, knowledge of opportunities and access to transport were all important factors in enabling some individuals and families to move over considerable distances. Thus, in a study of migration to Liverpool in the nineteenth century, it has been shown that the volume of movement was clearly related to distance, that access by sea enabled some migrants to move over longer distances, and that longer distance migrants tended to be both older and more skilled than those moving from places closer to Merseyside. Many longer distance migrants were attracted by the specific economic attractions of the expanding port city, and long distance migrants also tended to come from urban areas with labour markets similar to the destination. Only short distance movement from the surrounding region was rural to urban. Evidence of children's birthplaces suggests that some stepwise movement took place, though the static nature of census sources makes it difficult to assess the length of residence in any particular location (Pooley, 1978, pp. 58–99).

Ready availability of nineteenth-century census data has had some negative effects on research into migration in the past. It can be suggested that too many studies have focused on the years for which census data are available – thus neglecting the crucial periods of urban and industrial growth before 1841 (but see Clark and Souden, 1987) – and alternative sources of information for the study of migration have been relatively neglected. However, some studies have moved beyond the census, and in recent years researchers have become increasingly imaginative in their use of a wide range of sources to study migration.

Records generated by the operation of the Poor Law, apprenticeship registers, trade union records, the criminal justice system and, in some instances, parish registers can all give some information on mobility before the mid-nineteenth century, although in most cases they relate only to specific groups of the population who were moving for particular reasons.

Large cities such as London and Edinburgh attracted apprentices from considerable distances in the eighteenth century, and analysis of such sources enables some conclusions to be drawn about the distance and directional components of migration. In a study of movement to Edinburgh, Lovett *et al.* (1985) concluded that there was a strong distance-decay effect of apprenticeship movement in the eighteenth century, but that the migration field was not regular and that there were clearly defined flows from particular areas even after the effects of distance and population size of origin had been allowed for. The strong directional component in migration to eighteenth century Edinburgh was explained by factors such as the communication network, existing trading contacts, regional variations in levels of urbanization, and cultural differences. As other urban centres, notably Glasgow, became more important, so the area from which Edinburgh drew its apprentices changed to take into account these alternative opportunities. Although the directional component of apprenticeship migration to Edinburgh was more marked than in other studies, similar patterns have been demonstrated for London and other towns (Buckatzsch, 1949–50; Patten, 1976; Waring, 1980; Holman, 1980; Yarborough, 1980).

Other studies of mobility in late-eighteenth century rural areas have emphasized the importance of short distance circulation between places within a locality, rather than rural to urban movement or longer distance migration away from rural communities (Constant, 1948; Holderness, 1970; Schofield, 1987; Escott, 1988). Thus, following a study of especially detailed parish registers for the Plain of York from the 1770s to the 1820s, Holderness concludes that:

> The picture which emerges is of village communities regularly refreshed by influx from similar settlements lying at a comparatively short distance away. The ebb and flow of people affected each of the principal status groups but migration was most pronounced among labouring families. (Holderness, 1970, p. 452)

It was not until the later nineteenth century that depopulation significantly affected such rural districts and, even then, local circulation continued to occur despite increasing rural to urban movement.

Redford pioneered the use of Poor Law records for the study of migration in 1926, but more recent research has also utilized these sources (Redford, 1926, pp. 81–96; Parton, 1987; Taylor, 1989). They demonstrate not only the extent to which many poor families moved over quite long distances in search of work in the eighteenth and nineteenth centuries, but also the way in which implementation of the Laws of Settlement by local authorities who refused to accept responsibility for destitute migrants could fundamentally alter the lives of individuals. Thus a migrant to Preston from North Lancashire who fell on hard times was temporarily removed by the

Preston overseers to the parish in which he had served his apprenticeship, causing great hardship and splitting his family:

> I was advised to apply to Preston for relief, so my wife got 5 shilling or some trifle, & they removed us to Ellel, the town in which I lived when I was at dolphinholme, we left 2 lads at in preston with my sister, that was in work, & thought that they would keep themselves nerly, & sleep at home, for we did not sell our house-hold goods. (Shaw life history)

Green (1995) has used Poor Law and other records to examine aspects of mobility and community change in nineteenth-century London and, at a more aggregate level, Withers and Watson (1991) have used Scottish Poor Law records to analyze the nature of stepwise migration from the Highlands to Glasgow. They describe 'patterns of hierarchical-cum-spatial stepped movement' whereby migrants moved through a sequence of settlements enroute to Glasgow, and they speculate on the extent to which such movement aided acculturation for the rural to urban migrant. Undoubtedly some of the most disruptive migration is that which is forced upon individuals and families. The effects of Irish famine migration are well known but, in a second paper, Withers (1988) also highlights the migration experiences of Highland dwellers displaced by famine in the 1840s, and forced to move through directed labour mobility schemes. He argues that such schemes, developed as famine relief, did not fundamentally alter the constituent social relations of the destitute areas. Much of the burden of the schemes was placed on landlords, and a long tradition of temporary Highland–Lowland mobility meant that such movement was not unduly disruptive to local communities. Out-movement came principally from the North West, especially from Skye, and most migrants were males moving to railway work in Fife and East Lothian.

In historical research it is often the poorest members of society who are most invisible in the written records. Just as the destitute appear when they seek relief from the authorities, others were recorded when they brushed with the law. Most recent studies of crime in the nineteenth century have suggested that the majority of those arrested and convicted were a cross-section of ordinary working people, who temporarily resorted to petty crime in times of hardship (Emsley, 1987). Criminal records can provide some insight into the mobility of such people, including highly mobile vagrants. Relatively few studies have directly addressed the mobility of criminals, but Nicholas and Shergold (1987) have examined the mobility in Britain of those eventually transported in the early-nineteenth century, and have found patterns of movement very similar to those recorded from other studies. Most migration was short distance and longer distance migration was most likely to be undertaken by those with skills. A study of criminals in late-nineteenth century Cumbria, emphasizes the very frequent movement of many, and the way in which those recorded as tramping often moved through familiar places in what appeared to be a regular cycle. Over time, some such migrants covered long distances, but most movement was again within the local area (Pooley, 1994).

Not all of those who tramped around the country in search of work were especially poor or classed as vagrants. In the early-nineteenth century, in particular, the tramping artisan was a common feature, as skilled working men moved between settlements in search of suitable employment (Hobsbawm, 1951). Such migration was usually undertaken in a well-structured way, moving to places where work was known to be available and with the trade unions of skilled workers providing accommodation and support in each town. Such movement was almost entirely male, and married men would leave their family, sending money regularly and returning home at intervals. Southall (1991a, 1991b) has examined the migration paths of such skilled artisans, emphasizing the relatively long distance and high frequency of such mobility. Although movement of the tramping artisan was in no sense typical of wider migration patterns in the early-nineteenth century, it does emphasize the way in which those with relevant skills and contacts could move easily over long distances.

Some of the most revealing information about migration has come from studies which have utilized the evidence of diaries, letters, autobiographies or oral reminiscences. Although they may not represent a cross-section of society, such sources not only reveal the detailed pattern of movement, but also begin to hint at the reasons for and impacts of migration. Such sources have been most extensively used in studies of emigration, where letters home and accounts of a new life in the colonies have often survived in family archives (Erickson, 1972; O'Farrell, 1984; Richards, 1991). However, there are also an increasing number of studies which utilize diaries and autobiographies (Lawton and Pooley, 1975; Parton, 1980; Southall, 1991b; Pooley and D'Cruze 1994; Pooley and Turnbull, 1997a) or oral information (Jones, 1981; Perks, 1984: Bartholomew, 1991) to reveal more about the everyday experiences associated with short distance migration within Britain. Such studies emphasize the human side to migration decision making, including such aspects as the role of kin, which can never be revealed through census and similar sources. Thus an oral respondent who left Wales in the early-twentieth century recalls:

> My brothers paid for all my nursing, medical and surgery books. My grandmother and aunts paid for training school dresses. An uncle – my railway fare. My mother's eldest sister in North Wales contacted her brother-in-law who lived in Portsmouth, who met me at the station (Bartholomew, 1991, p. 182)

Leaving Wales to train as a nurse in England was clearly a major undertaking, but one which was eased by the involvement of all the family.

Personal accounts from diaries or oral evidence also emphasize the dynamic and continuous nature of migration experiences. Each migration event is not a separate entity, but one experience will feed into the next, and the impact and meaning of migration for individuals can only be understood as a longitudinal process which interacts with other life events. There have recently been some attempts to utilize static census sources to try to build up longitudinal profiles of a large number of migrants (Doherty, 1986; Pooley and Doherty, 1991). By linking individuals from one census to the next, it is possible to at least partially reconstruct the migration history of a more representative sample of people. Such longitudinal studies focus attention on family migration at different stages of the life-cycle, and in the study of migration from Wales to English towns some sixty-four per cent of the linked migrants had moved as part of a family group. Other studies have also begun to stress the importance of family migration, questioning the assumptions of Ravenstein and others that most movement was undertaken by young single migrants (Darroch, 1981; White, 1988; Anderson, 1990; Schurer, 1991).

This brief review of population migration in the past has concentrated mainly on studies relating to the eighteenth and nineteenth centuries. For the twentieth century, oral evidence can be combined with census data and other surveys, though the lack of individual census enumerators' books restricts the amount of research that can be conducted at the individual level. The twentieth century has seen two major geographical trends in population migration in Britain. First, the shift of population from older industrial areas of northern England to the South East and, secondly, the increasing importance of counter-urbanization as the more affluent left large cities and moved to smaller settlements. Well established migration characteristics, with young adults and the more skilled moving over longer distances, continued to be evident, though the movement of specific groups such as retirement migrants and the movement of students to higher education became increasingly important in the British migration system (Johnson, Salt and Wood, 1974; Coleman and Salt, 1992; Stillwell, Rees and Boden, 1992).

Recent migration research has thus refined some of the conclusions of nineteenth-century statisticians and has delved more deeply into the cultural and social meanings associated with the migration process, but it has not substantially altered many of the findings of authors such as Welton, Ravenstein and Redford. Throughout the period, short distance movement was the most common experience for most people. In the eighteenth century rural circulation between adjacent parishes was common but, increasingly, rural to urban moves became dominant although some studies emphasize the continued importance of rural circulation. By the mid-nineteenth century, as more than 50 per cent of the population lived in towns, movement between towns, both up the urban hierarchy and longer

17

distance moves between large centres, became more important with rural depopulation becoming marked in many areas. Despite improved transport, mobility in the twentieth century was still dominated by short distance moves, with intra-urban mobility and suburbanization increasingly import-ant, whilst longer distance moves shifted population towards the South East. Where studies have directly addressed the issues, there has usually been some evidence of stepwise migration and of distinct migration differ-entials, with longer distance moves more likely to be undertaken by those with skills (who were also more likely to be male). Some recent studies have also suggested that family migration was important, and most qualitat-ive work stresses the importance of kinship and friendship networks in both promoting migration and aiding the migrant to settle in a new environment.

It may seem that we already know a great deal about migration in the past, but there are still many areas where statements made from aggregate census data are unproven at the individual level, or where our inter-pretation is based on little more than speculation. In particular, we need to know much more about the life-time migration experiences of individuals and families, particularly the ways in which migration events were linked together and interacted with other life-cycle events such as marriage, child-birth, retirement, illness or death of a partner. Only by analyzing a large and representative sample of life-time residential histories can such aspects of the migration experience be examined. We also need to know much more about the ways in which the pattern and process of migration changed over time. Existing studies are heavily concentrated in the mid-nineteenth cen-tury, for which census data are available, and studies for earlier or later periods rarely use directly comparable sources or methodologies. By study-ing migration from the mid-eighteenth to the mid-twentieth centuries a clearer picture of temporal change should emerge.

Due to the large volume of data contained in the census and other sources, most existing migration studies focus only on one or two regions, or deal with the whole country at an aggregate level. We need to know much more about the extent to which there were variations in the patterns and process of migration between British regions. For instance, North-East Scotland and North-West England were economically, socially and cultur-ally very different in the nineteenth century. We do not know the extent to which such differences were reflected in the migration systems which oper-ated within and between regions. By taking a sample of life-time residential histories from all parts of the country it is possible to draw direct compar-isons between British regions.

Most studies of migration use only speculative or partial information on the reasons for migration and the census gives no such details. We need to know much more about why people moved, and about the ways in which the decision-making process interacted with other members of the house-hold. Although still imperfect, this study sets out to collect much more reliable and comprehensive data on reasons for moving in the past. The process of migration also needs to be related to the familial, social and economic context in which it took place. Most single-source studies do not make this possible: for instance, in most studies we do not know the family composition during migration. Data collected from multiple sources, and compiled into life-time residential histories, allows the event of migration to be set within a broader context. We should be able to answer questions about who moved together, about housing and employment change following migration, and about the ways in which migration decisions related to broader economic and social processes operating in particular regions.

Approaches to the study of migration in the past — *effects way we perceive people.*

Migration theory is relatively poorly developed. Most studies of migration in the past and the present are largely empirical and, to the extent that any theory is used, it tends to be drawn from quantitative and behavioural work (originating in the 1960s) which attempts to model spatial choice (Willis, 1974; White and Woods, 1980, pp. 1–56; Jones, 1981, 189–278; Lewis, 1982; Woods, 1985; Champion and Fielding, 1992). There have been numerous attempts to apply distance decay models of varying complexity to migration data, although it is difficult to account for the effects of time and ease of travel rather than simple linear distance (Zipf, 1946; Olsson, 1963; Gale, 1973; Willis, 1975). Other models have introduced concepts of intervening opportunities (Stouffer, 1940, 1960), and set the migration decision-making process within a simple push–pull framework, related to the opportunities available at one or a number of destinations (Lee, 1966). Much analysis specifically devoted to the longitudinal study of migration careers has also been largely quantitative, and ignores the human dimension of mobility (Davies, 1991; Davies and Flowerdew, 1992). However, recent work on event history analysis has added important new dimensions to the historical analysis of longitudinal data (Courgeau and Lelièvre, 1992; Courgeau, 1995). Behavioural work has focused on establishing decision-making frameworks for residential choice as it applies both to intra-urban mobility and to longer distance migration (Rossi, 1955; Wolpert, 1965; Brown and Moore, 1970; Golledge, 1980; Cadwallader, 1989). Some such frameworks appear quite elegant, but they are hard to operationalize given the level of data available to most researchers. A more satisfactory way to view the migration process is in the context of choice and constraint. For most people constraints to movement are as important as opportunities, and some studies which emphasize the constraints on migration decisions have drawn on Marxian theory (Harvey, 1985). Much migration theory neglects the cultural aspects of movement: the ways in which the migration decision, and reactions to migration, are embedded in the cultural values and meanings of individuals, groups and wider society. There have been recent calls to pay more attention to this cultural dimension, and to focus on the meanings associated with migration, but as yet relatively few studies exist (Fielding, 1992).

The approach which perhaps provides the best framework for explanation, and which will be utilized where appropriate in this volume, is that of structuration theory (Giddens, 1984; Gregson, 1987; Bryant and Jary, 1991). This focuses on the intersection between individual actions and wider structural processes, emphasizing the importance of the everyday social practices which actors adopt to make sense of the society in which they live. Whilst most actions are affected and constrained by structures, through specific social practices individual agency can also influence and change some structures. Through the concept of 'space–time distantiation' the theory also emphasizes the ways in which events and decisions taken in one place and time period may be affected by or distanced from processes operating elsewhere. It is clear that the decision to migrate is one which has immense personal ramifications but which is also fundamentally related to the social, economic and cultural context in which it is situated. As such, structuration theory provides an ideal framework in which to study and explain the migration process. Many of the ideas embedded in structuration theory also borrow quite heavily from the concepts of 'time geography' developed by Hagerstrand (1978). This work emphasizes the importance of individual life-paths, which bring actors into contact with places, people and ideas, and which cumulatively give meaning to everyday decisions such as those relating to employment change or migration. In an historical context, these ideas have been particularly developed by Pred (1990), who applies them to the study of nineteenth-century Swedish communities and, more generally, by Gregory (1982). Given that the focus of this research is on the creation of longitudinal residential histories, which can be related to other life-time events, it seems appropriate to utilize a time geography approach and, where relevant, present migration histories as life-paths which illustrate particular aspects of the migration process.

There has been some recent debate in the migration literature about the need to adopt what is termed a 'biographical' approach to the study of migration (Halfacree and Boyle, 1993; Skeldon, 1995). This is essentially a reaction against the largely quantitative work which dominates migration research, and is an attempt to introduce a more human element to migration studies. This clearly links closely with the concept of life-paths used in time geography. Such an approach is not entirely new. Often due to the lack of other data, studies of migration in developing countries have used a more ethnographic approach to the migration process than has been common in Europe or the USA (Pryor, 1979; Skeldon, 1990; Chant, 1992). This has typically meant the creation of detailed profiles of individual migration experiences. It is this rich ethnographic literature which could form the basis for future biographical research but, unfortunately, it is not possible to conduct true ethnographic studies in the past. However, the detailed life-time residential histories used in this research are a close surrogate, and a biographical or ethnographic approach to the study of migration is one which is relevant to both the data used and questions asked in this volume. In common with postmodern rejection of grand unifying theories, migration researchers have also, recently, combined a wider range of approaches – quantitative, behavioural, ethnographic – in explicitly multi-method approaches to migration research. This philosophy is also adopted in this volume.

The structure of the book

The remainder of this book focuses specifically on the results of one re-search project which has collected and analyzed a large number of indi-vidual residential histories, but draws additional material from relevant published work and other sources, and relates research findings back to some of the issues outlined above. Chapter 2 provides a detailed account of how and why the study was undertaken, paying particular attention to the characteristics of the data set and assessing its validity against comparative census data. Inevitable biases in the data set are highlighted and their implica-tions for the interpretation of results are discussed.

Chapters 3 and 4 focus primarily on the spatial and temporal dimen-sions of migration in Britain from the late-eighteenth century to the mid-twentieth century, providing an overview of the way in which the pattern and process of migration varied over space and time. Whereas Chapter 3 focuses on the simple question of where people moved at different periods of time, and the extent to which different people and places generated distinctive migration patterns, Chapter 4 focuses more specifically on the role of towns in the migration system. As outlined above, it is usually assumed that rural to urban migration dominated the migration system for much of the nineteenth century. This view is challenged and data from the longitudinal residential histories are used to demonstrate the complexity of the migration system, with rural to urban moves being one of many differ-ent migration experiences.

Subsequent chapters focus in detail on specific reasons for migration, examining the relative importance of particular motives and relating the migration process to broader economic, social and cultural trends within society. Topics considered include migration and the labour market, the role of the family in migration decisions, migration and the housing market, the significance of particular crises in stimulating migration, and the process of emigration to other countries. In each chapter aggregate patterns of spa-tial and temporal change are backed up by individual examples drawn from diaries and other records generated by family historians.

The two concluding chapters focus, first, on changes over time and space, using the unique longitudinal data set to examine the ways in which migration processes were linked into broader structures of economic, social and cultural change and, second, on migration theory, assessing the extent to which results presented in the book have challenged or substantiated existing knowledge about migration. In Chapter 10 the impact of different types of migration experience on both individuals and wider society is assessed, whilst in Chapter 11 the continuing relevance of the work of Ravenstein, Zelinsky and others is appraised, and British results are placed in a wider European context. Whilst the book is essentially an empirical analysis of migration over time and space, it is suggested that the evidence presented can contribute significantly to both migration theory, and to the development of new methodologies for the analysis of migration.

CHAPTER 2

How to study migration in the past

Conventional ways of studying migration in the past

As shown in Chapter 1, most of our current knowledge about migration in the past comes from census birthplace data, supplemented by sources such as Poor Law certificates, apprenticeship registers, diaries and oral reminiscences. Having reviewed the current state of knowledge about migration in the past derived from such sources, this chapter focuses in detail on the strengths and limitations of different classes of data for the analysis of migration. It should be stressed that this study is largely dependent on the same sorts of sources that other researchers have used, and thus the limitations of such sources need to be fully appreciated. However, the range of material utilized and the methods used to collect individual longitudinal data in this study overcome some of the previous difficulties. These methods, and the characteristics and biases of the data produced, are explained in the second half of this chapter.

It is possible to classify sources used in the study of migration according to a number of criteria. These include the extent to which the source gives direct or surrogate information on migration, the spatial and temporal coverage of the evidence, the extent to which the total population is covered, and the amount of explanatory and contextual information about migration that is included. In Britain, there are no sources which were designed to collect direct information on population migration. All data must be culled indirectly from sources which were used for other purposes. Even if continuous population registers were available (as in some continental European countries) this would not solve all problems of migration analysis. Although a requirement to register every change of address should give a continuous record of when and where people moved, such registers give very little contextual information about why people moved or who they moved with. Only the census of population gives a complete spatial coverage of information for the whole of Britain, although the nature and quality of information varies (Baines, 1972, 1978). Thus precise place of birth was only included for people born within the country of enumeration (England and Wales, Scotland), otherwise only country of birth was recorded. For instance, migrants from England to Scotland were simply recorded as having a birthplace in England. Likewise, only the census attempts to include the entire population, all other sources providing information on some sub-set of the total British population. All sources have a restricted temporal coverage, and there is no single source of information which can provide comparable data on migration from the eighteenth to the twentieth centuries. Because all sources were designed for purposes other than research on migration, the amount of explanatory and contextual evidence relating to migration is always limited. Those producing the sources were simply not concerned to explain why people moved. Only diaries, letters and oral evidence can be expected to provide reasonable explanatory details, and even this information must be treated with considerable caution.

The impact of migration on past society can be viewed from a number of perspectives. A strictly demographic approach focuses attention on the net impact of migration in changing the distribution of population and the size of settlements. Using aggregate information from the census (on total population change) and from the Annual Reports of the Registrar General (on births and deaths), the difference between total population change and change due to natural increase or decrease (the difference between births and deaths) can be easily calculated for registration districts and subdistricts from 1841 (Lawton, 1977). Such analysis clearly demonstrates the extent to which population increase or decrease in particular regions was due to migrational change or natural change, but concentration on the net balance of migration flows seriously underestimates the total impact of migration. Thus, a settlement may have received an inflow of (say) 1,000 migrants between 1841 and 1851, and experienced an outflow of 900 migrants over the same period. The net demographic impact on the population of the town was thus small (a surplus of 100 people), but a total of 1,900 individuals moved and this volume of migration could have had a significant impact on both the areas which gained and lost population and on the people involved. The social, economic, and cultural impacts of migration must therefore be studied in the context of gross migration flows, rather than the net demographic effects of population movement.

The published British census of population does not collect direct information on migration but, for much of its history, has recorded place of birth and place of residence on census night. It thus records the gross movement of people from their place of birth to their residence at an arbitrary point of their life on census night. Quite limited birthplace information was recorded in the census of 1841, and it is from 1851 that detailed data were collected. In 1961 the census of England and Wales dropped the detailed birthplace question for internal migrants, but since 1961 has recorded where respondents lived either one or five years previously. This again presents an arbitrary snapshot of gross movement rather than a more complete picture of life-time migration. Thus although the published census records changing population distributions, and enables an analysis of (say) the numbers of Irish-born or Yorkshire-born living in London on a given date (prior to 1961), it misses many intervening moves and presents only a partial picture of the migration system. However, tables in the published census volumes are accessible and easy to work with and, despite their limitations, they form the starting point for most studies of migration.

Evidence contained in the census enumerators' books (currently available from 1841 to 1891 in England and Wales and to 1901 in Scotland) suffers from the same problems as the published census, in that it only contains responses to a question on place of birth. But the census enumerators' returns do offer additional potential for local and regional studies (Mills and Pearce, 1989). In theory, precise place of birth is recorded within the country of enumeration from 1851 onwards, and thus detailed migration fields can be plotted. The census also contains some contextual information about the characteristics of individuals and their families. Although there is no direct information on when, why or with whom people moved, it is often possible to infer more contextual details from the available evidence. Thus examination of the ages and birthplaces of spouses and children may enable a more complete picture of longitudinal migration histories to be constructed. There are, however, limits to the use of such data. First, there are frequent obvious errors in the recording of precise birthplaces (for instance towns attributed to an incorrect county) with no way of knowing whether the town or county is the more accurate information, and studies which have linked families from one census to the next have frequently found discrepancies in the birthplace information recorded, with no way of knowing which is the correct data. Whilst the census enumerators' books are a key source, and have provided much valuable information on population movement, on their own they do suffer from a series of quite severe limitations (Tillot, 1972; Higgs, 1989).

There are a large number of sources which, whilst compiled for quite different purposes, provide some information on migration of particular groups of the population. Unfortunately, such sources are rarely available for long time periods or the whole country, and it is difficult to assess the typicality of the groups for whom data are available. The nature and usefulness of the sources depend very much on how they were generated and the purpose for which they were intended. At the local level parish and township records relating to the indenture of apprentices, the provision of settlement certificates, and the removal of paupers to their parish of settlement often contain quite detailed information on current and previous mobility. Thus a removal order produced in 1866 by the Overseers of the Poor of Kirkby Lonsdale (Westmorland, now Cumbria) to the Churchwardens and Overseers of the Poor of the neighbouring parish of Whittington (Lancashire) provided a detailed life history of the movements of George and Anne Wilkinson and their four children in justification of their removal from Kirkby Lonsdale to Whittington. Other Parish sources are rarely so detailed, but apprenticeship indentures, and registers of parish apprentices, normally recorded the place of residence of the apprentice and the name, address and trade of the person to whom they were bound. Quarter Sessions Examinations under the Laws of Settlement can also provide quite detailed information on the life history and previous mobility of individuals, as the court attempted to establish the Parish of Settlement of vagrants and other individuals brought before the Justices (Whyte and Whyte, 1986; Taylor, 1989).

However, all such records, together with other information generated by the legal system (for instance police and prison records) or by trade unions and businesses, have in common the fact that they only recorded information on particular sub-sets of the population (vagrants, paupers, apprentices etc.); that information was only recorded at points in which individuals came into contact with the administrative system established at the time; and that the only details of mobility recorded were those deemed at the time to be relevant to the case in hand. It is thus very likely that many moves were not recorded, but there is no way of knowing what is missing or of assessing the typicality and representativeness of information which is recorded. Such sources are clearly of great value to the study of migration, but they should be used with caution and, on their own, provide only a partial picture.

The only effective way to create life-time residential histories is to undertake large-scale record linkage at the individual level. In addition to the above-mentioned sources there are a large number of other classes of records which can yield cross-sectional information on place of residence or migration behaviour. On their own they are not especially illuminating, but when linked together they may allow a much more detailed picture of life-time mobility to be constructed. Sources which commonly yield relevant data include wills and inventories; parish records of births, marriages and deaths; rate books; electoral registers, trade directories and newspapers. Sources, such as directories, electoral registers and rate books which, in some cases, were compiled on an annual basis can provide good quality longitudinal information which allows individuals to be traced through a series of residences. Such sources have been used moderately successfully to examine intra-urban residential mobility (Pritchard, 1976, Dennis, 1977a; Pooley, 1979). However, all such sources again suffer from biases and omissions. Directories and electoral registers will, by definition, include only a sub-set of the population: those eligible to vote at a particular time or those whom it was thought appropriate to include in a trade directory. By and large such sources excluded the poorer members of society (Shaw and Tipper, 1988). Although rate books should be more complete, their survival is uneven and, where the owner was responsible for payment of rates, information on the occupant of a property may not be reliable. The names and addresses of individuals may be recorded in newspapers for a myriad of different reasons. In some cases these may be directly related to migration – such as lists of people arriving and departing on ships from a particular port – but more often chance entries have to be linked with other data.

The main difficulty with using any of the above sources for record linkage in historical research is the extremely time consuming nature of the work, and the large amount of effort that has to be expended for only limited returns. Thus, even when linking individuals through user friendly, indexed sources such as trade directories, the time taken to build up a significant sample of individuals can be considerable. Problems are further compounded by the fact that many sources provide only limited information on the characteristics of individuals. Thus record linkage may be difficult or impossible when the only identifier is a commonly occurring name. There have been a number of attempts to automate the process of record linkage, thus reducing the amount of labour involved and standardizing the decision-making process of whether or not a link has been made. In most cases, however, such computer based systems have met with only limited success, and depend on both good quality sources and a reliable range of contextual information with which to identify individuals. Even when large scale record linkage exercises have been successfully completed, we usually still only have information on the distance and direction of moves for a portion of an individual's life. Data on the reasons for and impacts of migration are much more elusive (Wrigley, 1973; Wojciechowska, 1988).

The only sources that can regularly provide some information on aspects of the migration process such as why and how people moved, or on the impacts of migration on the lives of individuals are diaries, accounts, letters and oral reminiscences. However, even these sources present problems. There is obviously no way of assessing the typicality of respondents, letter writers or diarists, and it can be suggested that those who left a written record of their life, or who respond to requests for oral informants, are by definition atypical. Care must certainly be used when drawing wider inferences from such data. Moreover, the amount of information on, and recall of, the circumstances surrounding migration can often be sparse and unreliable. Many diarists simply recorded the fact that they moved with no additional explanation. This may have been because the event was of limited significance in an individual's life, because the reasons for moving were so obvious at the time that there seemed little point in recording them, or because the purpose of the diary was simply to record other aspects of an individual's life. Thus the degree of information relating to migration can vary greatly between similar sources (Burnett, 1974, 1982; Vincent, 1981; Burnett, Vincent and Mayall, 1984–7).

Although information recorded in diaries, or even in life histories written from diaries and earlier notes, may be assumed to be reasonably accurate

– if filtered through the perceptions of the writer and the circumstances of the time – oral testimonies are much more difficult to evaluate (Humphries, 1984; Lummis, 1987; Thompson, 1988; Perks.1992). Short distance or unexceptional mobility may not be a significant event in the respondent's mind, in some cases a wife or child may have only partial knowledge of why a move was undertaken if they were not fully involved in the decision-making process, and there may be considerable post-move rationalization of the reasons for migration. This is likely to be especially problematic where moves were forced for reasons of poverty, redundancy or similar circumstances. At the time the family may have had no choice in whether to move or not, and may have been faced with a very limited number of alternative destinations. But many years later a respondent may rationalize a move by emphasizing the positive aspects and playing down the negative reasons for migration. Although oral evidence is of great value in understanding migration in the recent past, it should not be assumed that the evidence collected is necessarily less problematic than data collected for earlier periods.

Family historians, genealogists and historical research

Evidence presented in this volume is drawn mainly from data collected by family historians and genealogists. It is thus necessary to assess the objectives, sources and methods of family historians, and evaluate the biases that these may create. For many people interested in tracing the history of their family migration is not the central focus of interest. Long distance movement or emigration may excite interest, but frequent short distance moves are seen more as a problem than a significant aspect of the family history. Frequent moves make tracing individuals more difficult and thus often frustrate the research of family historians. Thus, very often, family historians acquire information on migration by default, because they need to know where people lived to discover other aspects of their lives. In the course of the project several family historians commented that they had not really thought very much about migration previously, but that our request for information had both stimulated their interest and encouraged them to seek additional data about residential change.

It should be stressed that family historians and genealogists do not have access to any special types of source. For the most part they use the same range of sources as all historians, and they encounter all the problems outlined above. This project is thus not using any new sources, but it is utilizing a large labour force who have certain advantages with respect to the range of sources they use and the way they tackle historical research. The importance of information collected by family historians and genealogists for broader historical research is now being recognized by the academic community (Gold, 1994). There are a small number of studies in

Britain and elsewhere that have used genealogical material for historical research (Courgeau and Lelievre, 1989; Dupaquier and Kessler, 1992; Prandy, 1997), but equally importantly an increasing number of family historians are becoming interested in historical research beyond the remit of their own family, and are beginning to apply their skills and knowledge to broader historical questions (Drake, 1994, Hurley and Steel, 1994). The divisions between family history, local history and social history are becoming increasingly blurred and each group has a good deal to contribute to the other.

The main advantages that family historians and genealogists have over most academic researchers interested in migration relate to the objectives of their research, the range of source materials used, the process through which they conduct research, and the time at their disposal. By definition, family historians and genealogists are interested in individuals. The focus of their attention is thus the life history of a series of people to whom they are related. Their objective is to build up a small number of longitudinal case studies which should encompass most aspects of an individual's life, rather than the snapshot approach adopted by most historians. Because family historians focus on only a small number of people, they are able to utilize a much wider range of sources than most academic historians. Thus, whereas most theses and research projects are based on one or two sources (most commonly the census backed up, perhaps, by information from directories or rate-books), family historians will search for their ancestors in every available documentary source. They will thus link individuals through parish registers, censuses, criminal records, poor law records, wills and many other pieces of evidence to build up a life history which is as complete as possible.

Family historians, genealogists and those involved in one-name studies also adopt a rather different research process to that of the academic historian. Typically the family historian will begin with oral evidence collected from parents and grandparents, often supplemented by family papers which would never find their way into a record office. From this skeleton of remembered events and supporting evidence of why events occurred, most family historians are able to go to specific sources to corroborate oral testimonies, add details, and delve back into the more distant past. Such research is, however, highly directed, has a strong element of cross-checking evidence between different sources, and by combining a wide range of documentary evidence, family papers and oral evidence can produce a much more rounded picture than that typically created from the research of academic historians. Because they are not interested *per se* in specific individuals, most academic historians adopt a more blanket approach to research and may, for instance, record details of all apprentices to a particular town in the late-eighteenth century. They will not, however, be able to relate this information to any other data about the individuals involved, and any individual examples cited may be from readily available diaries or accounts which may not relate directly to the aggregate data set used.

29

Because family historians are undertaking their research as a hobby, with no particular deadlines, they are also able to expend more time on their research than the majority of academic historians who have to complete work in a limited time and meet thesis or publication deadlines. Although family historians are amateurs in the sense that they undertake the research for pleasure, in their own time, this does not mean that they are unskilled. Most family historians have quite a high level of general education and can apply skills acquired in their own employment to family history research. For instance, many family historians have a much higher level of computer skill than some academic historians. Increasingly, family history societies and other bodies are also organizing a wide range of courses for people involved in tracing their ancestors. These are often run by skilled historians, and many family historians have acquired an impressive knowledge of sources and research methods from a combination of courses and private study. There is thus good reason to believe that the information produced by family historians will be carefully researched, at least as reliable as that produced by academic historians, and because of its longitudinal nature more suitable for the study of migration than snapshot data culled from a single source.

However, there are inevitable biases in the data produced by family historians and genealogists. These, too, are related to the objectives of their research, the methods used and the sources that are available. There are a number of demographic biases in the population researched by family historians. Because genealogists, family historians and, especially, one-name researchers are primarily interested in following a name down the male line, there tends to be a strong bias towards male ancestors in the research of family historians. This is further compounded by the relative invisibility of women in many historical sources, and the difficulties involved in tracing women who changed their name on marriage. For obvious reasons, family historians are also primarily interested in tracing ancestors who married and had children (and thus produced a family line): there is thus a bias towards those who eventually married and towards those who completed a family and lived to old age. Children who died young are of little interest to family historians and, in any case, they are less likely to appear in accessible sources. It can also be argued that family historians are most likely to research the history of ancestors who appear interesting or unusual (and such people may be more visible in the sources), and that there may be a temptation to infer information and elaborate a family story rather than basing it on hard information.

Other biases will be much the same as those found in the research of any historians. All researchers may make errors in transcribing documents or linking individuals, though the personal commitment of most family historians may mean that they make fewer such errors than the more detached professional historian. The fact that they know the context of a life history means that family historians are also more likely to spot and correct errors which produce results that do not tally with other evidence. The nature of all historical sources means that there may be an under-representation of the poor – who tend to be less visible unless they brushed with the law or the Overseers of the Poor – and that it may be more difficult to trace individuals in large and rapidly-changing urban areas. Family historians may thus be tempted to take the relatively easier route of tracing their more affluent ancestors who lived in small towns and rural districts. Given that many family historians are not primarily interested in residential change, and the problems of identifying frequent short-distance mobility from available sources, it is likely that some such moves may be missed. It can also be suggested that as most family historians are drawn from the middle and professional classes, this may be reflected in the social background of their ancestors. The extent to which these biases are evident in the data collected from family historians, and their implications for subsequent analysis, are outlined below.

The collection of longitudinal residential histories

The original idea to utilize data collected by genealogists and family historians emerged from a number of conversations with family and local historians following conferences at which papers on migration had been presented. It was clear that many family historians and genealogists had good quality information about their ancestors, were aware of the processes of academic historical research, and were keen to co-operate with a university-based research project. However, the total number of family historians with relevant information, and the numbers of ancestors for whom they had details, were unknown. It was clearly impossible to construct any form of controlled sample: the research would have to rely on responses from family historians to a general call for information. Inevitable biases in the sample would then have to be assessed against census data and controlled for in subsequent interpretation. Methods of data collection were first used in a pilot project which collected data only from the Lancashire and Cumbria Family History Societies during 1992. During this phase of research the methods of data collection were refined and improved before collecting information nationwide during 1994 and 1995. Although there were some changes in the data collection form between the pilot project and the main survey, the information collected is essentially compatible and for most analyses presented in this volume data from the pilot and main surveys are combined.

A total of 80 family history and genealogy societies were contacted and all bar 17 small and randomly distributed societies agreed to publicize the research in their magazines or newsletters (Appendix 2). Co-operating societies included national organizations such as the Society of Genealogists with some 12,000 members, large regional societies such as the Birmingham

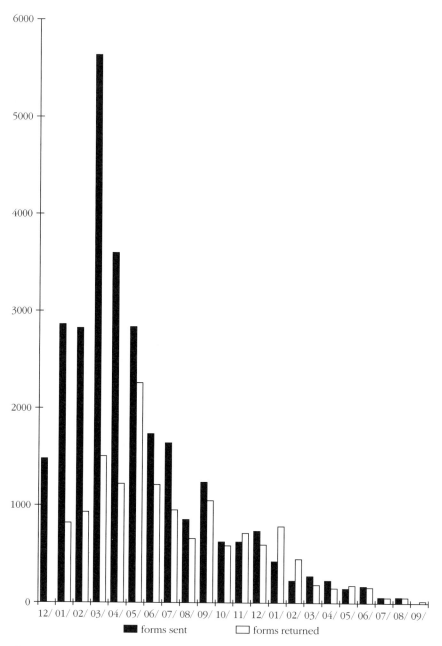

Figure 2.1 Graph of data collection: forms sent and received December 1993 to September 1995

and Midland Society for Genealogy and Heraldry (approximately 4,200 members), and smaller local societies such as the Huddersfield Family History Society with some 700 members. Most societies agreed to co-operate immediately and were happy to include details of the research in their journals. In some cases we also wrote longer articles to publicize the project and generate interest from family historians. There was no obvious reason why some societies did not co-operate, and because many people belong to more than one family history society (and receive several different newsletters) we actually received some responses from members of societies which did not themselves advertise the project. Societies from all parts of Scotland, England and Wales co-operated with the project, but we made no attempt to contact family history societies in Ireland. In retrospect this was, perhaps, a mistake as migrants from Ireland to England were under-represented in the final data set (see below). The publicity included in family history and genealogy magazines gave brief details of the project and simply asked family historians who had information on the residential histories of their ancestors to contact the researchers at Lancaster, stating the number of people for whom they had information.

All those who responded were sent the appropriate number of data entry forms (Appendix 3) and detailed instructions about what information was required. The data entry form had been revised several times following initial piloting with local family historians and in the light of responses from the pilot project with societies from Lancashire and Cumbria. The form required information on the full residential histories of ancestors born between 1750 and 1930 who had lived for at least part of their lives in Britain, and in addition requested details of all occupations and places of work, family structure at the time of migration, housing change, reasons for migration and other significant life course events (for instance marriage, childbirth or the death of a spouse). Including the pilot project, 2,420 individuals requested a total of 28,698 forms. By our deadline of 30th September 1995, 1,388 family historians and genealogists had returned 17,161 completed forms. Thus 57.4 per cent of those requesting forms returned information on at least one ancestor and 59.8 per cent of forms were returned. Figure 2.1 summarizes the flow of information from family historians during the main phase of data collection from December 1993 to September 1995. The pattern of requests for forms obviously reflects the dates at which relevant family history magazines were published with the peak months being March and April 1994. The graph of forms returned lags some two months behind requests with most forms returned during May 1994.

As forms were returned to Lancaster they were carefully checked for completeness and internal consistency, and inadequate forms were discarded. Our basic requirement was information on a complete residential history together with some additional information. We accepted forms where there was some missing data, for instance with regard to reasons for moves or occupations, but rejected all those where there were obvious gaps in the residential history, or where the information lacked internal consistency. In total 1,070 forms were not used (only 6.2 per cent of those returned). Overall, the quality of data returned by family historians and genealogists was very high, and it seems likely that the fairly complex and possibly daunting nature of the data entry forms had the desired effect of discouraging family historians with partial or poorly researched information. Those who completed forms clearly had a large amount of detailed information on their ancestors. During the project we did enter into extended correspondence with some respondents, and it is clear that a number of family historians undertook new research specifically to answer the questions posed on the forms. Some respondents commented that the project had required them to interrogate their data from a new angle, for which they were grateful. We also asked respondents to state the sources they used to produce their information, thus enabling us to check that the information provided could reasonably have been gleaned from the sources used. For instance, it was reassuring to find that reasons for migration were derived from a single documentary source in only 19.1 per cent of cases. Over 80 per cent of reasons for migration were deduced from oral evidence, family papers and correspondence or a combination of documentary sources (see below).

Between February 1994 and October 1995 information on 16,091 individuals who had made 73,864 moves was entered into three separate databases by a small team of research assistants. All data were entered into the computer in the form in which it was provided (but with standardization of addresses etc.), and in addition some codes were added to aid later analysis. Data entry was carried out on standard PCs using Microsoft Works (because it was widely available, easy to use and already known to our data-entry staff), but all files were then moved to the relational database INGRES, supported on the Lancaster University Sequent Symmetry computer. All places of residence and places of work within mainland Britain were assigned a six-figure Ordnance Survey grid reference (giving 1 km accuracy), and other places were given a coded identifier. The population sizes of all identified settlements within Britain containing more than 5,000 people were also recorded for four census dates (1801, 1851, 1891, 1951) and this information was added to the database. Initially the addition of grid references and town sizes had to be done manually using census volumes and gazetteers (a gazetteer published in 1972 was used to ensure the correct identification of places in relation to pre-1974 county boundaries), but we quickly established a series of updating files containing relevant data for all places encountered so that information could be added automatically for all locations apart from those new to the data set. Places which occurred for the first time were added manually to the updating file.

Most locations could be identified relatively easily, but there were a few very small settlements which were difficult to identify, and where the same place name occurred more than once within a county there was the potential for confusion. In such cases decisions were made in the context of other information on the form. When ascribing population sizes to settlements care was taken to use the appropriate administrative definition and, in large cities, adjoining suburbs which were not administratively part of the town were separately identified and coded so they could be linked to their parent city. London also posed particular problems. It was decided to use the 1891 County of London definition for most purposes, but we also identified all those settlements in surrounding counties which later became part of Greater London to allow us to reconstitute a larger definition of the metropolis. In addition to grid references and settlement sizes, separate updating files were also created for much of the coded information – so that codes could be added automatically – and straight-line distances between all origins and destinations in the database were calculated in INGRES and added to the database.

The process of data handling and establishing appropriate linked databases within INGRES was quite complicated: the basic structures of the procedures and resulting information are summarized in Figure 2.2. Three separate data tables were initially created within INGRES: a 'Name Table' which contained the attributes of all individuals for whom we had information; a 'Move Table' which contained information on all the residential moves undertaken by individuals for whom we had data (with each move entered as a separate case); and an 'Occupation Table' which contained data on the life-time employment history of all individuals who entered the workforce. Full details of the fields in all three tables are given in Appendix 4. Once created, the three tables were linked and numerous additional tables created to allow specific analyses to be undertaken (for instance if we wished we could create a table showing the distances of all moves from settlements of under 5,000 population undertaken by females born in Scotland between 1800 and 1820). The tables could be used to generate extremely complex queries and identify sub-sets of the data. Given the size of the data set (the largest single table contained 73,864 cases and 87 variables) it was essential to ensure efficient file management. For the most part we encountered few serious computational problems.

We used three separate but linked approaches to data analysis. First, much of the basic information required could be generated from queries within the relational database programme INGRES. However, most relational databases have only a limited capacity to handle statistical queries and thus most numerical variables were exported to the statistical package SPSS. This was used to produce more complex cross-tabulations and statistical analysis. As spatial analysis of locational data was a key component of the research it was also necessary to link the data tables in INGRES to a geographical information system (GIS) which could undertake both basic mapping and more sophisticated spatial analysis. ARC/INFO was used for all spatial analysis. For the purposes of this volume we have focused mainly

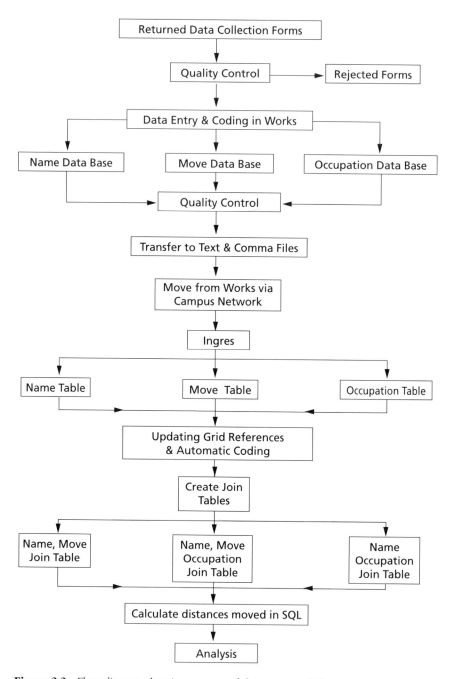

Figure 2.2   Flow diagram showing process of data entry and file structure

on relatively simple and straightforward analytic procedures. The way in which information was collected means that the data cannot be treated as a statistically representative sample (in terms of moves undertaken we have an unknown proportion of an unidentifiable total population). Thus, although we can make some judgements about biases in the data set, it is not appropriate to calculate levels of statistical significance. Because of the sheer volume of data, and the complexity of linking a series of attributes (moves) to individuals, we have concentrated on presenting clear tabular information which hopefully illustrates the key results of the research. Although there is some potential for using the longitudinal data for more sophisticated multivariate analysis and modelling of time series (Davies and Flowerdew, 1992; Courgeau, 1995), this approach has not been pursued in this volume.

It should be stressed that the data sets assembled from family historians, and on which most analysis reported in this volume is based, are truly unique. Although they cannot replace data derived from census and vital registration statistics for the calculation of net migration flows and the movement of population between large administrative units, they provide insights into the process of migration, and its impacts on the lives of individuals, which are otherwise only available for rather small and mostly atypical subsets of the population. Although we have been unable to conform to statistically valid sampling procedures, the data should be viewed as the historical equivalent of a very large questionnaire survey in which some 17,000 people were interviewed about all the residential moves they made in their lives. Although much of the information provided by family historians is necessarily secondhand, in that it is culled from documents not originally designed to yield data on issues such as the reasons for migration or household composition when a residential move took place, it does pertain directly to the lives of individuals, and it has been collected by family researchers who have often had access to very detailed family papers and related documents as well as oral evidence. As the analysis presented in the remainder of this volume should show, we believe that there is considerable potential for fruitful collaboration between academic researchers and family historians.

There is one major danger inherent in amassing a large volume of data: the analysis of aggregate characteristics may mask the voices of individual people. The great value of this data set is that, whilst information can be aggregated to produce trends over time and space, it is composed of the life histories of 16,091 separate individuals. Even within the confines of the data forms on which family historians were asked to record their ancestors' residential histories, many vivid and illuminating individual stories emerge from the aggregate data. Throughout the volume extensive use is made of individual examples, selected from many that could be chosen, to illustrate particular themes or events. These case studies add flesh to the bare bones of aggregate analysis. In choosing particular individuals to illustrate a point we are not arguing that these people are necessarily statistically representative of a larger sample, rather that their personal stories exemplify specific commonly occurring themes amongst the residential histories. All the individual life histories cited have been anonymized, though some readers may recognize their ancestors in the following pages.

During the project we have also acquired twelve detailed diaries and life histories written at various times in the past. These have been very kindly provided by family historians with whom we have corresponded (Appendix 1). Such documents provide insights into the past which cannot be achieved through other sources. All the diaries and life histories studied contain detailed personal information about residential moves of the individuals involved and (sometimes) of their relatives. Diarists often recorded their detailed reasons for moving, their fears when leaving a familiar area, the process of establishing a new home, and their thoughts about places left behind. The wealth of information provided in such sources means that they can be given only passing reference in this volume, and some merit separate detailed studies in their own right (Pooley and Turnbull, 1997a), but used selectively they can add immeasurably to the interpretation of other evidence on migration in the past. Where appropriate, evidence drawn from the diaries and life histories provided by family historians (together with some other similar sources previously used by the authors) is woven into the narrative to aid interpretation and illustrate specific themes. The authors of life histories have not been anonymized and explicit permission to quote from unpublished manuscripts has been sought in each case. We are certainly not claiming that such diarists are necessarily typical. By virtue of the fact that they wrote their diaries or life histories, often in an age when many were illiterate, they were almost certainly not typical. But this does not negate or devalue the individual experiences of migration and mobility which come through from the pages of such sources and, collectively, they enrich the analysis and interpretation of residential change in the past.

Assessing the representativeness of the data set

As this is the first time that data provided by family historians and genealogists have been used in a large scale academic research project, it is essential that any biases within the data set are fully assessed. As outlined above the data were provided by a self-selected group of respondents who themselves chose who to provide information on. We have thus had no control over either the selection of the sample or the historical research which has generated the data. All we can do is attempt to assess the representativeness of the population for which we have information by comparing their characteristics with census evidence. Some of the likely biases arising from the motives of family historians and the limitations of historical sources were outlined above. This section critically examines the data set, demonstrating

Table 2.1 Distribution of migration data by birth cohort

Date of birth	Number of people who moved	Number of people who never moved	Total number of people	Total number of moves
1750–59	280	98	378	699
1760–69	325	82	407	859
1770–79	410	94	504	1,074
1780–89	573	66	639	1,861
1790–99	805	81	886	2,605
1800–09	929	75	1,004	3,252
1810–19	1,071	55	1,126	4,093
1820–29	1,252	54	1,306	5,258
1830–39	1,279	29	1,308	5,625
1840–49	1,177	34	1,211	5,455
1850–59	1,170	18	1,188	5,534
1860–69	1,260	18	1,278	6,607
1870–79	1,048	17	1,065	5,768
1880–89	979	15	994	5,642
1890–99	1,060	13	1,073	7,084
1900–09	829	5	834	5,613
1910–19	448	3	451	3,300
1920–30	437	2	439	3,535
Total number	15,332	759	16,091	73,864

Data source: 16,091 life histories provided by family historians.

the extent of any biases, and assessing the impact this may have on future analysis and interpretation. It should be stressed that in highlighting these biases we are in no way criticizing the quality of information provided by family historians. Most biases are an inevitable outcome of the sources available and the objectives of family historians and genealogists. None of the biases invalidate the data set, but they do influence interpretation.

A problem common to all research which spans a period of over 200 years from the eighteenth century to the present is the fact that the quantity and quality of data available for the recent past is much better than that for more distant time periods. This is reflected in the distribution of data for different periods of time (Table 2.1). Although the data set contains a considerable number of people born in the period 1750–1819, there is a larger concentration of information for individuals born in the mid- and late-nineteenth century when a wider range of sources (especially census and vital registration data) are available to aid the family historian. Interestingly, relatively fewer people in the data set were born in the period 1890 to 1930. This cannot be due to a lack of information, but probably reflects family historians' greater interest in more distant relatives and a possible reluctance to provide data on those still living.

To assess the representativeness of the data set four cross-sections were identified: all those alive in 1801, 1851, 1891 and 1951. This yielded sub-sets of the data set containing 2,363 people in 1801, 6,607 in 1851, 7,421 in 1891 and 4,061 in 1951. These samples were then compared with the characteristics of the whole population of Britain derived from the relevant censuses. It should be noted that there are bound to be some discrepancies because of the cut-off points imposed by our own data collection procedures. We collected information only on those born between 1750 and 1930: thus in 1801 it is not possible for our sample to include anyone older than 51 and in 1951 the sample cannot contain any individuals younger than 21 years. Selected characteristics of the four cross-sections are compared with census data in Table 2.2.

There is a clear gender bias in the data set which becomes less marked in the twentieth century. Thus only 21.8 per cent of those in our sample who were alive in 1801 were female, whereas in 1951 44.2 per cent were female. This was anticipated, and is due both to the difficulty of identifying women in many historical sources (especially in the more distant past) and the fact that there is a tendency for family historians to follow surnames down the male line. The fact that we have information on a much higher proportion of women in the twentieth century, when oral evidence is available, suggests that the male dominance owes more to source problems than to a male bias among family historians. The fact that approximately 56 per cent of our respondents were female may partially explain the higher number of women returned in the twentieth century.

There are also discrepancies in the age profiles of the sample and census populations, though because detailed age structures were not recorded in the 1801 census comparisons are only made at three census dates. Concentrating on comparisons in 1851 and 1891 (which are not affected by the fact that we only collected data on those born 1750–1930), the sample from family historians has a markedly older age profile than the population of Britain as a whole. Although the proportion in the sample aged 20–39 years is quite close to that recorded in the census, in 1851 only 32.9 per cent were aged under 20 compared with 45.1 per cent in the census; whereas 33.8 per cent were over 40 compared to only 24.0 per cent in the census. There is a simple explanation for these differences. Nineteenth-century Britain had a predominantly young age structure, but with very high rates of infant and child mortality (infant mortality was around 150 per thousand in the 1850s (Woods, 1992, p. 29)). Thus many children failed to survive to adulthood, though those that did had a reasonable chance of living to old age. Because children who died young probably appeared in few historical sources (apart from birth and death registers), and because family historians are (by definition) primarily interested in those who survived to adulthood and had children to create a family line, we received very little information on those that died young. It is also likely that even when family historians had relevant

Table 2.2 Comparison of census and sample data by gender, age, marital status, region of residence, town size and principal town of residence (%)

Population characteristics	1801 Census	1801 Sample	1851 Census	1851 Sample	1891 Census	1891 Sample	1951 Census	1951 Sample
Gender								
Females	52.5	21.8	51.3	26.9	51.6	33.8	52.0	44.2
Age								
<20	–	55.6	45.1	32.9	45.3	26.3	28.7	–
20–39	–	31.8	30.9	33.3	30.3	32.9	28.8	21.2
40–59	–	12.6	16.7	22.8	17.0	26.1	26.8	42.6
60+	–	0.0	7.3	11.0	7.4	14.7	15.7	36.2
Marital Status								
Single males	–	57.2	63.0	31.1	62.6	27.0	20.4	5.3
Single females	–	55.8	60.4	28.8	60.0	29.3	19.6	9.0
Married males	–	40.0	33.2	64.6	34.0	68.6	74.0	88.0
Married females	–	33.4	32.2	62.3	32.4	62.8	65.8	71.0
Region of residence								
London	9.1	3.7	11.4	8.3	12.8	11.7	6.9	7.6
South East	8.4	9.4	7.8	8.4	10.2	8.2	11.8	12.3
South Midland	6.8	8.7	5.9	6.4	3.9	5.5	9.4	11.9
Eastern	6.6	8.2	5.4	6.8	4.9	5.4	6.2	6.5
South West	10.5	10.3	8.7	9.1	5.6	5.4	4.9	4.1
West Midland	10.4	9.2	10.2	7.5	9.8	7.7	10.8	8.0
North Midland	6.2	8.4	5.8	7.3	5.6	7.0	6.2	6.7
North West	8.3	9.0	12.0	13.2	14.3	15.4	13.3	16.1
Yorkshire	8.1	10.3	8.6	9.1	9.7	10.2	8.4	9.0
Northern	4.6	7.5	4.8	7.2	5.6	7.5	6.4	5.5
Wales	5.7	4.6	5.7	4.9	5.4	5.2	5.3	4.7
Scotland	15.3	10.7	13.7	11.8	12.2	10.8	10.4	7.6

Table 2.2 (cont'd)

Population characteristics	1801 Census	1801 Sample	1851 Census	1851 Sample	1891 Census	1891 Sample	1951 Census	1951 Sample
Settlement size of residence								
<5000	77.0	85.1	59.9	60.1	43.1	37.3	22.1	30.3
5000–9999	4.8	4.1	5.6	7.1	6.3	7.2	4.6	4.2
10000–19999	4.0	3.1	4.1	5.9	7.9	7.4	7.4	6.1
20000–39999	2.5	1.9	5.9	5.5	9.6	7.8	9.4	8.6
40000–59999	0.4	0.1	2.4	2.1	5.9	4.4	8.2	6.6
60000–79999	1.9	1.8	2.5	3.3	2.0	2.3	6.3	5.8
80000–99999	1.5	0.2	0.4	1.2	2.2	2.2	3.6	3.2
100000+	7.9	3.7	19.2	14.8	23.0	31.4	38.4	35.2
Principal town of residence								
Birmingham	0.6	0.5	0.8	0.4	0.7	0.7	2.3	1.6
Bristol	0.4	0.6	0.3	0.4	0.3	0.8	0.9	0.5
Bradford	0.1	0.1	0.3	0.2	0.2	0.6	0.6	0.4
Glasgow	0.7	0.0	1.7	0.7	1.9	1.1	2.2	1.2
Liverpool	0.7	0.4	1.2	1.3	1.5	1.6	1.6	1.3
Nottingham	0.3	0.4	0.3	0.6	0.5	0.7	0.6	0.3
Sheffield	0.3	0.2	0.4	0.4	0.3	1.2	1.0	0.9
Leeds	0.3	0.1	0.5	0.5	0.5	0.8	1.0	1.0
Edinburgh	0.8	0.2	0.9	0.5	0.8	1.2	1.0	1.3
Sample size		2,363		6,607		7,421		3,249

The characteristics of data collected from family historians are affected by the year of birth stipulated for people to be included in the sample. Data were collected for people born 1750–1930 thus particularly affecting sample characteristics in 1801 and 1951. Only limited comparative data are available from the 1801 census of Great Britain.

Data source: 16,091 life histories provided by family historians.

Table 2.3 Mean age of death (years) of sample by birth cohort

Date of birth	1750–1819	1820–49	1850–89	1890–1930	All
Age at death	69.9	68.6	72.6	74.4	71.3

Data source: 16,091 life histories provided by family historians.

Table 2.4 Comparison of census and sample data by marital status for those age over 20 years

Population characteristics	1801 Census	1801 Sample	1851 Census	1851 Sample	1891 Census	1891 Sample	1951 Census	1951 Sample
Single male	–	27.4	30.9	17.2	29.8	14.2	26.9	4.8
Single female	–	12.6	29.5	14.9	28.9	12.9	27.9	7.5
Married male	–	69.2	62.0	77.0	63.8	80.2	67.3	87.1
Married female	–	84.7	57.3	73.8	57.6	76.9	59.6	66.3
Widowed male	–	3.4	7.1	5.8	6.4	5.6	5.8	8.1
Widowed female	–	2.7	13.2	11.3	13.5	10.2	12.5	26.1

Data source: 16,091 life histories provided by family historians.

Table 2.5 Percentage of sample never married by birth cohort

Date of birth	1750–1819	1820–49	1850–89	1890–1930	All
Percentage never moved	1.8	2.7	3.5	5.4	3.1

Data source: 16,091 life histories provided by family historians.

information, they decided not to send data on those who died as children or young adults because they (perhaps wrongly) deemed such short lives to reveal little of interest. The bias towards those that survived is confirmed by the average age of death of the sample populations (Table 2.3), with a mean age of death of about 70 years for those born in all time periods. These figures can be compared with Victorian life tables which reveal a mean life expectancy at birth of only around 40 years in the mid-nineteenth century (Woods, 1992, p. 29).

The data provided by family historians are also biased towards those who were married and (to a much lesser extent) towards widows (Table 2.2). Whereas in 1891 60.0 per cent of females in Britain were unmarried only 29.3 per cent of our sample were single. This discrepancy is mainly an inevitable consequence of the age structure of the sample, as the majority of the unmarried population were children. However, there is a smaller bias towards married men and women amongst the adult population (Table 2.4). The proportion of the sample who never married is also less than it should be, with only 2.7 per cent of the sample born 1820–49 never marrying (Table 2.5) compared to around 12 per cent of the total population of

Table 2.6 Comparison of census and sample data by country of birth (%)

Country of birth	1801 Census	1801 Sample	1851 Census	1851 Sample	1891 Census	1891 Sample	1951 Census	1951 Sample
England	–	82.1	77.2	81.1	80.2	80.5	89.7	83.1
Scotland	–	12.0	13.2	12.1	12.0	11.9	0.9	4.7
Wales	–	4.5	5.0	4.6	4.5	5.1	6.6	9.5
Ireland	–	0.6	3.5	1.0	2.0	1.1	1.0	0.6
Island off Britain	–	0.1	0.6	0.1	0.1	0.1	0.1	0.2
Overseas	–	0.7	0.5	1.1	1.2	1.3	1.7	1.9

Data source: 16,091 life histories provided by family historians.

Britain who never married in the mid-nineteenth century (Wrigley and Schofield, 1981, p. 437). It can be suggested that this bias is again due to family historians concentrating on those ancestors who married and had children to form a family line. Unmarried relatives were ancestral dead-ends who were less likely to be researched in detail.

The geographical distribution of the sample population according to the census region of residence, and region of birth, is quite close to that revealed by the census, as too is the distribution between major provincial towns (Tables 2.2, 2.6). However, those born in Ireland are under-represented in the sample. This is probably due to four main factors: first, our decision to exclude Irish family history societies; second, the fact that we were not recording details of movement within Ireland may have deterred some family historians from sending details of those who moved from Ireland to Britain at some time in their lives; third, difficulties with Irish sources (particularly the lack of nineteenth-century census enumerators' books (Graham and Proudfoot, 1993)) may have deterred some family historians from tracing their Irish ancestors; and, fourth, the fact that many Irish migrants were poor meant that they would have been harder to trace within Britain (see below). Smaller discrepancies within Britain include an under-representation of those living in London in 1801, an over-representation of those from Northern and (to a lesser extent) North-West England, and an under-representation of those living in Scotland (though those born in Scotland are over-represented in 1951). People living in the South Midlands were also over-represented in 1951. Some of these variations are random fluctuations inevitable with the relatively small sample sizes for some regions and time periods, but others can be explained. The over-representation of Northern and North-West England is probably due to the fact that we had more contact with societies close to Lancaster:, the pilot project focused on these areas and thus family historians had a longer period in which to return information. The under-representation of London may relate to the difficulty of tracing people in a very large city (especially in the eighteenth and early-nineteenth centuries), and other variations reflect differences in response rates from particular societies. Thus although we had a very good response from the Aberdeen society we received relatively little data from much of Highland Scotland.

Even though the sample was broadly representative of the regional distribution of the British population, it is possible that the sample was biased towards rural or small-town populations. It can be argued that it is easier to trace ancestors in small communities. However, with the exception of the under-representation of London in 1801 (and to a lesser extent in 1851) outlined above, the settlement size distribution of the sample population was quite close to that in the census. In 1891 large towns were actually over-represented in the sample (for no obvious reason) and overall the proportion living in small places of under 5,000 people was close to the census figure (Table 2.2). Both in terms of regional characteristics, and the distribution of population through the urban hierarchy, the sample provides a good representation of the population as a whole.

It is more difficult to assess the extent to which the sample represents the socio-economic characteristics of the total population as, although nineteenth-century censuses provided detailed occupational information, these were based on industrial classes which each contained a variety of socio-economic groups, and which can only be compared for broad industrial categories (Armstrong, 1972, Bellamy, 1978). However, in the twentieth century a standard socio-economic profile can be applied (based on the 1951 Office of Population Censuses and Surveys classification). As shown in Table 2.7, although the sample contains people from all socio-economic categories, there is a bias towards those in higher status occupations. Although the proportion in skilled manual and non-manual occupations is largely representative, there is a large over-representation of those in professional and higher managerial categories and an equally large under-representation of the unskilled. This bias was anticipated and is primarily a function of the difficulty of tracing poor people in historical sources. Those who owned property, left wills, had obituaries written about them and were literate were more likely to appear in official documents and leave written records of their lives. It can also be suggested that family historians are themselves primarily drawn from professional and middle-class groups. Although many of them will have had ancestors drawn from all classes of society, this may account for the particularly large over-representation of professional occupations in 1951.

In addition to biases that can be assessed by comparing sample and census data, there are a number of other potential problems that can only be ascertained more subjectively. The relative paucity of information for the eighteenth and early-nineteenth centuries almost certainly means that some less visible short-distance moves were never discovered for the more distant past. The extent and implications of this are explored in more detail in Chapter 3. Undoubtedly the most difficult area in which to collect accurate

Table 2.7 Comparison of census and sample data by selected occupational and socio-economic groups (%)

Occupational group	1851 Census (over 20 yrs)	1851 Sample (over 20 yrs)	1891 Census (over 10 yrs)	1891 Sample (over 10 yrs)
Public service/professional	1.5	4.3	3.7	7.1
Defence	0.8	0.8	0.4	1.3
Iron/steel	2.2	1.6	1.8	2.0
Agriculture/fishing	16.1	14.1	6.3	7.2
Textiles	6.8	3.0	5.3	2.2
Domestic service	6.5	1.5	6.9	3.1
Sample size		6,607		7,421

Registrar General's Group	1931 Census (employed males)	1931 Sample (employed males)	1951 Census (employed males)	1951 Sample (employed males)
1 Professional	2.2	9.4	3.1	15.9
2 Intermediate	12.9	19.6	13.7	19.5
3 Skilled	48.9	48.8	52.6	46.5
4 Semi-skilled	18.2	15.0	16.0	13.2
5 Unskilled	17.8	7.2	14.6	4.9
Sample size		2,637		1,822

Data source: 16,091 life histories provided by family historians.

Table 2.8 Summary of sources used by respondents to determine reasons for migration (%)

Reason for move	Standard single documentary source	Oral evidence/ personal recollection	Family papers/ work records	Combination of sources
Work	21.5	26.0	7.6	44.9
Marriage	33.8	16.0	3.4	46.8
Family	15.2	41.0	5.8	38.0
Housing	14.0	41.5	4.1	40.4
Crises	16.0	35.4	6.2	42.4
Retirement	10.6	40.0	8.3	41.1
Other	9.0	37.0	12.4	41.6
Total %	19.1	30.5	7.0	43.4
Total number	9,025	14,415	3,326	20,517

Data source: 16,091 life histories provided by family historians.

information relates to the reasons why people moved. Ascribing motives to migration is difficult even in the present (Champion and Fielding, 1992), and it is reasonable to assume that at least some of the reasons stated by family historians for the pre-oral past were based as much on conjecture as on firm evidence. As noted above, we did ask respondents to indicate the sources they used to ascertain different items of information, and this allows us to make some assessment of the extent to which data on reasons for migration could reasonably have been deduced from the evidence available. For instance, if the only evidence was from the census enumerators' books it is reasonable to assume that the stated reason was little more than conjecture based on circumstantial evidence. However, if a combination of sources were used, especially family papers or letters, it is reasonable to assume that the stated reasons were based on firmer evidence. Table 2.8 shows that overall only 19.1 per cent of reasons for migration were deduced from standard single sources (such as the census or vital registration documents). Moreover, the reason which was most likely to be stated based on a single source was movement for marriage which can reasonably be ascertained from a marriage certificate. For all stated reasons a majority of family historians used a variety of sources, and a combination of oral information passed through the generations and family papers provided evidence for over one third of all stated reasons (especially movement for family, housing and retirement reasons). Although it is impossible to know with certainty why most people moved in the past, there are strong reasons to suggest that the data provided by family historians are based on the best available evidence.

There are, of course, some aspects of the migration process which it is impossible to deduce from any sources. For instance, although family historians may be able to state reasons why someone moved with a fair degree of certainty, we rarely if ever know how that decision was taken, the range of alternative locations that were considered, and the grounds on which potential destinations were accepted or rejected. In other words, as with much other historical research, our evidence relates primarily to outcomes – the fact that someone moved from one place to another to gain a different job – but we know little of the processes underlying particular outcomes. Even in the present it is difficult to research and understand the ways in which residential choice takes place. In the context of migration and mobility in the past this is an issue on which we can rarely say anything with complete certainty. Some individual migration histories, and especially diaries and life histories, provide some clues, but all too often the evidence is silent on processes of motivation, decision making and choice. Moreover, there is rarely direct evidence on the meaning of migration for individuals. This issue is returned to in Chapter 10.

Problems of interpretation

In conclusion, it is necessary to assess the implications of the biases outlined above for analysis and interpretation of the data set. To what extent do the biases create problems and invalidate conclusions? As stated earlier, the data could never be a statistically representative sample of the British population, and the data are not being used for inferential purposes or for the calculation of migration rates. Thus we have no intention of using the sample to make statements such as 'x' per cent of the population migrated from A to B in the 1850s. At a broad level of aggregation such evidence is, in any case, already available from published census tabulations. The data were never intended to be used in this way and it would be misleading to try to do so. The strengths of the data set are in the evidence they provide on the characteristics of migrants, on migration differentials, on the ways in which migration relates to the life-course, on spatial and temporal variations in such characteristics and on the reasons why people moved over long and short distances. In this context there are two important aspects of representativeness to be assessed. First, are the samples of particular sub-groups of the population large enough to make meaningful statements about migration characteristics and behaviour within those sub-groups and, second, do those excluded from the sample have migration characteristics which were significantly different from those included? Thus, it does not matter unduly if the proportion of women in the sample is 33 per cent instead of 50 per cent, but it does matter if the total number of women in particular categories is very small, and if those women missing from the sample come disproportionately from particular areas or have distinctive characteristics which affected their migration behaviour.

It can be suggested that the under-representation of women in the sample is not a major problem. In total we have information on 5,158 women who made 25,304 residential moves. This is a very large sample – particularly considering the difficulties of gaining data on women in the past – and there is no evidence that the distribution of women between areas, family circumstances or social groups is any different from the distribution for men. We can analyze the characteristics of female migrants with confidence, aways bearing in mind the biases outlined above which affect both men and women. The significance of biases in the age profile of the sample population is more difficult to assess. Because all our data relate to life-time residential histories we have a large amount of information on moves made by people of all ages (almost everyone who lived to be 70 also moved as a child or young adult). Thus, in total, 15,726 moves were made by those under 20. However, we do not know the extent to which moves made in childhood by people who eventually lived to an old age may have been different from moves made in childhood by those who died young. The data may reflect socio-economic and locational biases, in that mortality rates were higher amongst the poor in central urban areas, but in other respects there is no reason to believe that the childhood mobility of those who lived was any different from that of those who died young. In most families from all classes some children died whilst others lived, and it is reasonable to assume that – insofar as the mobility experience of any two people is similar – the adult mobility of those who died young would have mirrored that of their siblings who survived.

Of more consequence is the bias towards those who ever married. Although the longitudinal evidence provides information on a very large sample of people who at some time in their lives were single and moved alone (21,976 moves by single people and 11,780 moves undertaken alone), the majority of these people did eventually marry. The sample of those who remained single all their lives is much smaller (448 people), and given the importance of family and life-course events in structuring mobility it is likely that the mobility patterns of this group were different from those of the majority who married. However, although the sample is small and cannot easily be broken down into sub-categories, we do have some data on this group. The mobility of those who remained single is explored in more detail in Chapter 6 and compared to the migration patterns of those in different family circumstances. The same assumptions apply to the biases by socio-economic group. Although those in unskilled industrial occupations are under-represented in the data set, this group did make a total of 1,750 moves. Although the sample is too small to break down into many sub-categories, the mobility characteristics of different occupational and social groups can be compared (Chapter 5).

Although there are relatively few biases in the geographical distribution of the data set, there are severe limits to the extent to which the sample can be sub-divided into small geographical areas and time periods. Thus, although (for instance) the number of people in the sample born in Lanca- shire and Cheshire in the period 1750 to 1819 is quite large (2,044), when sub-divided into much smaller geographical units or time periods the num- bers become too few to be useful. Thus it is not possible to use the data to explore migration and mobility to and within particular villages and/or short time periods as the number of relevant people in the sample will be tiny. Thus the sample contains only 37 people born in the town of Lancaster in the period 1750–1819. Because most family historians returned information on several ancestors who may have been spatially clustered, and probably undertook some moves together, it is likely that in small places many of the people for whom we have information will have been related and thus their moves may have had shared characteristics. For these reasons, despite the very large total sample (16,091 people and 73,864 moves), spatial and tem- poral analysis is carried out within quite large aggregated units. However, local detail and individual perspectives are added through the use of diaries and selected life histories.

We do not claim that the data on which analysis in this volume is based are free from bias, or from the inherent problems of almost all historical sources. However, we are able to demonstrate the extent and nature of the biases, to assess the extent to which they may influence later analysis, and to take these factors into account in later interpretation. Given that all migration decisions in the past or present are based on a complex web of factors which come together in a particular time and place in ways which may be poorly understood even at the time, we must be aware that any information on migration in the past is one of several possible views of what occurred, why it occurred, and how it affected the lives of people involved. We do not claim that the data we have analyzed present the only, or even the most important, view of migration in the past. However, we do believe that the voices from the past provided by family historians deserve to be heard, and that collectively they provide unique insights into migra- tion and mobility in the past.

CHAPTER 3

Where people moved: the spatial and temporal pattern of internal migration in Britain

Introduction

Despite the large number of migration studies outlined in Chapter 1, there remains a basic lack of information about the ways in which the direction and characteristics of migration changed over time and space from the late-eighteenth century. This chapter, therefore, provides an introductory description of the pattern of migration in Britain investigating a series of questions, including the following. What were the characteristics of migrants and non-migrants? Where did people move from and to? How frequently did individuals move home? What distance did people move? Why did individuals and families decide to move house? The chapter investigates the extent to which these features varied over time, between British regions, and from one migrant to another depending on personal characteristics. Later chapters will focus in more detail on the process of migration, in an attempt to explain these patterns and investigate the impact of migration on people and its significance for wider society.

Analysis presented in this chapter is based almost entirely on the data collected from family historians. The methods used to collect this information, its strengths and limitations, have been described in Chapter 2. Despite the very large data set (16,091 individual migrants who made a total of 73,864 separate moves), when the data are broken down into regions, time periods and migrant characteristics, some sub-groupings can become quite small and problems of representativeness may occur (see Chapter 2). The main value of this information is thus at the regional scale, where broad trends can be identified within reasonably large time periods. In later chapters, specific information is selectively examined at the local scale and for particular periods of time (such as during war time) when there were unusual societal influences on migration.

The spatial framework used for analysis in this chapter consists of the 20 census regions used in the 1891 census. Temporal analysis can be undertaken either in relation to the date on which a move was undertaken, or the date

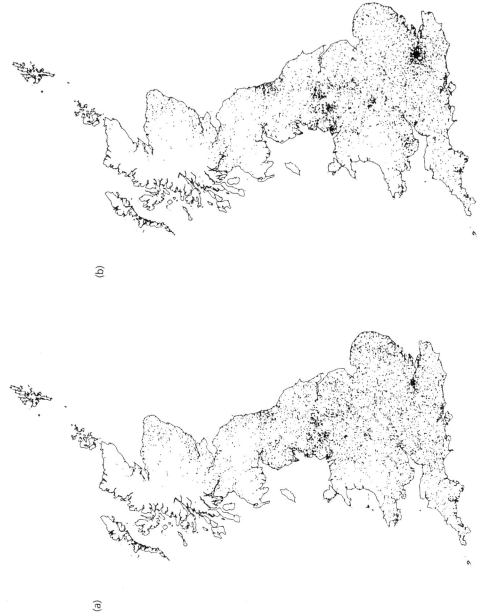

(b)

(a)

Figure 3.1 All migration origins, (a) 1750–1879, (b) 1880–1994

of birth of the migrant. There are arguments for and against analysis by both migrant birth cohorts and date of move. As the main focus of this chapter (and most of the book) is on the way in which mobility changed over time, most analysis is undertaken by dividing moves into four broad time periods which, it can be suggested, coincide with key periods of social, economic, demographic and cultural change in Britain and which, in turn, had implications for the pattern and process of migration. These are a period of sustained industrialization and urbanization in the early phase of the demographic transition from 1750–1839; a mid-Victorian period from 1840–79 which saw sustained economic growth, in which urban living became the dominant feature of life, and much concern was expressed about living conditions and high mortality in towns; a period from 1880–1919 which saw the demographic transition towards lower mortality and fertility, the beginnings of economic restructuring as older industry declined and newer sectors emerged, and the growth of new levels of social and political awareness; and a final period from 1920 which experienced more complete economic restructuring, massive increases in personal mobility and the emergence of a more complete system of welfare provision (Mathias, 1983; Evans, 1983; Robbins, 1983; Lawton and Pooley, 1992; Hoppit and Wrigley, 1994). In addition, certain analyses, which relate to the life-time experience of individual migrants, are more sensibly examined in relation to the birth cohort of the migrant. For this purpose migrants have simply been divided into four groups of approximately equal size: those born 1750–1819; 1820–49; 1850–89; 1890–1930. Although it is possible to argue about the appropriateness of any of the above dates, the analyses are extremely robust, and small changes in the time periods used make little difference to the results.

The national migration system: where did people move?

The data set of 73,864 separate moves can be analyzed in a large number of different ways. We begin by simply evaluating the spatial pattern of moves generated, before exploring more of the characteristics of the data later in the chapter. Explanation of how and why people moved is largely left to later chapters. Not surprisingly, and rather reassuringly, a map of all the origins or destinations occurring in the data base is very similar to a map of population distribution in Britain (Figure 3.1). Any attempt to map all migration flows simply produces a large inkblot but, in doing so, demonstrates the extent to which the migration experience covered by the data covers all parts of the country. Not surprisingly, most migration occurred where most people lived, and the characteristics of migration will inevitably reflect the make-up of the areas in which migrants were located.

If maps of the total migration system are divided into a series of time periods, with shorter time spans for those periods for which we have most data and during which there was likely to have been rapid social and economic change, it is possible to say something about the way in which the British migration system changed over time and space (Figures 3.2–3.5). These maps obviously do not represent the total migration system. They can only show longer distance moves (a minority of all moves), and short distance intra-urban mobility cannot be represented at this scale. The maps also do not indicate the direction of each move. The features that emerge most clearly are the relative paucity of long distance moves in the eighteenth century, the increasing development of large urban areas as the foci for migration activity, the dominant role of London in the British migration system in all periods, the declining importance of movement from rural areas as rural depopulation began to take effect, the increasing significance of long distance interchange of population between large urban centres, and the particular importance of local circulation at all time periods and in all regions. Overall, all 16 maps are quite similar, suggesting that migration patterns were quite stable over time.

The maps also suggest a complex pattern of local and national population movement, with all parts of the country to some extent related to each other, but with significant local migration networks. Changes over time also suggest ways in which population dynamics, including the volume and direction of migration, may have been associated with national and regional patterns of economic and social change. Such features, hinted at in the national maps, will be explored in more detail in subsequent sections. Although, at one level, such maps merely confirm the pattern of inter-county flows demonstrated by previous census-based analyses (Smith, 1951; Lawton, 1958; Friedlander and Roshier, 1966), they add to such studies a much more complex pattern of local and sub-regional movement which cannot be revealed from census data. It is the inter-play between such local and national migration systems which forms the focus for much of the subsequent analysis.

The national migration system: how often did people move?

It is impossible to assess the extent to which the data provided by family historians captured every move undertaken by an individual. However, there are a number of common-sense checks and balances which need to be taken into account when interpreting information on the frequency with which people moved in the past. First, it seems reasonable to assume that data relating to mobility in the twentieth century are reasonably complete. Information is drawn mainly from oral evidence or recent memory, and it is most likely that all residential moves have been recorded. Second, it is equally obvious that information drawn from pre-oral records is likely to

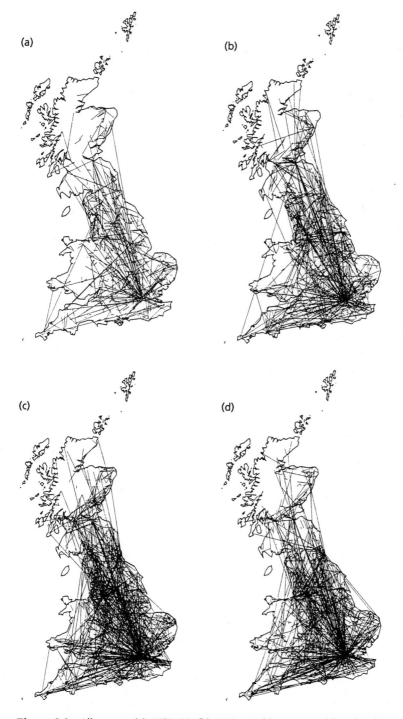

Figure 3.2 All moves, (a) 1750–99; (b) 1800–19; (c) 1820–39; (d) 1840–49

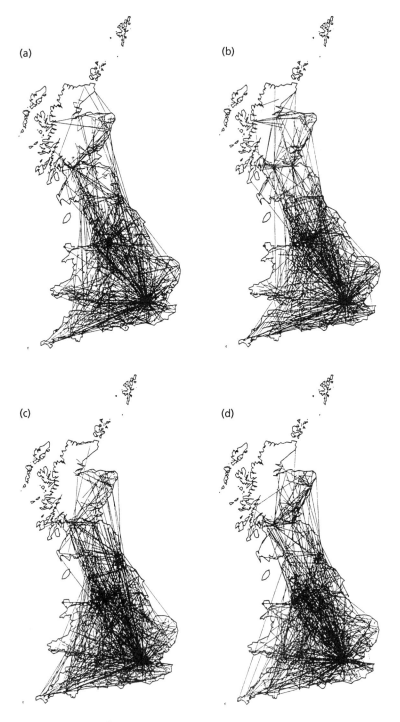

Figure 3.3 All moves, (a) 1850–59; (b) 1860–69; (c) 1870–79; (d) 1880–89

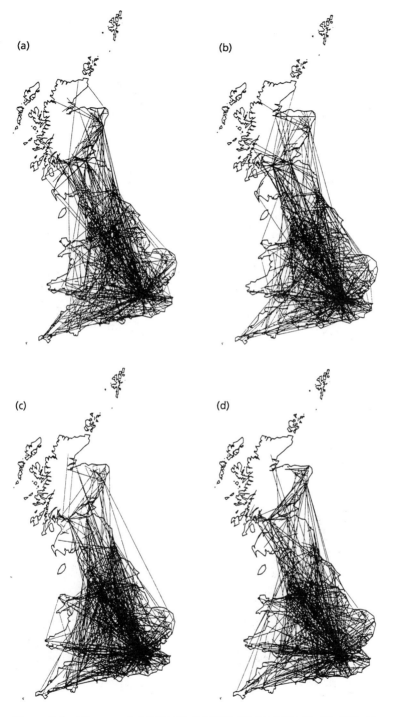

Figure 3.4   All moves, (a) 1890–99; (b) 1900–09; (c) 1910–19; (d) 1920–29

Figure 3.5 All moves, (a) 1930–39; (b) 1940–49; (c) 1950–59; (d) 1960–94

Table 3.1 Mean number of moves and mean length of residence by gender and birth cohort

	1750–1819	1820–49	1850–89	1890–1930	All moves
Mean number of moves					
Male	3.3	4.5	5.2	7.1	4.7
Female	3.2	4.2	5.4	7.0	5.0
All moves	3.3	4.4	5.3	7.0	4.8
Mean length of residence					
Male	13.5	10.9	10.7	8.0	10.7
Female	14.2	11.6	11.2	8.8	10.9
All moves	13.6	11.1	10.9	8.4	10.8

Data source: 16,091 life histories provided by family historians.

understate mobility. Family historians piece together their ancestors' movements from a variety of sources which are mostly cross-sectional in nature. Although diaries and family papers may fill in some gaps, these do not exist for all people, and even diaries may be incomplete. It is likely that most longer distance moves, and moves to addresses at which the individual remained for several years will have been recorded. However, it is probable that some short distance moves within settlements, or addresses at which a family lived for only a short time will have been omitted. Third, it is probable that the degree of under-representation of short distance and short-term mobility will have been greater for the eighteenth and early-nineteenth century than for later periods. The availability of census enumerators' books, reliable trade directories and similar sources make it more likely that individuals were traced at all their addresses. Nominal sources that allow reliable record linkage are less commonly available in the eighteenth century. Assessment of the sources family historians used in tracing their ancestors thus suggests that the quality of information on the frequency of mobility will improve with time and that, for the eighteenth century especially, short distance intra-settlement moves and moves of short duration are under-represented.

The mean number of moves undertaken by all individuals in the data set during the course of their life-time was 4.8, with a mean length of residence of 10.8 years at any one address. There was little difference between males and females, but there is a marked increase in the frequency of mobility over time. Thus those born 1750–1819 moved on average 3.3 times during their life-time whereas those born 1890–1930 moved 7.0 times on average. Mean length of residence at an address fell from 13.6 years for the first cohort to 8.4 years for the cohort born after 1890 (Table 3.1). These data are thus consistent with the biases that are assumed to exist in the data set, but there may also be good historical reasons why the frequency with which people moved changed over time. These are discussed below. Mean figures obscure considerable variations between individuals, and the data set contain a range of mobility experiences from those who remained in the same house all their lives to others who moved twenty or more times during their life-time (Table 3.2). The data returned by family historians has certainly picked up very frequent mobility for some individuals.

One way in which the validity of the longitudinal records provided by family historians can be assessed is by comparing them with the small number of diaries which are available, and which are known to give a complete record of either life-time mobility or, at least, of movement during the period of the diary. Such evidence as is available suggests that whilst some individuals experienced very high levels of mobility, others moved only infrequently. At one end of the spectrum, Henry Jaques moved 25 times during his life in and around London between 1842 and 1907 and David Brindley moved home 12 times in just eight years in Liverpool in the 1880s. Such very frequent short distance intra-urban mobility is only likely to be detected from diaries or oral evidence, and is almost certainly under-represented in the evidence provided by family historians, especially for the eighteenth and early-nineteenth centuries. However, not all families moved frequently. Joseph Yates moved only three times in his life-time from 1826 to 1896: once when he left home to take an apprenticeship, once when he married, and once later in life as his family circumstances changed. Similarly, although Benjamin Shaw had moved five times with his parents when a child, once married he remained in the same house in Preston for 46 years from 1795–1841, interrupted only by a short interlude in 1808 when he was removed to his Parish of Settlement by the Overseers of the Poor. What these diaries suggest is that individual experiences were extremely varied: although some people moved very frequently others could stay in the same place for long periods. It should not be assumed that variations over time shown in the data set are due only to the sources used; there may also have been some real changes in the frequency of mobility from the eighteenth to the twentieth centuries.

There are, indeed, good reasons for supposing that at least some of the apparent rise in mobility over time was real. From the late-nineteenth century an increasingly diversified housing market, improved public and personal transport, wider job opportunities and broadening personal horizons linked both to better education and the dislocating effects of two world wars increased the opportunities for mobility. On the other hand, it can also be suggested that the growth of home ownership and acquisition of a larger number of personal possessions acted as a brake on mobility: in the nineteenth century it was much easier to move sparse personal possessions between nearby rented property than it is to move in the late-twentieth century. In the twentieth century the relative difficulty of moving even short distances, including the large number of organizations that have to be

Table 3.2 Frequency distribution of number of moves by birth cohort

Number of moves undertaken by each individual	Number of people born 1750–1819	Number of people born 1820–49	Number of people born 1850–89	Number of people born 1880–1930	All people born 1750–1930
1–3 moves	2,442	1,367	1,195	412	5,416
4–6 moves	1,120	1,292	1,605	858	4,875
7–9 moves	279	444	846	697	2,266
10–12 moves	44	130	261	376	811
13+ moves	18	43	108	233	402
Never moved	551	117	68	23	759
Total number	4,454	3,393	4,083	2,599	14,529

Data source: 14,529 life histories provided by family historians (data excludes pilot study).

informed of a change of address, may mean that short distance mobility now has greater significance and meaning for the individual migrant than it did in the past. It is impossible to fully resolve the extent to which the increase in mobility demonstrated in the data is due to real changes in migration propensity or is due to changing availability of information. It seems most likely that a combination of factors should be taken into account. It would be surprising if data for the eighteenth and nineteenth century did indeed record every short distance move, and we should assume that there is some under-representation of this category. However, it also seems possible that a doubling of the frequency with which people moved from the eighteenth to the twentieth centuries also represents some real increase in the rate of mobility, though it is likely that the margin of change was not as great as the raw data suggest.

Although data on changing levels of mobility over time are difficult to interpret, variations between different categories of migrant are likely to be more reliable. There is no reason to suggest that any bias in the sources relating to the frequency of mobility disproportionately affected particular groups of the population, thus the relationships between groups should be reliable even if the absolute levels of mobility recorded may be open to question. Table 3.3 shows the mean length of residence at an address broken down according to the characteristics of the migrants at the time they moved to that address. Most variations are predictable and easily explained. Those over 60 at the time they moved had the shortest mean residency period (obviously due to mortality), widows and widowers also had a short mean length of residence at the address they moved to for the same reason. Those moving alone also tended to have a short mean length of residence at the address they moved to, whilst married couples and nuclear families stayed in one place for slightly longer than average. This can be explained by a tendency towards stability in the family-building stage of the life-cycle as small children acted as a brake on mobility, and by the fact that those moving alone tended either to be relatively old or young. The former were affected by high mortality whilst the latter moved more frequently due to both career and family adjustments.

Variations by occupation are relatively small, but there is some evidence that farmers were particularly stable – often remaining in the family farm all their lives – and unskilled agricultural workers also had a high degree of stability in the eighteenth century. This, however, may be partly due to the difficulty of tracing this group who might have moved frequently from farm to farm within a small area. Those who moved most frequently included domestic servants, those in the armed services and skilled non-manual workers. For the first two groups their mobility is clearly related to the nature of their work, a factor which is explored in more detail in Chapter 5. It is not immediately apparent why skilled non-manual workers (clerks, shop assistants and the like) should move more frequently than

Table 3.3 Mean length of residence (years) by age, marital status, household composition and occupation after move by birth cohort

Migrant characteristics	1750–1819	1820–49	1850–89	1890–1930
Age				
<20	16.1	11.0	10.3	7.4
20–39	15.0	10.7	10.9	8.2
40–59	12.9	12.7	12.8	11.1
60+	7.5	7.5	7.4	7.4
Marital status				
single	14.0	10.1	9.3	6.5
married	14.2	11.5	11.8	10.0
other	8.5	8.9	8.2	6.5
Companions				
alone	11.7	8.4	8.6	5.0
couple	16.8	11.7	11.6	9.4
nuclear family	13.5	11.5	11.5	9.7
extended family	10.6	10.9	10.0	9.2
other	10.9	10.9	9.7	5.2
Position in family				
child	15.8	11.2	10.2	8.1
male head	13.2	10.7	10.8	8.3
wife	15.5	12.1	12.5	10.8
female head	11.4	9.9	9.0	5.2
other	9.1	9.6	8.1	5.1
Occupational group				
professional	11.5	9.1	9.4	7.3
farmer	19.7	16.6	16.2	14.4
intermediate	14.0	12.5	12.9	9.9
skilled non-manual	9.2	9.1	10.0	7.7
skilled manual	13.3	10.7	11.3	9.1
skilled/semi-skilled agricultural	13.4	12.2	10.8	8.9
semi-skilled manual	12.5	10.1	11.3	9.2
unskilled manual	12.6	10.3	11.1	8.8
agricultural labourer	16.4	10.6	11.9	7.3
domestic service	13.2	9.0	8.9	6.2
armed forces	8.2	6.5	5.4	3.6
unpaid household	14.5	11.2	12.1	10.0
Total number	14,443	16,338	23,551	19,532

Data source: 16,091 life histories provided by family historians.

other occupational groups, though it may be that such people combined both an aspiration for self-improvement and a reasonably reliable income, and thus were able to move frequently in response to changing circumstances. However, overall, the frequency of mobility, and the mean length of residence at a particular address was quite stable for different groups of the population, suggesting that the factors which influenced mobility decisions were widely applicable to most ages and social groups.

The national migration system: how far did people move?

In many respects this is not the most appropriate question to ask. For most people the absolute distance over which movement took place was not as important as the travel time involved. Ease and speed of travel affected the decision to move, the ease with which migration took place and the extent to which migrants were able to maintain links with family and friends after migration. Unfortunately it is impossible to collect systematic information on the actual journey times involved in particular migrations, so linear distance has to suffice as a reasonable surrogate. The analysis of distance must, however, be interpreted within the context of changing transport systems and travel times.

The decision to migrate is likely to have been influenced as much by perceived distance as it was by actual distances. As contemporary studies have shown, perceptions of distance can be affected by many factors, including mode of transport, cost of transport, levels of information about different places and social, cultural and physical differences between regions (Gold, 1980; Gould and White, 1986). Such factors were likely to have been more important in the past than they are in the present. Late-twentieth century global communications, and the availability of instant news and information through computer networks and the media, has led to the shrinking of distance (Harvey, 1989; Thrift, 1990; Allen and Hamnett, 1995). Concepts of distance have changed radically since the eighteenth century as today most people think nothing of travelling to parts of the world which many people were scarcely aware of 200 years ago. However, even in the late-twentieth century, not everyone has equal access to global information and rapid international travel. Although concepts of distance have changed most radically over time, and have altered particularly quickly in the last quarter century, there are also persistent variations between different groups of the population. Thus attitudes to and opportunities for travel and movement may vary by gender, age, occupation, education and social group. In the past, as in the present, the rich could travel much more easily than the poor, and attitudes towards long distance movement may vary as much between groups as it does over time.

Changing transport technology from the mid-eighteenth century caused a gradual reduction in average travel times within Britain. In the mid-eighteenth century, before widespread turnpiking of major roads, coach travel was slow, hazardous and expensive. Average journey times from London to Manchester were four and a half days in the 1750s: this was clearly a journey that was not undertaken lightly and one which was only available to the wealthy. For most people the majority of journeys were undertaken on foot and, hence, the area over which people had information and through which they could easily move was restricted. By the early-nineteenth century turnpiking had reduced travel times by road (30 hours by coach

Table 3.4 Distance moved by year of migration

Distance	1750–1839	1840–79	1880–1919	1920–94
Mean distance (km)	37.7	33.7	38.4	55.5
Distance band (km)				
<1	24.5	40.5	43.7	35.5
1–4.9	14.7	13.2	11.8	9.9
5–9.9	14.1	9.3	7.9	8.5
10–19.9	12.3	8.3	7.0	8.3
20–49.9	15.4	11.0	9.5	10.1
50–99.9	8.6	7.2	7.7	9.0
100–199.9	5.8	6.1	6.8	9.2
200+	4.6	4.4	5.6	9.5
Total number	8,199	19,656	20,934	17,864

Data source: 16,091 life histories provided by family historians.

from London to Manchester in the 1820s), the frequency of coaches had increased, but travel remained expensive. Development of the railway network from the 1840s significantly reduced travel times – 7 hours 45 minutes from London to Manchester in 1844 – but fares remained relatively high and for most ordinary working people rail travel was a major and only occasional expense until at least the 1890s, by which period travel time from London to Manchester had further reduced to around 4 hours 15 minutes. Only in the twentieth century with rising real incomes and, especially, development of the motor car did personal travel throughout Britain become more common (Dyos and Aldcroft, 1969; Aldcroft and Freeman, 1983; Freeman, 1986; Thrift, 1990; Simmons, 1990). By the 1950s there were over four million licensed vehicles on the roads of Britain, a figure which doubled during the next decade, although even in the late-twentieth century there are substantial portions of the population – women, children, the elderly – who have poor access to public or private transport and whose travel behaviour is restricted (Whitelegg, 1997). Concurrent with changes in transport, other forms of communication within Britain were also revolutionized as changes to the postal service, development of the telegraph, expansion of newspaper circulation and, in the twentieth century, development of the radio and then television enabled information to spread rapidly from one part of the country to another (Gregory, 1987; Daunton, 1985; Thrift, 1990).

Analysis of changes in the distances moved within Britain for all migrants in the data set reveals two overwhelming trends. First, in all periods, migration was dominated by short distance mobility and, second, the average distance over which people moved increased substantially in the twentieth century, having remained relatively constant over the previous 150 years (Table 3.4). More than 50 per cent of all moves were less than ten kilometres in each time period and the largest single category was migration over less than one kilometre (and therefore impossible to represent on Figures 3.2–3.5). Most people were clearly moving very short distances within settlements or between adjacent settlements. The mean distance moved by all migrants was within the range 33.7–38.4 km for the three time periods before 1919, but in the twentieth century shot up to 55.5 km. This is caused mainly by an increase in very long distance moves of over 100 km in comparison to the earlier periods. In the period 1920–1994 most people still moved over short distances (35.5 per cent less than one kilometre), but an increasing minority were moving over much greater distances within Britain. In the period 1750–1839 the mean distance moved was slightly higher than in the period 1840–79, and the proportion moving less than one kilometre is the lowest in any category. This confirms the interpretation outlined above that very short distance moves are under-represented in data for the eighteenth century, though such migration still makes up the largest single category and clearly family historians were returning data on a large number of very short distance moves. It should be emphasized that all these data refer to moves within Britain: movement overseas is examined in Chapter 9.

Although mean distances migrated clearly rose in the twentieth century, it is pertinent to ask whether such increases matched the changes in travel time and availability of information that occurred over the same period. For instance, it could be that although absolute distances increased; in relation to the ease and cost of moving the distances over which people moved have relatively decreased. There was very little change in the mean distances moved in the nineteenth century at a time when travel times by the fastest mode of transport were decreasing substantially, first through turnpiking of the roads and, second, through the spread of the railways. Given this apparent shrinking of distance it is, perhaps, surprising that the average distance over which people moved remained almost unchanged. Most significantly, the proportion of people moving over 100 km increased only marginally from 1750–1880, and this is a category of move for which information should be equally reliable in each time period. These trends suggest that throughout the nineteenth century most people were still constrained by the high cost of public transport, that their horizons were limited, and that their perceived work opportunities were still constrained within local and regional labour markets. It can be argued that by the late-nineteenth century, relative to travel time by the fastest mode of transport, people were actually migrating over shorter distances than they had been a century earlier.

Mean distances moved increased by some 20 kilometres in the twentieth century, and the proportion moving over 100 km almost doubled, but it can still be argued that this lags behind changes in travel time and ease of movement for those with access to a private car. What the data suggest, is that changes in income and access to transport are more important than changes in fastest travel times, which may be irrelevant to the majority of the population who cannot afford that mode of transport. Thus, although relatively fast rail transport was available in the second half of the nine-teenth century, relatively few people used it routinely. In the twentieth century although travel times by train fell only slowly, with rising real in-comes more people were able to use the railways. However, private cars and, especially, internal air travel remained the preserve of the relatively affluent until the late-twentieth century, and thus distance moved did not increase in line with the theoretical compression of distance. Clearly other factors also affected the decision to migrate, including attachment to a par-ticular community which may have constrained people to move over a short distance despite the ability to move further. Such issues will be ex-plored in more detail in later chapters. But relating mobility to the compres-sion of travel time has emphasized the extent to which short distance movement dominated the migration system and the fact that, in relative terms, migration distances may have been getting shorter rather than longer. This has implications for the application of social theory which assumes time–space compression and time–space distanciation in the process of globalization (Harvey, 1989; Giddens, 1990; Allen and Hamnett, 1995). It is argued not only that the process of communication is speeded up in a shrinking world, but also that remote interaction across space has become as important as face-to-face contact. Whilst this is undoubtedly true for some with access to modern communications technology, many remain excluded, and in terms of residential migration short distance moves to familiar neighbourhoods remained important throughout the study period.

These trends are further illustrated if distance moved is related to the characteristics of migrants (Table 3.5). There is no evidence to back up Ravenstein's (1885/1889) assertion that females were more migratory than males over short distances, but that males were more likely to move over long distances. For the whole data set around 38 per cent of both males and females moved less than 1 km and around 13 per cent moved over 100 km. The mean distance moved by females is actually slightly greater than that for men, but this is mainly due to differences in the twentieth cen-tury. In the period 1750–1839 females do tend to move less distance than men, but from 1840 to 1920 (including the period covered by Ravenstein's analysis), the mean distances moved are almost identical for men and women. Differences in distances moved by age were also relatively small. Short distance moves dominated all age groups, but those over 40 had slightly lower mean-distances and a higher proportion of short distance moves. This trend is evident in the first three time periods, but after 1920 there is little difference in the distances moved at particular ages. Given that age is closely related to life-cycle stage, this suggests that constraints on mobility for those over about 40 were much more severe in the past, possibly relating to a combination of family commitments, reducing incomes related to a less competitive position in the labour market, and attachment to a known area where there would be friends or kin to provide care in case of infirmity or

Table 3.5 Mean distance moved (km) by year of migration by migrant characteristics

Migrant characteristics	1750–1839	1840–79	1880–1919	1920–94	All moves
Gender					
male	39.5	34.1	38.5	54.0	40.8
female	32.3	32.7	38.2	57.3	42.9
Age					
<20	40.5	36.8	42.0	57.7	41.9
20–39	38.3	36.8	41.9	58.2	43.5
40–59	29.9	26.9	30.6	51.3	36.6
60+	19.7	18.0	24.4	54.4	40.4
Marital status					
single	52.3	42.6	52.1	72.2	52.8
married	28.6	28.3	30.7	50.3	35.6
other	35.3	27.9	31.4	52.8	41.4
Companions					
alone	60.2	61.5	83.6	84.5	75.6
couple	24.9	30.7	31.1	52.9	37.3
nuclear family	32.7	29.3	31.2	44.7	33.8
extended family	35.4	24.4	25.4	53.6	35.6
other	118.1	43.4	57.3	73.9	61.9
Position in family					
child	35.4	30.7	32.0	36.7	32.9
male head	36.9	34.4	39.3	54.9	41.4
wife	26.1	30.6	32.2	50.8	37.5
female head	41.4	34.2	62.7	70.7	60.8
other	110.1	49.1	57.9	71.7	64.9
Occupational group					
professional	85.9	81.1	83.8	74.4	79.3
farmer	19.4	23.5	27.4	22.8	22.7
intermediate	35.0	25.8	27.3	48.5	33.1
skilled non-manual	97.8	44.6	40.9	48.7	47.6
skilled manual	32.5	28.3	25.9	35.6	29.4
skilled/semi-skilled agricultural	38.3	32.2	36.5	35.6	34.9
semi-skilled manual	36.8	26.8	23.2	29.3	26.9
unskilled manual	20.3	21.2	28.1	28.7	24.2
agricultural labourer	16.7	15.0	16.3	45.0	16.7
domestic service	41.9	42.7	60.3	69.0	56.0
armed forces	128.5	124.9	129.5	155.3	138.7
unpaid household	27.8	31.3	32.4	57.0	41.3
Total number	8,199	19,656	20,934	17,864	66,653

Data source: 16,091 life histories provided by family historians.

incapacity. In the twentieth century the elderly are not only often fitter and more affluent, but it is relatively easy to move long distances to live with or near relatives if necessary. It would seem that migration patterns have changed most for those over 40, and especially those over 60, in the past 200 years. During, and immediately after, the industrial revolution the young moved further than the old, but in the twentieth century these differences have largely disappeared.

Age, marital status, and position in the household are obviously related, but it would seem that it is life-cycle factors related to family responsibilities that were the main determinants of distance moved, rather than age *per se*. Single migrants moved over longer distances than those that were married or widowed; those moving alone moved further than other groups; and, especially after 1880, female-headed households (mostly single females) moved further than the average (though not further than single males). Thus those groups that were previously shown to be the most mobile in terms of the frequency of their moves, also tended to move over the longest distances. These trends are easily explained. Single people, whether male or female, moving alone without family or other responsibilities, were clearly less constrained in their migration choices. Such people were also likely to be young, possibly more adventurous, unencumbered by children, property or large quantities of possessions, and in a position to move readily to a completely different part of the country. However, none of this should obscure the fact that for all groups and time periods short-distance migration was the most common experience.

The ability to move over long distances depends not only on personal initiative and family ties, but also on the amount of information available, the ability to pay relatively high costs of long distance travel, and the extent to which the migrant had transferable skills which gave access to employment in a new labour market. Links between migration and employment are explored in detail in Chapter 5, but it is evident that migrants in certain occupations were much more likely to move over long distances than others. Particularly high mean distances, caused mainly by a large amount of movement over 100 km, were found amongst migrants in the professions and those in the armed forces. Thus 28.9 per cent of moves by migrants in professional occupations were over 100 kms and no less than 45.9 per cent for those in the armed forces. These patterns are apparent in each time period and were clearly related to the combination of a high income and national labour market for those in the professions, and to the peculiar demands of life in the militia or other armed services, which demanded movement around the country as men were transferred from camp to camp (Skelley, 1977).

Skilled non-manual workers (clerks, shop assistants etc.) moved over relatively long distances before 1880, possibly due to a relatively high income and a willingness of parents to place apprentices to such trades in businesses some way from the parental home, and domestic servants recorded above average migration distances, especially after 1880. This was possibly due to a relative shortage of servants in the late-nineteenth and twentieth centuries leading affluent households to recruit servants over longer distances, though it could also be explained by servants moving with professional households. This trend may have become more noticeable after

69

about 1880 as servants were harder to recruit and households were keen to retain a good servant rather than having to recruit locally after they had moved (Horn, 1975). Farmers were the least mobile group, and they were also more likely than most migrants to move over relatively short distances. Other groups with short mean distances were unskilled manual workers and agricultural labourers. Both of these would have been constrained by low wages and the operation of local and regional labour markets. Even in the late-twentieth century the unskilled have difficulty gaining information about distant job opportunities (Saunders, 1986; Johnson and Salt, 1990).

Some of these themes, and especially the problems involved in moving long distances, are illustrated by accounts of moving house provided in the diaries supplied by family historians. Joseph Yates moved only three times in his life, but in 1838 at the age of 12 years, he left his parental home in Battersea, London, to be apprenticed as a clerk to a firm in Leominster, Herefordshire, a distance of some 250 kms. His diary entry for October 23rd 1838 reads:

> Left home at Battersea at 5 a.m. with Aunt Davies and Mr. John Townsend. Coach '??' left 'Bull and Mouth' Inn, Oxford Street at 7 a.m., arrived Oxford at 1 p.m., stayed an hour for dinner and reached Hereford at 10 p.m. Slept at the Green Dragon Hotel. Rose at 6 a.m. on the 24th, breakfasted, started by coach at 7 a.m. and arrived at Hope Gate. Thence to Mr. Green, Hampton Mill.

The journey was clearly uneventful. As a 12 year old leaving home for the first time he had few personal possessions to transport, and he was accompanied by people he knew; but travelling by coach from London to Herefordshire was a significant undertaking which necessitated a 15 hour journey prior to an overnight stop before reaching his destination the following day. It is, perhaps, significant that he did not return home again until the New Year holiday over four years later, leaving Leominster at 3 pm on 30th December 1842 and arriving in London 24 hours later for a two-week holiday. The impact of a four year absence from home on both Joseph Yates and his parents can only be speculated upon.

As today, not all removals were straightforward, and the life history of Henry Jaques recounts an occasion when, at the age of 15 in 1857, he accompanied his Aunt's furniture on her removal from the Islington area of London to Shorne, between Gravesend and Rochester in Kent. Although only a distance of some 45 kms the journey was not easy:

> ... it was thought that one large van would take all the goods and accordingly they were subsequently packed into one vehicle piled up to a great height. Soon after the start disaster came. We had not proceeded 200 yards when one of the springs gave way, and most of the goods were toppled over into the road. The services of a wheelwright

were called into requisition, the breakage made good, a small two wheel cart obtained which was filled with goods thus easing the van, and after a delay of three hours, and about 12 o/c midday a fresh start was made for our 30 miles journey.

One thing noted by me on the journey was the frequent stoppages at the road-side Inns and the condition of the driver in consequence. Progress was slow but at 1 o/c the following morning we entered the quiet village of Shorne, all was quiet, not a soul to be seen. The exact locality of the house was not known so we had to wake up some of the sleeping villagers to guide us to the Ridgeway. So about two o/c we drew up at the house . . . The unloading was all over by 5 o/c by which time the village was all alive and we were regailed with a crowd of small children with open mouths wondering what it all meant.

If moving less than 50 kms was so difficult for a relatively affluent lady, it is not surprising that long distance moves were undertaken relatively rarely.

The national migration system: why did people move?

Any explanation of where and how far people moved must be related to an understanding of why they moved. The main reasons why people moved will be explored in more detail in subsequent chapters, which focus on movement related to such factors as employment change, family circumstances, housing and personal crises. This chapter simply provides an overview of the reasons why people moved, and relates this to the spatial pattern of migration outlined above. As discussed in Chapter 2, stated reasons for migration are probably the least reliable aspect of the data set. Even present-day studies find it difficult to analyze the reasons for migration, as such decisions are made for a variety of reasons and people may forget or rationalize the real reasons quite soon after the event. Family historians were asked to state the main reason for moving, where it was known from oral or written evidence, or where it could reasonably be inferred from other sources. The information seems quite reliable when aggregated into broad categories, though there is a large category of 'other' and multiple reasons that defy simple classification, and it is likely that information will be least reliable for the more distant past. All the problems of deducing migration motives in the past must be borne in mind when interpreting these results.

In each of the four time periods used, movement for work-related reasons was the single most common experience, but the frequency with which this occurred declined from 47.8 per cent of all moves in the period 1750–1839 to 26.3 per cent 1920–94. In the period before 1839 marriage was the second most important stated reason for migration (26.5 per cent) but after

Table 3.6 Reasons for moving by year of migration (%)

Reason	1750–1839	1840–79	1880–1919	1920–94
Work	47.8	47.4	38.8	26.3
Marriage	26.5	18.8	14.7	9.5
Housing	4.3	9.5	16.0	23.6
Family	1.4	2.9	2.9	5.7
Crisis	5.1	6.6	7.2	9.3
Emigration	1.3	2.3	1.9	1.0
War service	3.1	2.2	6.3	4.9
Retirement	0.7	1.7	2.7	6.0
Other	9.8	8.6	9.5	13.7
Total number	6,258	13,923	17,972	18,445

Data source: 16,091 life histories provided by family historians.

1920 this had also declined to only 9.5 per cent. In contrast, moves for housing related reasons increased from 4.3 per cent in the first time period to 23.6 per cent after 1920, and moves for family reasons, personal crises and retirement all increased by smaller amounts. It would appear that, in relative terms, the key factors of work and marriage became less important from around the 1880s, and that people increasingly moved for a wider range of reasons (Table 3.6).

However, these data should be interpreted with considerable care. It can certainly be argued that increased choice within the housing market, greater affluence, increased personal mobility, smaller families and increased longevity into a retirement that was supported by the welfare state could all contribute to a trend in which people moved more often for a wide range of reasons, rather than moving only on marriage and to change their employment. However, it can also be suggested that the observed trends are, at least in part, an artefact of the data set. As outlined in Chapter 2, the range of sources on which family historians can draw is more limited in the eighteenth- and early-nineteenth centuries. It has already been noted that some short distance moves may be missing for the earlier periods. It is just these moves that would be likely to occur for housing and some other non-work reasons, and thus it is possible that in the earlier time periods family historians picked up moves for major events such as marriage and job change (for which it is more likely that records survive), but missed moves undertaken to improve housing or for family reasons. It is most likely that both sets of explanation hold some truth: that part of the trend is accounted for by constraints of the data, but that there was also a real change towards mobility being undertaken for a wider range of reasons. Some of these reasons are explored in more detail in later chapters.

Tables 3.7a–d examine the stated reasons for migration broken down by the characteristics of migrants and by time periods. Before 1840 women

Table 3.7a Reasons for moving by migrant characteristics, 1750–1839 (%)

Migrant characteristics	Work	Marriage	Housing	Family	Crisis	Emigration	War service	Retirement	Other	Total number
Gender										
male	51.5	20.7	4.7	1.3	4.8	1.4	3.8	0.6	11.2	4,720
female	36.6	44.5	3.2	1.8	5.8	1.2	0.8	0.7	5.4	1,538
Age										
<20	54.9	12.3	3.9	1.5	7.3	1.5	3.9	0.4	14.3	1,632
20–39	44.4	36.2	3.6	1.0	2.8	1.0	3.0	0.1	7.9	3,884
40–59	54.3	7.7	9.6	1.9	10.3	3.5	1.9	1.9	8.9	593
60+	22.0	2.0	10.0	11.0	19.0	1.0	0.0	18.0	17.0	100
Marital status										
single	56.9	9.2	3.2	1.6	6.8	1.4	5.8	0.4	14.7	2,074
married	40.3	39.9	4.8	1.0	2.9	1.2	1.9	0.8	7.2	3,211
other	18.3	8.4	6.9	6.9	39.7	1.5	0.8	1.5	16.0	131
Companions										
alone	50.2	16.0	0.7	2.3	5.0	1.0	6.0	0.3	18.5	1,173
couple	17.5	72.1	1.2	0.9	0.7	0.3	0.5	0.6	6.2	1,761
nuclear family	67.2	5.3	8.0	1.4	7.1	1.7	1.0	0.9	7.4	2,770
extended family	43.8	14.1	7.4	0.8	15.7	5.8	0.0	0.0	12.4	121
other	35.1	2.5	2.1	1.7	7.5	3.4	32.6	0.0	15.1	239
Position in family										
child	67.9	3.4	6.7	1.9	9.7	1.7	0.4	0.6	7.7	993
male head	48.9	25.9	4.4	1.2	3.4	1.3	2.6	0.8	11.5	3,609
wife	32.0	55.7	3.5	0.9	1.8	0.7	1.1	0.8	3.5	1,012
female head	25.5	39.6	0.5	4.7	19.3	1.0	0.0	0.0	9.4	212
other	34.6	4.0	0.8	2.4	11.2	2.8	30.1	0.0	14.1	249
Occupational group										
professional	68.3	12.1	6.8	1.6	2.5	1.6	0.3	–	6.8	322
farmer	56.8	21.8	2.0	0.9	2.4	4.9	0.0	–	11.2	454
intermediate	55.5	19.2	6.1	1.1	5.0	0.3	0.0	–	12.8	360
skilled non-manual	78.4	10.2	1.7	1.1	3.4	0.6	0.0	–	4.6	176
skilled manual	47.5	33.3	6.2	1.3	2.9	0.4	0.4	–	8.0	995
skilled/semi-skilled	58.3	25.1	6.1	1.6	2.4	0.4	0.0	–	6.1	247
agricultural										
semi-skilled manual	55.8	31.4	3.3	0.5	1.9	0.5	0.0	–	6.6	213
unskilled manual	50.0	28.6	10.3	1.5	4.4	1.5	0.0	–	3.7	136
agricultural labourer	55.9	29.9	2.4	0.8	3.3	0.2	0.2	–	7.3	481
domestic service	83.7	7.2	2.0	1.0	3.1	1.0	1.0	–	1.0	98
armed forces	8.7	2.4	0.0	0.0	2.4	1.0	83.6	–	1.9	207

The table should be read as follows: 51.5% of moves by men between 1750 and 1839 were for work-related reasons and 20.7% for marriage.
Data source: 16,091 life histories provided by family historians.

Table 3.7b Reasons for moving by migrant characteristics, 1840–79 (%)

Migrant characteristics	Work	Marriage	Housing	Family	Crisis	Emigration	War service	Retirement	Other	Total number
Gender										
male	49.9	16.1	9.8	2.5	5.9	2.3	2.7	1.9	8.9	9,856
female	41.4	25.6	8.9	4.2	8.1	2.4	0.7	1.2	7.5	4,067
Age										
<20	53.9	7.5	9.4	3.0	9.0	2.0	3.0	0.5	11.7	3,456
20–39	44.9	30.2	7.8	1.7	3.2	2.4	2.1	0.2	7.5	7,479
40–59	55.2	4.0	14.3	2.0	8.8	2.8	1.8	3.2	7.9	2,154
60+	20.2	3.4	10.9	16.9	22.1	1.1	0.0	17.7	7.7	752
Marital status										
single	55.7	6.0	8.2	3.1	8.4	2.1	3.5	0.7	12.3	4,269
married	43.5	28.5	10.8	1.8	3.2	1.7	1.6	2.0	6.9	7,282
other	20.4	4.9	5.0	14.7	37.9	0.8	1.0	6.6	8.7	618
Companions										
alone	50.5	12.1	0.7	6.3	9.0	2.7	4.3	1.0	13.4	2,119
couple	17.4	66.8	2.6	1.7	1.2	0.9	0.6	2.4	6.4	2,816
nuclear family	60.4	4.5	14.4	1.9	6.5	2.2	1.2	1.5	7.4	7,257
extended family	33.9	13.1	14.1	6.8	15.1	5.3	0.0	2.9	8.8	717
other	38.0	5.1	12.0	3.0	10.8	4.0	13.4	1.2	12.5	734
Position in family										
child	57.9	3.0	14.1	2.5	10.0	2.2	1.1	1.1	8.1	2,483
male head	49.2	20.0	9.1	2.3	4.4	2.0	1.9	2.2	8.9	7,504
wife	39.5	36.6	9.7	1.8	2.3	2.0	0.9	1.3	5.9	2,304
female head	34.9	19.6	1.9	10.9	20.3	2.6	0.0	0.7	9.1	690
other	29.2	9.1	4.9	7.2	15.9	5.0	14.2	1.3	13.2	684
Occupational group										
professional	70.6	8.5	9.1	0.8	2.2	1.5	0.4	0.6	6.3	729
farmer	64.4	12.0	3.2	1.1	3.9	5.6	0.0	0.7	9.1	464
intermediate	55.3	15.9	12.3	1.8	5.7	1.1	0.1	0.4	7.4	1,142
skilled non-manual	58.2	17.0	11.5	1.1	4.7	0.5	0.2	1.0	5.8	619
skilled manual	44.8	26.7	11.5	2.9	4.4	2.0	0.2	0.2	7.3	2,389
skilled/semi-skilled agricultural	54.5	20.3	9.4	1.3	4.6	0.6	0.0	0.4	9.0	479
semi-skilled manual	48.8	24.6	11.1	1.6	6.6	1.2	0.0	0.2	6.0	679
unskilled manual	48.3	25.3	10.1	1.6	5.9	1.3	0.0	0.3	7.2	387
agricultural labourer	64.7	19.8	4.3	1.9	3.3	0.8	0.0	0.3	4.9	672
domestic service	75.6	5.1	2.8	1.5	5.9	1.0	0.5	0.5	7.1	393
armed forces	10.7	4.1	0.4	0.7	1.7	0.0	81.3	0.4	0.7	289

Data source: 16,091 life histories provided by family historians.

Table 3.7c Reasons for moving by migrant characteristics, 1880–1919 (%)

Migrant characteristics	Work	Marriage	Housing	Family	Crisis	Emigration	War service	Retirement	Other	Total number
Gender										
male	39.8	12.5	15.2	2.5	6.1	2.1	9.2	3.0	9.6	11,566
female	36.9	18.4	17.5	3.8	9.1	1.6	1.1	2.2	9.4	6,406
Age										
<20	43.6	4.2	17.5	2.7	9.0	1.5	7.7	0.9	12.9	4,703
20–39	37.6	25.4	12.8	1.7	4.3	2.3	7.8	0.5	7.6	9,093
40–59	43.1	3.7	24.3	2.3	9.2	1.7	2.0	3.7	10.0	2,958
60+	16.7	2.2	13.6	15.8	16.7	0.9	0.1	24.7	9.3	1,171
Marital status										
single	46.5	3.2	13.6	2.6	8.1	2.0	11.1	1.1	11.8	6,626
married	35.3	24.2	18.1	2.0	4.2	1.4	3.4	3.6	7.8	9,030
alone	13.9	3.4	7.9	16.3	36.9	0.8	0.9	7.1	12.8	776
Companions										
alone	45.2	5.9	1.4	5.7	8.1	3.3	16.2	1.0	13.2	3,489
couple	15.2	65.2	5.1	2.0	1.1	0.8	0.6	4.4	5.6	3,108
nuclear family	47.1	3.1	25.7	1.9	7.7	1.5	1.4	2.6	8.8	9,302
extended family	24.7	11.1	20.8	7.2	16.4	1.9	0.3	4.8	12.8	806
other	22.8	2.3	8.4	1.9	9.3	3.6	38.5	1.3	11.7	1,056
Position in family										
child	47.3	2.1	25.5	1.9	9.6	1.3	1.2	1.8	9.3	3,630
male head	39.6	16.3	14.1	2.4	5.3	2.1	7.3	3.4	9.5	8,533
wife	33.3	30.9	19.1	1.8	3.6	1.2	0.6	2.6	6.9	3,254
female head	38.9	8.8	4.7	9.3	17.5	2.4	2.3	1.0	15.1	1,238
other	19.4	4.3	5.3	5.6	12.3	3.5	36.9	2.0	10.7	1,114
Occupational group										
professional	64.9	8.2	11.6	1.0	3.0	1.3	1.5	0.7	7.8	1,107
farmer	58.6	9.5	4.2	1.9	5.3	4.9	0.0	1.5	14.1	263
intermediate	50.6	12.5	19.5	1.0	4.8	1.0	0.2	0.6	9.8	1,256
skilled non-manual	45.9	17.9	17.3	1.9	4.9	1.2	0.6	0.8	9.3	1,406
skilled manual	39.7	22.9	19.6	2.0	5.5	2.6	0.4	0.3	6.8	2,374
skilled/semi-skilled agricultural	57.9	13.7	11.9	1.3	4.1	2.6	1.8	0.8	5.9	388
semi-skilled manual	38.1	25.8	17.6	2.1	6.8	1.3	1.3	0.2	6.8	825
unskilled manual	39.0	18.5	20.2	4.3	8.2	2.8	0.5	0.2	6.1	392
agricultural labourer	56.9	16.8	6.7	4.8	5.3	4.8	0.9	1.4	2.4	209
domestic service	74.5	5.6	5.7	2.5	5.5	1.3	0.5	0.5	3.9	634
armed forces	4.2	1.5	0.5	0.0	1.7	0.1	89.8	0.5	1.7	1,121

Data source: 16,091 life histories provided by family historians.

Table 3.7d Reasons for moving by migrant characteristics, 1920–94 (%)

Migrant characteristics	Work	Marriage	Housing	Family	Crisis	Emigration	War service	Retirement	Other	Total number
Gender										
male	29.0	9.3	23.4	3.9	6.9	1.2	6.9	6.4	13.0	10,432
female	22.8	9.8	23.9	8.1	12.3	0.7	2.4	5.4	14.6	8,013
Age										
<20	30.4	2.3	25.2	3.7	7.1	1.0	9.4	0.6	20.3	1,804
20–39	31.9	18.9	22.1	2.7	4.1	1.2	7.5	0.4	11.2	8,151
40–59	32.3	2.4	29.4	2.9	8.5	1.0	2.9	5.0	15.6	4,448
60+	6.3	1.7	19.4	15.9	21.6	0.5	0.2	20.6	13.8	4,021
Marital status										
single	38.1	2.6	17.8	3.8	7.3	1.3	10.2	1.6	17.3	3,839
married	25.7	13.2	27.3	3.6	5.4	0.8	3.6	7.7	12.7	11,287
other	6.8	1.9	16.6	19.7	36.4	0.7	0.5	3.7	13.7	2,033
Companions										
alone	30.0	3.7	7.8	11.4	17.9	1.5	10.2	1.8	15.7	3,809
couple	14.5	32.0	18.2	5.3	4.1	0.7	1.4	14.2	9.6	4,313
nuclear family	32.3	1.6	35.4	3.1	6.7	0.8	2.2	3.7	14.2	8,160
extended family	17.1	7.4	22.8	8.2	17.3	0.9	0.6	6.5	19.2	1,205
other	24.8	1.2	12.8	3.5	9.4	1.5	29.9	3.6	13.3	856
Position in family										
child	32.0	1.8	35.4	4.2	9.1	0.6	1.0	2.3	13.6	1,578
male head	29.3	10.4	23.5	3.8	6.4	1.2	5.3	7.2	12.9	8,955
wife	24.5	15.7	28.9	3.4	4.7	0.6	2.1	7.5	12.4	4,280
female head	17.3	3.1	14.5	16.5	24.7	0.7	2.6	2.2	18.4	2,282
other	20.2	3.6	9.8	9.2	16.9	1.7	20.1	3.3	15.0	1,256
Occupational group										
professional	46.7	8.0	22.6	1.7	3.7	1.0	1.2	1.6	13.5	1,864
farmer	53.0	9.7	5.9	1.5	10.5	3.7	0.8	3.7	11.2	134
intermediate	48.2	6.4	25.0	1.3	6.0	0.5	0.2	1.5	10.9	1,299
skilled non-manual	34.5	14.9	26.9	2.8	6.2	0.4	1.2	1.5	11.6	2,220
skilled manual	29.9	16.8	31.6	3.3	4.8	1.1	1.1	0.3	11.1	1,713
skilled/semi-skilled agricultural	43.1	13.4	13.4	1.8	7.2	4.8	0.5	1.0	14.8	209
semi-skilled manual	24.9	17.8	35.9	3.8	4.9	1.2	0.9	0.2	10.4	585
unskilled manual	23.4	16.8	37.6	4.0	6.6	1.0	1.5	0.0	9.1	197
agricultural labourer	49.2	4.8	7.9	6.4	6.4	14.3	1.5	0.0	9.5	63
domestic service	57.4	4.9	14.6	4.4	7.5	0.8	0.5	0.8	9.1	385
armed forces	5.5	3.4	3.7	0.6	0.7	0.1	82.0	0.5	3.5	857

Data source: 16,091 life histories provided by family historians.

were much less likely than men to move for work related reasons and rather more likely to move on marriage, suggesting that brides commonly moved to their future husband's home. However, from the mid-nineteenth century these differences largely disappear, though women were still slightly less likely than men to move for employment reasons, and women also moved more frequently for family reasons and personal crises. Variations by age are predictable, with those under 20 years old and those age 40–59 most likely to move for work-related reasons. Movement for marriage was most common for those age 20–39, whilst those aged 60 and over moved mainly for a combination of family reasons, crises and retirement. Until the 1880s a substantial proportion of the over 60s also moved for employment reasons, but in the twentieth century with earlier and more secure retirement, this changed markedly. Moves for housing reasons were undertaken mainly by those age 40–59, especially after 1880, as families were able to adjust their housing to changing circumstances in an increasingly diversified housing market.

Migration for work related reasons was not only important for those who were single and moved alone, often migrants moving to their first employment, but also for married couples moving as part of a nuclear family as people moved in later stages of their career cycle. Moves for family reasons were undertaken most frequently by single women and those who had been widowed as people adjusted to changed circumstances, and females without other responsibilities took on a caring role within the household. Variations in reasons moved by occupation were limited, but domestic servants and those in the professions were most likely to move for work related reasons before 1920. In summary, young and single migrants were most likely to move for employment, couples in their 20s most commonly moved due to marriage, nuclear families with more mature household heads moved most frequently for work and housing reasons, whilst movement of the over 60s was most often stimulated by retirement, personal crises (including ill health) and the desire or need to be near relatives. Women, especially single women, were also more likely to move for family reasons, and patterns of change over time suggest relative stability from 1750–1920, but with more significant changes in the twentieth century.

Analysis of the reasons for migration can be related to the spatial pattern of moves and the distance over which people migrated (Table 3.8). In every time period very short distance moves of under 1 km were mainly for housing reasons as people moved within a settlement in response to changed housing needs. Such short distance moves were also quite often stimulated by marriage and by crises such as ill-health, redundancy, or bereavement. Longer distance moves occurred for a variety of reasons, including work, marriage, family reasons and crises (with work the dominant reason in all distance bands over 10 kms in each time period). Although not numerically particularly important, movement with the armed services was over-represented in

Table 3.8 Reasons for migration by mean distance moved (km) by year of move

Reason	1750–1839	1840–79	1880–1919	1920–94
Work	48.3	55.4	67.0	101.5
Marriage	18.0	20.2	23.1	26.4
Housing	12.6	4.2	3.7	8.1
Family	64.7	42.9	62.9	87.4
Crisis	44.2	38.5	37.9	44.3
War service	138.9	145.9	150.8	176.4
Retirement	45.8	27.6	49.3	63.9
Total number	5,646	12,738	16,264	15,917

Data source: 16,091 life histories provided by family historians.

long distance moves of over 100 kms. Thus although, predictably, the short distance milling around of population which dominated the spatial pattern of migration in all periods was due mainly to non-work reasons, and longer distance moves were mainly stimulated by employment change, there was quite a wide distribution of migration motives in all distance bands. Detailed explanation of these trends is complex, and some of the issues involved will be explored in later chapters.

The changing significance of regions

It is often suggested that one effect of improved communications and high speed transport is not only the shrinking of distance but also the removal of differences between places (Anderson, Brook and Cochrane, 1995; Massey and Jess, 1995). It can be suggested, for instance, that in the eighteenth century the different regions of Britain were both clearly defined and poorly integrated with one another. As the nineteenth century progressed such differences might be expected to have disappeared as people, goods, information and ideas flowed more freely from place to place and few locations were truly marginal or isolated (Everitt, 1979; Hudson, 1989). In the twentieth century, European or global integration seems at first sight more relevant than issues of integration between regions in a country as small and relatively homogeneous as Britain.

However, such ideas are open to considerable debate. Whilst some writers have emphasized the extent to which the changes wrought by the processes of urbanization and industrialization broke down barriers between regions and led to a society that was defined more by class differences than by regional differentiation (Gregory, 1988a, 1988b, 1990), other authors have suggested that although communications between places improved this served to heighten awareness of differences and thus a new sense of regional identity and regional distinctiveness emerged in the nineteenth century (Langton, 1984, 1988; Freeman, 1984). In the twentieth century, although processes of globalization allow rapid movement of people, goods, capital, news and ideas around the world, not every person or place shares or benefits equally in this process. There have been globalization losers as well as winners, and for many people the late-twentieth century has been a period in which regional and cultural identity has assumed increasing importance. This has manifested itself in events such as ethnic conflict, movements for independence by small cultural groupings in larger nation states and the growth of local media (Featherstone, Lash and Robertson, 1995). The processes and effects of changing levels of regional integration are thus complex and are certainly worthy of further investigation.

One major problem in the analysis of regional identity and distinctiveness relates to the ways in which regions are defined. Most historical data are primarily available for administrative regions, but although individuals may identify with such units for some purposes (for instance local taxation or representation on local government), it can be suggested that for many purposes administrative regions have little real significance. In relation to the movement of people, goods and ideas between regions, physical barriers may have been of much greater significance in the past. These can be defined partly in terms of distance, but also relate to the extent to which a region is divided from other areas by topographic features such as a mountain range. Regions may also have had economic and associated social distinctiveness based on a particular way of life which meant that interaction was most likely to be with other regions with similar characteristics (Butlin, 1990; Lawton and Pooley, 1992, pp. 323–33). These and other factors are difficult to measure for past societies, though analyses of the impact of local government reorganizations in Britain have highlighted the extent to which people do have strong attachments to particular places even in the late twentieth century (Horne, 1984). In this study, data are analyzed within a county framework, but the problems of relating such administrative units to the everyday regions with which people identified need to be borne in mind.

There have been a number of attempts to analyze levels of regional interaction and integration in the past. Two examples will suffice to illustrate the complexity of the debate that surrounds such studies. Gregory (1987) has examined the circulation of information in early-nineteenth century England, focusing on the different modes of mail transport that were available, ranging from the Standard Mail, through the Privileged Post to clandestine mail carriers. He demonstrates that although the various mail systems tended to operate hierarchically, with major centres better connected than smaller places, and with some individuals having better access to rapid communications than others, overall there was a well-integrated network of postal communications operating in nineteenth-century England. This he argues demonstrates a high degree of regional integration within the English space-economy, and he goes on to argue that ease of communication

79

Table 3.9 Type of move by region and year of migration (%)

Region	1750–1879				1880–1994			
	Within region	Into region	Out of region	Total number	Within region	Into region	Out of region	Total number
London	57.2	25.5	17.3	3,981	46.2	22.0	31.8	5,024
South East	62.5	18.7	18.8	2,973	50.8	27.6	21.6	6,166
South Midland	55.0	20.5	24.5	2,011	44.2	30.8	25.0	4,626
Eastern	67.1	12.8	20.1	1,842	53.4	25.0	21.6	3,106
South West	69.0	11.6	19.4	2,358	51.2	24.1	24.7	2,460
West Midland	57.3	27.3	15.4	3,043	50.5	33.4	16.1	4,684
North Midland	71.9	12.5	15.6	2,110	62.9	18.3	18.8	3,043
North West	77.0	14.6	8.4	4,471	73.8	13.3	12.9	6,738
Yorkshire	71.4	13.9	14.7	2,931	68.1	15.5	16.4	4,368
Northern	76.2	12.1	11.7	2,535	70.6	12.9	16.5	2,796
North Wales	67.2	14.9	17.9	403	46.6	26.2	27.2	654
*South Wales	79.6	12.2	8.2	1,236	75.3	12.4	12.3	1,918
**Northern Scotland	83.9	5.8	10.3	2,073	72.9	11.6	15.5	2,061
***Southern Scotland	74.5	12.3	13.2	1,814	66.0	15.9	18.1	2,241

* Includes Monmouth and Glamorgan

** Includes Highland, North East Scotland and Scottish Midland census regions.

*** Includes South West Scotland, South East Scotland and Southern Scotland census regions.

Data source: 16,091 life histories provided by family historians.

and a low friction of distance also enabled the flow of ideas which were important to national political integration in the early-nineteenth century.

In his study of the movement of financial capital in early industrial England, Black (1989) has critically examined the Gregory–Langton debate about the nature of English regions, and has demonstrated the complexity of capital flows in the early industrial revolution. Rather than seeing integration and differentiation as polar opposites, he chooses to stress their inter-relatedness, suggesting that it was perfectly possible to have a 'regionally fragmented series of export-based economies bound together by metropol-itan dominated flows of finance capital and commercial information.' He further emphasizes that 'The spatial mobility of finance capital during the industrial revolution was remarkable both in its fluidity and its ability to integrate seemingly disparate production systems' (Black, 1989, p. 381). He thus concludes that both regional distinctiveness and a degree of national integration, dominated by the London finance market, were compatible and important components of early industrial growth in England.

The debate about the extent to which there was regional integration or regional distinctiveness in Britain is thus complex and can be tackled in a variety of different ways. Undoubtedly one of the major factors affecting the level of interaction between regions was the ease and frequency with which people moved from one locality to another. It can be suggested that a migration system in which most population flows were contained within their region of origin would have heightened regional differences, whereas a system which consisted of substantial flows between regions would be indicative of a more fully integrated space-economy in which labour, goods and ideas could have moved easily from one location to another. It can also be suggested that the traditional role of London, which dominated the Brit-ish space-economy from at least the seventeenth century, was also an im-portant element in the development of a national migration system (Wrigley, 1967; Finlay, 1981). Moreover, it can be argued that regions distant from London might be expected to show a higher degree of marginalization and distinctiveness than those closer to the metropolis. These ideas can be examined using the longitudinal migration data drawn from family historians.

Patterns of regional migration

Analysis of the spatial pattern of regional migration can be carried out within individual counties and by larger census regions. The latter frame-work provides a larger sample, and is used for most analyses, but individual counties are examined to illustrate particular points. One simple measure of the degree to which migration flows formed a series of regionally differen-tiated systems, or were part of a larger national pattern of population inter-change, is to examine the extent to which movement was contained within defined regions (Table 3.9). Using relatively large census regions and taking data for two time periods, within-region flows – that is moves that had both an origin and destination within the region – were the largest single category in all regions and formed more than half of all moves in every region before 1880, and all bar London, the South Midlands and North Wales after 1880. If the smaller units of individual counties are examined the same trends are apparent but, not surprisingly, the proportion of within-region moves decreases. Even so, over the whole time period within-region moves are the single most important category for all English counties except Buckinghamshire, Middlesex, Surrey, Worcestershire and Rutland. Whilst the data for Rutland can possibly be discounted as the county was much smaller than the other units and consists only of a very small sample (113 moves), it can be suggested that the other four counties were each situated close to a large and expanding conurbation thus accounting for the higher than expected movement across a county boundary. There are also a number of counties in Scotland and Wales with relatively low proportions of within-county movement but, like Rutland, these are mostly very small counties and/or regions for which the sample is rather small. It should, of course, be stressed that much movement outside a region could still be short distance as administrative boundaries did not necessarily have any significance for those who migrated (see below).

The same trends are true if the data are disaggregated into four time periods: within-region moves remain the single most important category in all census regions and time periods except London after 1920 when movement out of the metropolis becomes the dominant trend. There is also a general trend for within-region moves to become rather less important in the twentieth century, with such moves accounting for less than 50 per cent of all moves in half the census regions (London, South East, South Midlands, South West, West Midlands) in the period after 1920. It is clear that population interchange with other regions both increased with time, as transport became more widely available and opportunities for longer distance moves increased, and was most common in regions which contained or were adjacent to the larger centres of population which were undergoing significant population change. Even so, it should be stressed that within-region flows were dominant in most regions and time periods.

It is also worth examining the balance between movement into and out of the census regions. Migration into and out of the regions was very evenly balanced and the only region with a large difference between the two was the West Midlands, which experienced approximately twice as much in-migration as out-movement. Even when the data are broken down into smaller county units movement into and out of the regions remains surprisingly evenly balanced, and it is notable that the counties which comprise the West Midlands also then had balanced migration flows. Thus counties such as Gloucestershire, Herefordshire, Shropshire, Staffordshire, Worcestershire and Warwickshire must have been disproportionately attracting migrants from outside the West Midland region, but those leaving the counties must have stayed within the regional migration system. English counties where the in-flow exceeded the out-flow of migrants by about five per cent or more include Essex and Surrey, whilst those with a similar excess outflow include Cumberland, Norfolk, Wiltshire, the North Riding of Yorkshire and the very small county of Rutland. In Scotland and Wales county-level imbalances, particularly population losses, were rather greater with counties such as Bute, Clackmannan, East Lothian, Kinross, Ross and Cromarty, Sutherland, West Lothian, Wigtown, Anglesey, Cardigan, Carmarthen and Radnor all having out-flows which were substantially greater than inflows. This fits with what is known about rural depopulation in remoter districts (Saville, 1957; Flinn, 1977; Devine, 1992); but many of these counties are small in area and sample size, in most cases within-county movement remained the single most important component, and movement out of the county was not necessarily over a long distance (see below).

Variations in levels of in- and out-migration are more marked when the data are divided into four time periods (though sample sizes are smaller). The census regions can be classified according to their balance between population loss and population gain (bearing in mind that in almost every case local population redistribution remained the single most important component). Only the West Midlands had more in-migrants than out-migrants in all four time periods, emphasizing the extent to which this region experienced industrialization and urbanization relatively early, sustained this during the nineteenth century and successfully underwent a transition to a new set of industries in the twentieth century thus avoiding large-scale out-migration associated with the inter-war depression (Gill, 1952; Briggs, 1952; Sutcliffe and Smith, 1974). The only areas to have a net loss of population in all four time periods were both Northern and Southern Scotland, though in most time periods (and especially in Southern Scotland) the differences were small. Only in Northern Scotland between 1840 and 1919 did population loss markedly outstrip population gain. Three areas, London, North-West England and South Wales experienced significant population gain during the industrial revolution followed by population loss in the twentieth century. Both the North West and South Wales are classic industrial regions that have been unable to adjust successfully to changed economic conditions in the twentieth century (Walton, 1987; Singleton, 1991; Humphries, 1972; Cooke and Rees, 1981), whilst large-scale out-movement from the metropolis is well known, though it could be argued that London has, in effect, simply transferred its population to adjacent counties through the process of suburbanization (Hoggart and Green, 1992). The largest group of counties (South East, South Midlands, Eastern England, South West, North Midlands, North Wales) are those that lost population to more industrial areas during the industrial revolution, but which have gained population from the late-nineteenth century and, especially, in the twentieth century.

Finally Yorkshire and Northern England show a more mixed and overall quite stable pattern, with population losses and gains being relatively evenly balanced in each time period.

The spatial patterns associated with moves into, out of and within the British regions are summarized in maps presented in Appendix 5. Such maps clearly only show longer distance moves: moves of under 1 km cannot be mapped at this scale. The maps contain a rich detail which it is not possible to explore in depth for every region in this chapter. Rather, this section will describe and explain the general trends which emerge, drawing examples from selected regions. Some of the processes underlying these patterns are explored in later sections. First, the maps demonstrate the extent to which there was a relative shift from within-region moves in the eighteenth and early-nineteenth century, to inter-regional moves in the twentieth century. Although a large amount of migration remained within a regional network, there is evidence of increasing integration between the British regions from the late-nineteenth century. Second, it is clear that both the volume and pattern of in- and out-migration were quite closely matched for all regions. This confirms Ravenstein's (1885/1889) assertion that every flow produced a counterflow, and suggests that there were well-established migration routes over which interchange of population was a well-established feature. Patterns of in- and out-migration were quite stable over time, which suggests that although the volume of interaction between regions increased from the late-nineteenth century, the spatial networks over which such movement took place were of long standing. It is likely that, given this stability, flows of information about opportunities and links between friends and kin would have aided the migration process. These issues are explored in detail in later chapters. Third, the maps show that relatively few regions exchanged substantial numbers of migrants with all parts of Britain. The majority of inter-regional movement was with adjacent counties.

Thus, much of the inter-regional movement which occurred was short distance and did not remove migrants far from familiar territory. For instance, most movement out of the West Midlands was either pulled to the adjacent industrial counties of North-West England, or to expanding urban centres to the south. The main exception to this pattern was London and, increasingly in the twentieth century, the wider South East, which had a substantial interchange of migrants with all parts of the country. Thus, inter-regional flows to and from South-East and South-West Scotland (principally Edinburgh and Glasgow) were dominated by population interchange with London, and both South-West and North-West England also had well-established and substantial flows of migrants to and from London and the South East. These patterns of inter-regional migration seem to confirm Black's (1989), hypothesis about the parallel and interrelated process of regional integration and differentiation. Whilst much migration was short distance within regions or to adjacent counties, longer distance moves were dominated by flows to and from London and the South East. It can thus be suggested that throughout the period from the mid-eighteenth century there existed a series of well-defined regional migration systems, superimposed upon a national system which was dominated by population interchange with London. Although the relative importance of inter-regional movement increased over time, the basic pattern remained remarkably stable over two centuries.

It is also instructive to examine the pattern of mobility within the relatively large census regions used in this analysis. It is clear that within most regions there were well-defined sub-regional migration systems, usually focused on the major urban centres or reflecting the physical structure and pattern of communications which existed within larger regions. Such patterns are particularly obvious in areas such as South-West England and Eastern England where, in both cases, there were a series of only partially interconnected migration networks operating within each region. In the large Northern region, bisected by the Pennines, it is clear that North-East England and Cumbria actually formed two quite distinct migration systems with very little interchange of population across the Pennines. Indeed, inter-regional flows to London and the South-East from both regions were more significant than movement between the Eastern and Western parts of the Northern region.

Perhaps what is most striking about the maps produced for each of the British regions is their similarity. Despite very different social, economic and physical characteristics, patterns of migration within and between the British regions were remarkably similar. Movement within regions, or to adjacent regions, was of overwhelming importance in all areas, each area contained clearly defined local migration systems, usually focused on the main urban areas but with substantial amounts of rural circulation, and most longer-distance population interchange was dominated by flows to and from London and the South East. Location and marginality with respect to London had some effect so that, not surprisingly, the volume of interchange with London was much greater from Eastern England than from Southern Scotland. But, even in the case of most Scottish or Welsh regions, the single largest inter-regional flows were to and from London and South-East England.

In terms of the debate about the degree of differentiation or regional integration which existed in the past, these data send contradictory messages. There clearly was a national migration system, centred on London and reflected by the way in which the British transport network also focused on the metropolis (Dyos and Aldcroft, 1969; Aldcroft and Freeman, 1983). Superimposed on this, however, were a series of very strongly-defined regional and sub-regional migration systems. At one level this suggests a strong degree of regional differentiation, but the fact that the migration process seemed to operate in much the same way in every region suggests that the underlying processes producing these regional systems were remarkably similar. Thus, it can be suggested that despite the physical, social and economic differences between regions, all parts of Britain were being affected by the same sorts of processes, which were quite stable over time and between areas. The existence of well-defined regional migration systems can thus be attributed not to a high degree of regional differentiation *per se*, but to the local and regional outcomes of national processes which structured the nature and extent of population migration in the past. The extent to which such similarities between regions extended beyond the basic spatial patterns of population migration is explored below.

Regional variations in the characteristics of migration

The characteristics of migration within each of the British regions strongly reflect the national pattern outlined above. The data are presented in Tables 3.10–3.12 but only the key points to emerge from the analysis are discussed here. Variations over time are not examined in detail as sample sizes often become rather small when the data are broken down by region, migrant characteristics and time period. It should also be noted that the data cannot be directly compared with figures from the national data set, as in the regional analysis all intra-regional moves are counted twice, once for the region of origin and once for the region of destination.

The level of mobility as demonstrated by the mean number of moves per person and the average length of residence at one address was remarkably constant across the different parts of Britain (Table 3.10). There is some evidence that people moved slightly less frequently in some remoter rural areas in Wales and the Highland region of Scotland, and slightly more frequently in London, but the differences are small. As with the national data set there are also no significant differences between male and female migrants at the regional level. In terms of the propensity to move, migrants were behaving in similar ways in all British regions, and the possible biases and omissions in the data outlined above appear to apply equally to all parts of the country.

There are rather larger variations in the distances moved within and between regions (Table 3.11), but these can mostly be explained in relation to the size, level of urbanization and industrialization of the regions (and hence their ability to retain migrants), and the marginality of particular regions to the London-dominated British migration system. Taking all moves together (that is all moves which had an origin, destination or both within a particular region), the areas which experienced notably longer-distance migration within Britain were the Scottish Highlands, South-East Scotland and South-West England. Regions with predominantly short distance migration were North-West England and Monmouthshire. Whilst data for Monmouthshire alone are slightly suspect due to a relatively small sample size, and for most analyses it is amalgamated with other parts of South Wales, the

Table 3.10 Mean number of moves and mean length of residence (years) by region of birth

Region	Mean number of moves	Mean length of residence	Number of individuals	Total number of moves
London	5.8	9.3	1,220	7,053
South East	4.8	9.7	1,323	6,369
South Midland	4.4	10.3	1,083	4,766
Eastern	4.4	11.3	1,039	4,515
South West	4.1	10.8	1,235	5,116
West Midland	4.5	9.9	1,249	5,601
North Midland	4.3	11.4	1,180	5,048
North West	4.6	10.7	2,044	9,376
Yorkshire	4.3	10.9	1,557	6,692
Northern	4.5	10.9	1,141	5,111
Monmouth	6.0	10.8	108	649
North Wales	3.8	11.2	232	891
South Wales	4.0	11.5	209	831
Glamorgan	5.2	10.3	226	1168
Highland	3.8	12.4	145	550
North East Scotland	4.6	11.9	441	2,012
Scottish Midland	4.6	10.3	432	1,981
South West Scotland	5.1	9.3	326	1,653
South East Scotland	5.2	9.7	185	960
Southern Scotland	4.3	10.8	198	850

Data source: 16,091 life histories provided by family historians.

other figures can be quite easily interpreted. Areas with predominantly long distance migration were distant from London and or had relatively few industrializing and urbanizing areas close by. In contrast North-West England retained a high proportion of its migrants or exchanged population with adjacent industrial areas.

Short within-region moves were mainly (and obviously) associated with small regions, and with large urban areas such as London and Edinburgh where there was a large amount of short distance intra-urban movement. London was particularly notable for a large amount of short distance movement within the metropolis. In contrast migrants within the large and remote Highland and North-East regions of Scotland moved on average over much longer distances. Remoteness and distance from London also largely explain differences in the mean distances moved into and out of regions. The Highland region and North-East Scotland attracted migrants over very long distances when they moved into these regions and, together with South-West Scotland, South-East Scotland and Northern England, also had out-migrants who travelled over longer than average distances. These were all areas that experienced economic decline and rural depopulation and these factors, combined with their remoteness, account for the relatively long distance moves. Short inter-regional migration was associated particularly

Table 3.11 Mean distance moved (km) by type of move by region

Region	Within region moves	Into region moves	Out of region moves	All moves with an origin, a destination or both in the region
London	2.5	142.2	93.3	58.5
South East	15.8	135.7	140.5	71.2
South Midland	10.5	96.6	105.4	57.9
Eastern	12.2	129.7	145.3	64.3
South West	15.0	206.6	191.8	88.4
West Midland	10.1	143.2	132.0	70.6
North Midland	9.5	134.7	128.2	50.2
North West	7.9	154.9	163.7	45.5
Yorkshire	10.1	152.3	140.9	51.8
Northern	11.8	188.8	217.0	63.0
Monmouth	4.2	77.3	98.0	43.1
North Wales	12.3	124.3	134.6	65.7
South Wales	14.1	127.2	128.0	69.6
Glamorgan	8.2	130.8	131.8	55.5
Highland	31.1	261.6	276.0	153.7
North East Scotland	17.0	243.1	289.6	68.9
Scottish Midland	12.1	140.2	167.3	62.7
South West Scotland	7.0	188.1	207.1	78.0
South East Scotland	4.1	180.2	223.9	86.9
Southern Scotland	11.6	129.9	175.1	64.3

Data source: 16,091 life histories provided by family historians.

with the South Midlands (for moves into the region) and London (for moves out of the region). In both cases this relates to proximity to other centres of population and, in the case of London, represents short distance suburbanization to the Home Counties.

Detailed examination of variations in distances moved by region and a wide range of migrant characteristics confirms that the migration process was operating in very similar ways in all parts of Britain. In all cases distance moved increased over time and, with a few exceptions, females travelled slightly further than males (as in the national data set). However, in seven regions (Northern England, Monmouthshire, South Wales, Glamorganshire, North-East Scotland, South-West Scotland and Southern Scotland) there was a tendency for men to migrate slightly further than women. These were all areas which were relatively remote and distant from London, and in which there was a tendency for longer distance movement to be more common for both men and women. This suggests that although in most cases women migrated further than men, in those regions where long distance out-migration was due mainly to economic decline and marginality, men were prepared to move further than women to find work. Again, it is worth emphasizing that emigration is excluded from all these calculations.

As shown in the national data set variations in distance moved by age were small, although there was a tendency for young adults to move further than those in a later stage of the life-cycle, and single people and those moving alone tended to move over longer distances than married couples and family groups. The only real exception was the Highland region where older people, married couples and family groups were at least as likely to move over long distances. The distinctiveness of migration from and to the Scottish Highlands is further emphasized by the fact that this is the only region which departs significantly from the national pattern when distance moved is related to occupation. In almost all regions long distance movement was mainly undertaken by those in professional work and by domestic servants, with very short distance moves most common among the unskilled. However, in the Highlands the furthest distances were moved by skilled manual workers with unskilled migrants moving almost as far as those with professional occupations. It would seem that the particular locational, economic and social circumstances that affected the Scottish Highlands from the mid-eighteenth century, including the decline of rural industry and the clearance of crofters from the land, led to more sustained and widespread long distance migration than that experienced in any other British region (Richards, 1982–5; Craig, 1990; Whyte, 1991).

The reasons why people migrated also varied little between the British regions (Table 3.12). Movement for work related reasons was of most importance in all regions, but there was a tendency for work related reasons to be particularly important in remoter districts of Scotland and Wales. It was perhaps in such districts, where job opportunities were limited, that the need to migrate for work was felt most acutely. When reasons for migration are broken down by regions and migrant characteristics all regions are remarkably similar, displaying characteristics very much in line with the national database discussed previously. The only variable where there is any significant variation is age, where it seems that although nationally and in most districts movement due to work became less important with age, in some regions work related moves were as important for those in the 40–59 age group as they were for younger adults. This feature is seen most clearly in South-West England, perhaps a symptom of pre-retirement migration as people moved to jobs in a desirable part of the country in which they wished to retire, but similar trends are also seen (but to a lesser extent) in the West Midlands, North-West England, Yorkshire, North Wales and North-Central Scotland. These were perhaps regions where economic circumstances made career adjustments and associated mobility in middle age more likely (Marsden, Lowe and Whatmore, 1990; Marsden, 1993).

In summary, regional analysis of migrant characteristics and of the reasons for migration emphasizes the similarities between the migration experience in different parts of Britain. Most differences that do occur were clearly related to the size of the region and its proximity to London and

Table 3.12 Reasons for moving by region* (%)

Region	Work	Marriage	Housing	Family	Other	Total number
London	34.6	18.5	16.5	3.7	26.7	6,224
South East	39.0	12.2	12.6	5.0	31.2	7,227
South Midland	35.3	13.3	16.7	5.2	29.5	5,420
Eastern	38.5	15.3	13.3	4.5	28.4	3,847
South West	45.6	14.1	10.4	4.6	25.3	3,726
West Midland	47.1	13.7	11.0	4.1	24.1	6,067
North Midland	42.1	15.9	13.8	3.9	24.3	4,129
North West	39.2	16.6	18.1	4.4	21.7	8,214
Yorkshire	40.4	17.1	15.7	4.0	22.8	5,622
Northern	45.9	14.8	16.8	3.9	18.6	4,270
North Wales	46.6	12.9	8.1	4.4	28.0	918
**South Wales	49.4	13.3	11.6	4.0	21.7	2,646
***Northern Scotland	50.5	13.3	11.7	3.0	21.5	3,273
****Southern Scotland	45.2	12.8	16.2	3.4	22.4	3,128
Complete dataset	37.8	15.4	15.6	3.7	27.5	

* All moves with an origin, a destination or both in the region.
** Includes Monmouth and Glamorgan
*** Includes Highland, North-East Scotland and Scottish Midland census regions.
**** Includes South-West Scotland, South-East Scotland and Southern Scotland census regions.
Data source: 16,091 life histories provided by family historians.

other urbanized and industrialized regions. The most distinctive British region was undoubtedly Highland Scotland where a combination of remoteness, economic and social change leading to loss of employment for many, and a lack of alternative opportunities nearby led to a much higher proportion of very long distance movement for work, undertaken by all ages and by a much wider range of occupational groups than in most regions. Interestingly, Highland Scotland not only experienced a surfeit of long distance out-migration, but people moving into the Highlands also travelled over equally long distances. Thus, with the possible exception of the Highlands, it would seem that although there were well-defined regional migration systems, the processes operating in each region were remarkably similar, with migrants moving for the same reasons and demonstrating the same range of characteristics in all parts of Britain.

The aggregate analysis which largely forms the basis for this chapter does of course obscure the diversity of individual experiences of migration which occurred in all regions. Although at the aggregate level similarities between regions appear greater than differences, this does not deny the significance of local variations related to particular circumstances pertaining to a village or town. Such variations cannot be revealed at this scale of analysis. The analysis presented in this chapter also gives no indication of the meaning associated with migration for those involved. Though basic patterns varied little over time and space, the meaning ascribed to movement by individual migrants may have changed significantly. The effects of time–space compression and the forces of modernity may have induced different attitudes to similar patterns of movement. Thus, although the volume of inter-regional migration was relatively low in both the eighteenth and twentieth centuries, the meaning of such moves for individual migrants may have changed significantly, with longer distance moves more difficult and disruptive in the past. Such issues are returned to in later chapters when individual life histories are explored in more detail.

CHAPTER 4

The role of towns in the migration process

Introduction

The century and a half after 1750 was primarily associated with the process of urbanization (Weber, 1899; Law, 1967; Robson, 1973; Adams, 1978). In the mid-eighteenth century only London (population 675,000 in 1750) could claim to be a large urban centre. The metropolis outstripped the next largest towns by over half a million people with Edinburgh (57,000), Bristol (50,000), Norwich (36,000), Glasgow (32,000) and Newcastle (29,000) comprising the top five urban centres after London. By 1801, although the newer industrial cities of Manchester (89,000) and Liverpool (83,000) were now ranked second and third – and Bristol had sunk to seventh and Norwich eleventh in the urban hierarchy – only one third of the population of England and Wales and one fifth of the population of Scotland were classified as living in towns in the 1801 census. London (959,000) remained dominant as the only city with over 100,000 population. By 1851, however, over 50 per cent of the population of Britain lived in towns and ten cities had exceeded a population of 100,000. Proportionately, the first half of the nineteenth century saw the most rapid rates of urban growth throughout Britain. Although London continued to be at least five times as large as the next biggest city, by 1891 Manchester, Glasgow, Birmingham and Liverpool all exceeded half a million inhabitants, almost 40 per cent of the population of England and Wales lived in towns with over 100,000 inhabitants, and over 73 per cent of the British population lived in towns. By 1951 this had risen only marginally to 81 per cent, and the rate of growth for the largest towns had slowed markedly. However, almost 40 per cent of the British population now lived in cities of over half a million inhabitants, and Birmingham and Glasgow were both conurbations of over one million. Such urban growth had a massive impact on the lives of contemporaries and focused attention on the problems associated with urban living (Abrams and Wrigley, 1978; Corfield, 1982; Waller, 1983; Gordon and Dicks, 1983; Williamson, 1989).

Many contemporary observers linked high rates of urban growth to the observable process of rural to urban migration (see Chapter 1), and assumed that population migration was the most significant component in the rapid expansion of towns (Danson and Welton, 1859; Ravenstein, 1885/ 1889; Redford, 1926). For instance Danson and Welton commented on move-ment into the industrial Lancashire town of Preston:

> A stream of population constantly passes into Preston from the north. This we may reasonably suppose to consist, to a large extent, of persons born in the districts of Fylde, Garstang and Clitheroe, to whom such movement is not only obviously profitable, but also comparatively easy. (Danson and Welton, 1859, pp. 48–9)

Although rural to urban migration was the dominant demographic component of urban growth in pre-industrial towns (Finlay, 1981; Wrigley, 1987), most industrial cities grew more due to natural increase than by in-migration. Although some rapidly emerging towns such as resorts, some dormitory towns and a few new industrial centres owed more of their population growth to migration than natural increase for short periods, in most instances after 1850, natural growth was the most significant component (Welton, 1911; Cairncross, 1949; Lawton, 1983). This suggests that, if contemporary observations about the large numbers of rural to urban migrants were correct, then in-flows to towns were almost matched by out-flows. This is consistent with Ravenstein's observations that every flow had a counterflow, but goes against his own hypothesis that most migrants were drawn to the centres of industrial expansion, and is contrary to the observations of contemporaries about the significance and permanence of rural to urban movement. Thus the novelist Charles Dickens caught the flavour of contemporary opinion when he stated in Dombey and Son:

> Day after day, such travellers crept past, but always . . . in one direction – always towards the town. Swallowed up in one phase or other of its immensity, towards which they seemed impelled by a desperate fascination, they never returned. (Dickens, 1982, pp. 404–5)

These conflicting observations and interpretations of the role of towns in the nineteenth-century migration system require much more detailed investigation and, in particular, we know very little about the significance of rural to urban movement for urban growth before census and civil registration data are available from 1841.

This chapter examines a series of specific hypotheses about the extent, nature and characteristics of migration between places of different size. First, to what extent was rural to urban migration, and movement from small towns to large towns, a dominant and consistent process in all British regions in the period after 1750? Second, were flows of migrants towards urban centres matched by counterflows away from such places to other towns and rural areas? Third, to what extent did the selective characteristics of rural to urban migrants lead to the transfer of a younger population to towns and thus, it can be argued, the migration process effectively shifted rural fertility to towns and contributed to the high rates of natural increase in urban centres? Fourth, how important were migration experiences other than rural to urban moves, including processes of counter-urbanization, rural circulation and intra-urban residential mobility? By examining the totality of the migration experience over the 200 years from 1750, it should be possible to place assumptions about the significance of rural to urban moves in a wider context and resolve some of the apparent contradictions that emerge from earlier work.

One of the problems associated with studying the significance of rural to urban movement lies in the arbitrary nature of rural and urban definitions. Studies of urbanization in nineteenth century Britain have adopted criteria of population size, sometimes backed up by evidence on population density and degree of nucleation. It is argued that such measures can, to some extent, be used as surrogates for, otherwise unavailable, data on urban functions (Law, 1967; Robson, 1973). However, there is no ready agreement about the size threshold that should be used to determine an urban settlement (Carter, 1995, pp. 10–17), and concepts of rurality are as elusive as definitions of urban settlements (Cloke and Edwards, 1986; Cloke, 1994). In this analysis implications about the degree of rurality or the extent of urban functions associated with particular settlements are avoided. Movement between settlements is assessed purely on the basis of population size, to examine the degree to which movement was mainly from small to larger places or vice-versa.

Even the use of size categories is not straightforward in an analysis which spans 200 years. A settlement that was a village in 1800 may have been a substantial town in 1900, and a small settlement in 1800 may have delivered similar functions to a much larger place a century later. The extent to which people moved between places of different size must, obviously, also be related to the size distribution of settlements within a region. Thus, if there are no towns between 50,000 and 100,000 in a region, it is not possible for there to be intra-regional movement to towns in that size category. These considerations are particularly important when regional variations in the migration system are examined. To ensure reasonable sample sizes, the analysis of movement between settlements of different size is carried out within the four time periods used in Chapter 3, and town size is identified according to the 1801 census for moves undertaken in the period 1750–1839, the 1851 census for 1840–79, the 1891 census for 1880–1919, and the 1951 census for moves after 1919.

The same town size categories are used throughout the analysis although, as noted above, the relationship of places of a particular size to the delivery of urban functions will have varied over time. The eight size classes are: settlements under 5,000 people; those 5,000–9,999; 10,000–19,999; 20,000–39,999; 40,000–59,999; 60,000–79,999; 80,000–99,999; and settlements of 100,000 population and over. London was always identified as a separate category, and particular care was taken to identify emerging suburbs of large towns which, whilst not within the administrative framework of a town, were contiguous with it and could reasonably be argued to be part of the same settlement. Thus, in the 1891 census, Didsbury (Lancashire) was classified as a separate settlement, but in reality was a contiguous part of the Manchester urban area. Such suburbs have been separately identified in the analysis and added to their adjacent city. Thus, movement from an inner residential area of a conurbation to a contiguous suburb is not recorded as movement down the urban hierarchy, but can be correctly identified as intra-urban suburban movement whether or not the settlement concerned was administratively part of the conurbation.

Did people move from small places to larger towns?

It has been suggested that census-based studies of migration have placed undue emphasis on inter-regional movement at the expense of short distance migration and local circulation. It was also demonstrated in Chapter 3 that only London was able to attract migrants from all parts of Britain and that for most people movement within a local region was the most common experience. It can also be suggested that such short distance movement was relatively undisruptive to both individual families and local communities, and that it was long distance moves that had the greatest impact on people's lives. The pattern of movement between settlements in different size categories is given in Appendix 6 and summarized in Tables 4.1 and 4.2. Taking all the data together, the most common migration experience, and almost certainly the one which had least impact on people and places, was very short distance intra-settlement mobility. Overall, 41.7 per cent of all moves had an origin and destination within the same settlement and 64.9 per cent of moves within the same size category were intra-settlement moves. This confirms evidence from local studies which have emphasized the importance of intra-urban movement in both the past and the present (Dennis, 1977a; Pooley, 1979; Stillwell, Rees and Boden, 1992).

The relative significance of within-settlement moves varied both over time and between places of different size. Intra-settlement moves were least important in the first time period (1750–1839), accounting for 27.5 per cent of moves within the same size category; from 1840 to 1919 intra-settlement moves accounted for over 40 per cent of all moves, but after 1920 within-settlement moves declined in importance. It can be suggested that the figure for the first time period may be artificially low because, as noted in Chapter

Table 4.1 Summary of all moves by settlement size and year of migration (%)

Type of move	1750–1839	1840–79	1880–1919	1920–94	All moves
Within same settlement*	27.5	45.1	47.9	37.3	41.7
To new settlement in same size category**	49.8	25.9	16.9	16.9	23.6
To settlement in larger size category	13.8	16.4	17.8	20.4	17.6
To settlement in smaller size category	8.9	12.6	17.4	25.4	17.1
Total moves	8,198	19,658	20,935	17,873	66,664

* For towns over 100,000 population moves to contiguous suburbs have been classed as within-settlement moves.
** Population size categories as given in Table 4.2.
Data source: 16,091 life histories provided by family historians.

Table 4.2 Intra-settlement moves as a proportion of moves in each settlement size band by year of migration (%)

Size of settlement	1750–1839	1840–79	1880–1919	1920–94
<5,000	18.5	28.1	26.8	20.9
5,000–9,999	28.1	43.8	42.3	31.1
10,000–19,999	34.6	48.9	41.4	41.4
20,000–39,999	48.6	54.1	47.1	40.3
40,000–59,999	52.3	61.6	59.8	39.3
60,000–79,999	61.6	72.4	50.4	43.6
80,000–99,999	35.4	63.3	58.9	44.5
100,000+*	N/A	52.5	57.2	42.8
London	68.7	72.6	62.8	41.4
Total number	2,250	8,863	10,036	6,658

* excluding London.
Settlement sizes were calculated at four census dates: 1801, 1851, 1891, 1951.
Data source: 16,091 life histories provided by family historians.

2, the more restricted range of sources available for the eighteenth and early-nineteenth centuries may have led to the omission of some very short distance mobility. The figures in the twentieth century suggest two trends. Increased affluence and housing choice enabled families to move more frequently to meet their changing housing needs, but the ability to commute over longer distances has meant that there is relatively less short distance adjustment within a settlement. Not surprisingly, in all time periods intra-settlement moves formed a smaller proportion of moves within a size class in small places of under 5,000 people (Table 4.2). This simply reflects the fact that there was both less opportunity and need to adjust housing to other changes in employment and family circumstances. In settlements of over 10,000 people intra-urban moves formed in excess of 90 per cent of all moves within the same size category in almost all time periods and size bands (Appendix 6).

If movement within the same settlement was the most common migration experience, then removal to another settlement in the same size category was the next most frequent occurrence. In total around two thirds of all moves were either within one settlement or to another settlement in the same size band. Although such movement inevitably could involve movement to a slightly larger or smaller place, there is little evidence of large scale movement up or down the urban hierarchy. The proportion of all moves which had an origin and destination in the same size category did decline from the eighteenth to the twentieth century, and in all time periods such moves were most common from origins of under 5,000 people. But the overwhelming impression is of people moving either within a settlement or between places of similar size. This directly challenges Ravenstein's (1885, 1889) assertions about the importance of stepwise movements up the urban hierarchy. If people were moving in stages up the urban hierarchy they were moving in very small increments which rarely took them outside the size bands used in this analysis.

Even if movement between towns of different size was less common than the observations of contemporaries might have led us to expect, migration theories and contemporary observations would still predict that most moves to places of a different size would be up the urban hierarchy to progressively larger settlements. There is little evidence that this was indeed the case. Over the whole time period moves to larger settlements were almost equally matched by moves to smaller settlements. It should again be noted that suburban movement in large towns is excluded from such calculations because these have been treated as intra-settlement moves. In the first two time periods (1750–1879), coinciding with the era of most rapid urbanization and industrialization, there were more moves up the urban hierarchy than down it, though the differences were relatively small. From 1880–1919 moves up and down the hierarchy were almost evenly matched, and in the period after 1920 there are clear signs of early counter-urbanization as more people moved to smaller than to larger settlements (Fielding, 1982; Champion, 1989; Pooley and Turnbull, 1996).

Closer examination of the data by time period and size category shows that in the periods from 1750–1879, migrants leaving towns in all size categories over 5,000 people were more likely to move to a smaller place than to a larger one. This directly contradicts Ravenstein's (1885/1889) assertion about progressive movement up the urban hierarchy during the period of most rapid urbanization and industrialization. After 1880 the same trends are evident but there was marginally more movement up the urban hierarchy from places under 20,000 population. By definition, all those leaving settlements of under 5,000 population to move to a town in a different size category must have been moving up the urban hierarchy, and such migrants formed the vast majority of people moving to a larger settlement in all time periods. Although in the first time period, when all towns except London were relatively small, most movement from places under 5,000 population was to towns under 20,000 in the next two size categories, thereafter, movement to places of over 100,000 population (London, the principal provincial towns and their suburbs) was as significant as movement to smaller places. In each time period around one third of such migrants moved to very large cities (100,000 plus) and one third to places of under 20,000 people. Thus it would seem that in the late-eighteenth and early-nineteenth centuries those migrants who were drawn to larger settlements went mainly to places only marginally larger, that there is little evidence that they subsequently moved in a stepwise fashion up the urban hierarchy, and that after 1840 a substantial proportion of those leaving places of under 5,000 people went directly to very large cities.

What is equally interesting, is the fact that of those moving down the urban hierarchy from larger to smaller settlements a substantial proportion moved directly to a very small settlement of under 5,000 people. This holds true for all time periods and for migrants leaving towns in all size classes. Thus of those leaving cities of over 100,000 people for towns in a different size class (who by definition must have moved down the urban hierarchy) at least 48 per cent went directly to places of under 5,000 people. Thus although there was movement from small places to some large cities, there was an almost equal amount of migration in the other direction. Whilst this confirms Ravenstein's observations on the extent to which each migration stream had a counterflow, it runs against his argument that there was substantial net rural to urban movement with people being sucked into the large centres of industry and commerce.

These interpretations are backed by further analysis of the data. Table 4.3 examines the extent to which sequential moves took migrant families progressively up (or down) the urban hierarchy, focusing particularly on the extent to which population was retained in towns within a series of broad size bands. The data are organized according to the date of birth of the migrant and track moves sequentially through places of different size. Thus for those born 1750–1819, 79.1 per cent originated in a place of under 5,000 people. After their first move 70.7 per cent of such people remained within small settlements, 67.6 per cent after a second move and so on with 45.0 per cent remaining in small places after eleven or more moves. The tables show three main trends. First, not surprisingly, the proportion originating in small settlements declined substantially from the eighteenth to the twentieth centuries as the British population became increasingly urbanized. Second, for the first two time periods (1750–1849) there was a tendency for the proportion living in small places of under 5,000 people to

Table 4.3 Sequential moves by town size category and birth cohort (%)

Residential sequence	1750–1819		1820–49		1850–89		1890–1930	
	<5,000	≥100,000	<5,000	≥100,000	<5,000	≥100,000	<5,000	≥100,000
Origin settlement	79.1	4.2	60.4	12.5	42.0	33.9	25.1	41.0
Destination settlement 1	70.7	7.2	48.6	18.8	34.3	31.4	23.7	38.3
Destination settlement 2	67.6	8.8	46.1	20.2	31.2	33.4	24.5	34.8
Destination settlement 3	64.2	10.5	42.8	22.0	30.8	31.8	23.9	34.6
Destination settlement 4	61.8	11.0	42.8	22.7	31.9	31.0	24.8	34.1
Destination settlement 5	57.8	12.2	41.8	25.1	32.5	29.7	28.4	31.8
Destination settlement 6	55.8	14.0	40.1	26.2	33.8	28.8	27.8	31.9
Destination settlement 7	50.8	18.4	41.7	24.0	34.1	27.4	32.8	27.3
Destination settlement 8	49.1	21.1	37.9	28.0	34.8	27.7	31.8	26.6
Destination settlement 9	47.7	17.7	43.0	24.8	36.6	27.0	33.9	27.3
Destination settlement 10	44.9	20.3	36.4	31.6	41.3	22.8	32.9	26.7
Destination settlement 11	45.0	19.8	37.2	24.8	37.8	23.0	36.0	22.1
Total size of cohort	4,394		3,706		4,456		2,773	

The table should be read as follows: 79.1% of those born 1750–1819 originated in a settlement of under 5,000 population, and 4.2% in a settlement of 100,000+ population (including London). After their first residential move 70.7% of the cohort born 1750–1819 remained in a settlement of under 5,000 population and 7.2 lived in a settlement of 100,000+ population. Of those who made 10 moves, 44.9% were then in a settlement of under 5,000 population and 20.3% were in a settlement of 100,000+ population.

Data source: calculated from 16,091 residential life histories provided by family historians.

Table 4.4 Type of move by mean distance (km) and year of migration

Type of move	1750–1839	1840–79	1880–1919	1920–94	All moves
Within same settlement*	0.5	0.5	0.5	0.5	0.5
New settlement in same size category	33.7	41.4	54.7	86.4	51.8
New settlement in larger size category	94.3	83.6	88.4	95.4	86.3
New settlement in smaller size category	87.4	75.8	79.8	88.9	79.3

* Includes contiguous suburbs
Data source: 16,091 life histories provided by family historians.

decline steadily with progressive moves, but in the period after 1850, moves which came later in an individual's life history were more likely to be back to a small settlement, and the proportion living in small places began to rise. Third, in the first three time periods 1750–1889, the most significant drop in the proportion living in small places occurred after the first move undertaken by individual migrants. Thus there was a tendency for migrants to make an initial move to a larger place, but thereafter the proportion that remained in small settlements was quite stable.

It is also important to examine the distances over which moves between places of different size were undertaken. For instance, it could be argued that, although suburban moves have been treated as intra-urban movement, most moves from large to small places were very short distance moves of only a few kilometres to semi-rural villages which, whilst distinct settlements, were in effect suburbs of a large conurbation. However, Table 4.4 suggests that this was not the case. Although, as shown in Chapter 3, average distances moved increased over time, in all time periods the shortest moves were to places within the same size bands. As these also form the largest category of migrants in most time periods, we can confirm that most people were moving relatively short distances to settlements of a similar size within the same locality. Distances moved to larger places were only slightly longer than those to smaller settlements, with moves to smaller places averaging almost 80 kms. These were clearly substantial long distance moves rather than short distance migration to settlements close to a conurbation from which the migrant had come.

These data have important implications both for our understanding of the relative importance of migration and natural increase for the growth of towns in the period before about 1840, and for our interpretation of the impact of the migration process on people and places. As outlined in Chapter 1, we know that natural increase was a more important component of urban growth than migration in most towns after the mid-nineteenth century. There is also evidence that pre-industrial towns depended much more

Table 4.5 Type of move by reason and year of migration (%)

Reason	Within settlement move*	To new settlement in same size category	To settlement in larger size category	To settlement in smaller size category
1750–1839				
Work	28.8	56.4	54.2	57.6
Marriage	42.3	26.1	16.8	14.0
Housing	12.8	2.0	0.9	2.1
Family	0.9	1.5	2.3	2.5
Crisis	6.4	4.1	5.9	17.4
Retirement	0.5	0.7	0.1	1.7
Other	8.3	9.2	19.8	4.7
Total number	1,478	2,904	814	528
1840–79				
Work	29.0	65.1	57.9	58.2
Marriage	28.7	15.2	16.0	12.7
Housing	22.6	2.8	1.8	4.0
Family	2.4	3.2	2.8	4.3
Crisis	7.3	5.7	7.2	14.3
Retirement	2.1	1.3	1.4	2.0
Other	7.9	6.7	12.9	4.5
Total number	4,827	3,690	2,351	1,847
1880–1919				
Work	19.7	60.2	54.3	51.0
Marriage	23.0	10.9	11.8	11.8
Housing	35.8	6.2	4.9	6.2
Family	2.7	3.2	3.9	3.4
Crisis	8.2	7.1	6.6	17.9
Retirement	2.3	3.8	2.9	3.4
Other	8.3	8.6	15.6	6.3
Total number	6,605	2,665	3,089	2,932
1920–94				
Work	9.5	36.7	35.8	30.9
Marriage	14.8	6.7	7.6	8.1
Housing	47.3	13.4	11.7	15.8
Family	3.8	7.5	4.3	7.8
Crisis	10.5	9.0	11.1	5.5
Retirement	4.3	8.3	5.6	8.7
Other	9.8	18.4	23.9	23.2
Total number	6,177	2,479	3,500	4,251

* Includes contiguous suburbs.
Data source: 16,091 life histories provided by family historians.

heavily on in-migration to maintain their population in the face of very high mortality, and it has been suggested that in the eighteenth century cities such as Liverpool derived the greater part of their population growth from in-migration (Langton and Laxton, 1978). The data derived from individual residential histories provided by family historians suggests that in the century before 1850 the volume of migration from small to larger settlements was only slightly greater than the amount of movement from large places to small. Thus although there was some shifting of population from countryside to town, this could not have provided sufficient migrants to account for the very rapid expansion of urban areas in the first half of the nineteenth century. This is not out of line with what we know was happening later in the century, and is consistent with recent research on fertility trends in Britain from the mid-eighteenth century which has emphasized the importance of high and increasing fertility for population growth in both urban and rural areas (Wrigley and Schofield, 1981).

If most people were not moving up the settlement hierarchy, but rather were principally moving short distances within settlements and to settlements of a similar size, then this has implications for the impact of migration. The data suggest that most people were not moving to alien urban environments, but rather remained within the same town or moved to nearby and similar settlements. Whilst some people did move to large towns, and the biggest towns attracted the most migrants, almost as many moved in the other direction. Many of these would have been people who had begun their life in the countryside. Such return migration and its implications for people and places will be explored further below.

Such patterns of migration are clearly related to the reasons why people moved (Table 4.5). As with the full data set, migration for work related reasons was most important for almost all types of move, but within-settlement moves were much more likely to be undertaken for reasons of marriage (especially in the period 1750–1839), and housing change. By the twentieth century almost half of all within-settlement moves were for housing reasons, emphasizing that families often traded a longer journey to work for improved living conditions (see Chapter 5). For inter-settlement moves the stated reasons for migrating were quite stable regardless of whether migration was up or down the urban hierarchy, though before 1920 moves to smaller places were more likely to be stimulated by a crisis. This possibly represents people returning to settlements they had previously left when some form of crisis struck (unemployment, ill health, a family tragedy). This will be explored in more detail below.

The characteristics of migrants undertaking different types of move were also quite stable over time (Tables 4.6a–b). Within-settlement moves were dominated by older, married migrants with families, and moves by those with relatively low-skilled craft and industrial occupations were over-represented. Such movement reflects the short distance residential adjustment of

Table 4.6a Type of move by migrant characteristics, 1750–1879 (%)

Migrant characteristics	1750–1839				1840–79			
	Within settlement move	To new settlement in same size category	To settlement in larger size category	To settlement in smaller size category	Within settlement move	To new settlement in same size category	To settlement in larger size category	To settlement in smaller size category
Gender								
male	27.6	48.8	14.6	9.0	46.5	24.4	16.4	12.7
female	26.8	53.0	11.6	8.6	45.9	25.6	15.9	12.6
Age								
<20	23.4	51.8	16.5	8.3	43.8	25.4	18.6	12.0
20–39	27.8	50.3	13.0	8.9	44.0	24.6	17.4	14.0
40–59	34.4	42.9	12.5	10.2	53.0	23.3	12.8	10.9
60+	42.7	42.0	7.3	8.0	55.2	25.4	10.9	8.5
Marital status								
single	19.5	51.8	19.8	8.9	40.6	26.1	20.6	12.7
married	32.7	48.5	10.1	8.7	50.1	23.8	13.8	12.3
Companions								
alone	12.9	53.0	23.8	10.3	25.5	30.1	29.8	14.6
couple	35.2	47.6	9.3	7.9	47.8	23.5	16.5	12.2
nuclear family	29.0	50.8	11.3	8.9	49.3	24.2	14.0	12.5
extended family	29.2	44.6	10.9	15.3	55.7	21.7	10.8	11.8
other	19.0	36.6	32.5	11.9	45.2	26.9	15.5	12.4

Position in family								
child	26.5	51.9	13.1	8.5	49.2	23.1	15.2	12.5
male head	28.4	48.8	13.6	9.2	46.4	24.3	16.4	12.9
wife	30.6	51.8	9.2	8.4	47.2	26.0	14.2	12.6
female head	20.5	57.4	14.9	7.2	39.6	28.0	21.2	11.2
other	17.6	39.0	30.2	13.2	40.4	28.8	18.6	12.2
Occupational group								
professional	23.9	35.4	21.2	19.5	31.2	23.9	19.6	25.3
landed farmer	24.3	70.7	1.0	4.0	29.1	58.3	4.7	7.9
intermediate	41.4	29.9	19.1	9.6	56.9	14.5	14.4	14.2
skilled non-manual	22.6	31.5	29.2	16.7	47.0	16.5	22.2	14.3
skilled manual	33.9	40.8	15.0	10.3	52.5	19.7	16.1	11.7
skilled/semi skilled agricultural	26.4	61.8	7.9	3.9	40.9	37.5	11.5	10.1
semi-skilled manual	31.5	41.5	16.9	10.1	51.2	17.2	18.5	13.1
unskilled manual	23.3	60.5	9.2	7.0	53.5	19.0	20.2	7.3
agricultural labourer	21.6	75.0	1.3	2.1	38.9	51.3	4.6	5.2
domestic servant	12.1	59.5	15.5	12.9	33.3	33.9	23.1	9.7
armed services	14.5	16.5	40.8	28.2	13.2	14.9	37.2	34.7
Total number	2,248	4,086	1,134	730	9,106	4,868	3,204	2,480

The table should be read as follows: 27.6% of all males who moved between 1750 and 1839 made within settlement moves whilst 48.8% made moves to new settlements in the same size category; 38.9% of agricultural labourers who moved between 1840 and 1879 made within settlement moves whilst 51.3% made moves to new settlements in the same size category.

Data source: 16,091 life histories provided by family historians.

Table 4.6b Type of move by migrant characteristics, 1880–1994 (%)

Migrant characteristics	1880–1919				1920–94			
	Within settlement move	To new settlement in same size category	To settlement in larger size category	To settlement in smaller size category	Within settlement move	To new settlement in same size category	To settlement in larger size category	To settlement in smaller size category
Gender								
male	49.5	15.4	17.8	17.3	39.7	14.1	20.6	25.6
female	48.8	15.9	18.2	17.1	37.8	15.9	21.2	25.1
Age								
<20	45.6	16.6	20.5	17.3	39.3	13.5	22.3	24.9
20–39	48.5	14.8	18.3	18.4	40.0	13.8	21.3	24.9
40–59	54.6	15.6	14.4	15.4	39.7	15.3	19.6	25.4
60+	52.9	16.9	15.7	14.5	36.0	16.6	20.9	26.5
Marital status								
single	42.2	16.3	22.6	18.9	33.0	14.7	25.3	27.0
married	53.7	14.9	15.2	16.2	41.3	14.7	19.3	24.7
Companions								
alone	24.2	19.2	32.9	23.7	25.5	16.2	29.5	28.8
couple	53.0	14.4	16.2	16.4	40.2	14.6	19.3	25.9
nuclear family	53.0	15.4	15.5	16.1	43.4	14.4	18.4	23.8
extended family	58.7	13.1	13.8	14.4	40.4	14.8	20.4	24.4
other	41.0	14.8	19.8	24.4	34.0	14.8	21.6	29.6

Position in family								
child	52.2	15.5	16.3	16.0	48.2	12.6	16.7	22.5
male head	49.2	15.6	17.8	17.4	39.2	14.4	20.7	25.7
wife	52.7	15.3	15.5	16.5	40.2	15.6	19.4	24.8
female head	32.6	18.7	28.8	19.9	31.4	15.8	26.1	26.7
other	42.2	12.6	20.9	24.3	32.6	16.3	23.3	27.8
Occupational group								
professional	30.5	17.9	22.8	28.8	31.2	14.9	24.3	29.6
landed farmer	27.0	52.0	8.0	13.0	29.3	52.0	4.9	13.8
intermediate	56.6	11.6	14.6	17.2	44.3	12.7	17.8	25.2
skilled non-manual	52.5	8.6	21.7	17.2	39.0	11.4	22.8	26.8
skilled manual	59.7	11.2	15.2	13.9	51.8	8.9	18.2	21.1
skilled/semi skilled agricultural	37.9	32.3	16.3	13.5	35.3	37.7	12.3	14.7
semi-skilled manual	61.2	10.1	14.8	13.9	60.6	8.9	16.4	14.1
unskilled manual	59.2	10.3	18.4	12.1	60.8	10.0	12.0	17.2
agricultural labourer	31.9	46.9	10.7	10.5	9.9	63.9	9.8	16.4
domestic servant	33.9	22.7	28.0	15.4	32.9	15.6	25.5	26.0
armed services	15.2	11.9	30.3	42.6	13.6	19.5	26.5	40.4
Total number	10,312	3,266	3,747	3,610	6,949	2,677	3,731	4,546

Data source: 16,091 life histories provided by family historians.

families in a later stage of the life course and the constraints on those with low skills and incomes to remain within the same settlements. Movement to other settlements in the same size category was the single most important category and, not surprisingly, contained a wide cross-section of all types of migrants. The only group that were consistently over-represented were farmers and agricultural labourers who, obviously, moved from one rural community to another. Migrants moving up and down the urban hierarchy had very similar characteristics, with young, single migrants moving alone over-represented in such flows. However, it should be stressed that even for such movement the numerically largest categories were moves by married migrants and nuclear families, and the age differential for movement down the urban hierarchy was much less significant for the earlier time period. In terms of occupations, moves by professionals, skilled workers and domestic servants (mostly moving with their masters) were over-represented in moves both up and down the urban hierarchy. This confirms the extent to which longer distance moves were constrained by marketable skills, resources and knowledge of new labour market opportunities (see Chapter 5).

Regional variations in the importance of rural to urban migration

It can be suggested that national trends in the relative importance of rural to urban movement *vis-à-vis* other types of migration may obscure important regional variations. For instance, it seems likely that the more highly urbanized and industrialized regions of Britain would have had rather different migration patterns from those regions which were more rural and marginal to the metropolitan-dominated urban system. However, data analyzed in Chapter 3 suggest that in terms of inter-regional flows all the different parts of Britain experienced very similar migration trends. This section assesses the extent to which movement between settlements of different size was similar in each of the British regions, and relates any differences to the process of urbanization and industrialization that was taking place from the mid-eighteenth century.

Obviously, the extent to which movement between places of different size could occur within one region will depend in part on the urban hierarchy within that region. Although in the country as a whole there is a relatively even gradation of towns in different size bands, in some areas there were few or no towns in a particular size category. This must obviously be borne in mind in subsequent interpretation, but to lessen this effect the analysis looks simply at migration within four categories: those who stayed within the same settlement; those who moved to another settlement in the same size category; those who moved to a larger settlement and those who moved to a smaller settlement. As in Chapter 3 the regional analysis considers all moves with an origin or destination or both within the defined region, thus moves between regions are counted twice (in the region of origin and destination) and mean figures cannot be directly compared with those for the whole data set. Data for London are in some cases excluded as, by definition, all those moving from the conurbation must be moving to smaller settlements.

In most respects the regional data reflect the national picture with remarkably little variation from place to place and only limited changes over time (Table 4.7). This interpretation will concentrate only on those areas where significant variations occur. In all regions most moves were either within the same settlement or to a settlement in the same size category, but there were variations in the balance between these two categories. A high proportion of within-settlement moves occurred in North-West England, South-West Scotland and South-East Scotland (including Glasgow and Edinburgh), Yorkshire and, in the later time period especially, in North-East Scotland and Northern England. There was thus a clear tendency for population to be retained within the same settlements in the urban and industrial areas of Northern England and Southern Scotland, especially before 1880. In contrast, regions in which movement to other places in the same size category was dominant included the Highland and Border regions of Scotland, North and South Wales and, before 1880, Eastern England. These were all predominantly rural and relatively marginal regions and the vast majority of such movement was between small places of under 5,000 population.

As with the whole data set, movement up and down the urban hierarchy was quite evenly balanced in most regions. In the period 1750–1880 all regions did experience more movement up the urban hierarchy than down, thus contributing to the process of urbanization, though in some cases the differences were small. However, after 1880 the position was reversed with the majority of regions experiencing more movement down the urban hierarchy than up it. The only areas which did not experience these counter-urbanizing trends after 1880 were the more remote and marginal areas of North Midland and Northern England, South Wales (excluding Glamorgan), and the Highland, North East, South East and Border regions of Scotland. Regions which experienced the clearest urbanizing trends in the period before 1880 were the West Midlands and South-West Scotland whereas, in the period after 1880, movement to a larger settlement was most important in relation to all types of moves in the outer South East though even in these areas it was outstripped by movement to smaller places.

These trends are borne out by an analysis of longitudinal sequential moves by region (Table 4.8). In the period before 1850 in all regions the proportion of migrants living in small places at the end of their life-course was always smaller than it had been at the beginning. As with the whole

Table 4.7 Type of move by region by year of migration (%)

Region	1750–1879					1880–1994				
	Within settlement move	To new settlement in same size category	To settlement in larger size category	To settlement in smaller size category	Total number	Within settlement move	To new settlement in same size category	To settlement in larger size category	To settlement in smaller size category	Total number
London	69.1	1.0	–	29.9	3,094	67.8	5.0	–	27.2	4,189
South East	24.4	33.8	22.9	18.9	2,972	21.9	23.9	24.9	29.3	6,166
South Midland	26.6	32.8	22.5	18.1	2,011	24.7	24.6	23.0	27.7	4,623
Eastern	26.9	40.7	20.4	12.0	1,852	27.2	25.2	21.5	26.1	3,105
South West	27.5	36.4	22.3	13.8	2,362	26.0	25.6	24.1	24.3	2,456
West Midland	26.3	31.3	24.6	17.8	3,042	29.4	21.3	23.9	25.4	4,682
North Midland	28.9	39.3	19.7	12.1	2,108	37.7	19.6	21.8	20.9	3,040
North West	44.2	21.0	20.7	14.1	4,471	43.8	13.6	19.9	22.7	6,738
Yorkshire	36.1	27.3	21.1	15.5	2,933	44.3	14.3	20.0	21.4	4,368
Northern	30.6	40.2	16.4	12.8	2,534	41.1	19.7	20.0	19.2	2,796
Monmouth	15.9	56.4	15.9	11.8	264	35.0	25.6	19.4	20.0	547
North Wales	28.0	45.4	16.2	10.4	403	22.9	35.9	19.0	22.2	654
South Wales	28.0	43.2	16.3	12.5	971	31.5	34.1	17.1	17.4	1,371
Highland	19.3	61.9	12.9	5.9	171	17.3	41.3	22.2	19.2	104
North East Scotland	37.9	44.5	10.8	6.8	849	39.8	30.7	17.6	11.9	909
Scottish Midland	30.4	36.1	19.7	13.8	961	33.7	21.3	22.7	22.3	951
South West Scotland	42.0	19.2	22.9	15.9	840	43.8	17.4	19.5	19.3	1,151
South East Scotland	38.0	23.0	22.8	16.2	534	50.4	13.0	19.7	16.9	808
Southern Scotland	25.8	53.4	12.4	8.4	429	33.4	39.5	18.8	8.3	276

The table should be read as follows: 24.4% of all moves made between 1750 and 1879 which had an origin or destination in the South East were within settlement moves whilst 33.8% were to new settlements in the same size category.

Data source: 16,091 life histories provided by family historians.

Table 4.8 Sequential moves by region and birth cohort (% resident in settlement with a population under 5,000)

Region	Origin 1	Destination 1	Destination 2	Destination 3	Destination 4	Destination 5	Destination 6	Destination 7+	Total number
1750–1849									
London	41.7	21.8	19.3	22.0	25.3	22.4	23.1	30.3	1,177
South East	67.5	55.1	50.9	51.3	47.1	42.0	39.7	41.5	1,079
South Midland	69.8	54.6	51.0	51.7	47.1	45.3	37.8	43.0	859
Eastern	76.9	64.9	58.7	54.7	52.1	40.3	38.6	39.2	718
South West	73.2	59.4	56.0	47.2	51.1	46.4	45.7	42.4	889
West Midland	66.4	55.3	51.6	47.4	44.4	42.5	35.2	37.7	922
North Midland	77.7	63.6	56.6	52.4	47.8	46.0	38.6	36.3	761
North West	61.0	47.9	45.1	40.5	40.6	40.8	42.9	41.2	1,371
Yorkshire	64.6	53.1	47.6	41.9	40.1	38.2	39.5	36.9	1,067
Northern	70.4	63.0	58.2	54.5	55.2	50.0	58.6	52.5	817
North Wales	80.5	65.4	69.1	65.4	61.3	57.8	52.2	67.4	159
South Wales	80.6	71.9	69.0	63.2	62.1	54.4	46.6	50.6	345
Northern Scotland	82.6	71.9	66.1	63.2	59.3	60.2	56.5	42.5	581
Southern Scotland	71.3	59.3	51.4	50.0	49.2	47.0	47.5	37.0	464
1850–1930									
London	23.6	21.7	20.4	24.0	26.6	30.9	34.0	41.9	1,533
South East	33.7	32.5	34.4	36.2	39.8	40.7	45.3	48.9	1,728
South Midland	38.7	37.0	36.0	35.2	38.4	41.9	43.2	48.5	1,401
Eastern	40.3	38.0	36.6	40.2	41.4	43.4	44.0	49.9	935
South West	46.4	40.4	38.1	41.9	41.6	44.9	44.2	47.5	806
West Midland	41.0	37.0	35.3	38.5	36.4	38.5	36.5	43.3	1,116
North Midland	46.3	37.5	40.4	37.4	37.4	43.1	42.8	47.6	886
North West	32.0	27.9	29.1	27.7	30.6	32.5	33.4	41.2	1,644
Yorkshire	35.2	29.7	29.7	28.3	29.8	26.9	34.9	38.3	1,166
Northern	41.9	36.4	39.3	35.8	38.7	37.9	40.6	54.3	738
North Wales	47.6	54.1	46.9	55.6	51.9	55.3	52.8	61.5	231
South Wales	56.7	50.7	49.5	50.5	50.9	50.5	48.0	52.0	446
Northern Scotland	51.7	44.2	36.6	36.8	38.2	40.1	44.4	42.7	513
Southern Scotland	43.2	37.7	31.4	27.8	31.5	34.9	35.5	42.0	581

The table should be read as follows: 41.7% of people born between 1750 and 1849 who lived at some time in their life in London were born in a settlement with a population <5,000; after their first move 21.8% were still resident in a settlement with a population <5,000; after their sixth move the proportion living in a settlement with a population <5,000 was reduced to 23.1%.

Data source: 16,091 life histories provided by family historians.

data set, in all regions there was a tendency for movement to larger places to occur early in the life-course. However, in the period after 1850, in most regions the proportion living in places of under 5,000 people at the end of their life-course was larger than it had been at the beginning. The only regions which did not share this counter-urbanizing trend were South Wales (excluding Glamorgan), and the whole of Scotland. Thus it would seem that the process of counter-urbanization, usually thought of as a late-twentieth century phenomenon (Fielding, 1982; Champion, 1989), was beginning to have a significant effect from the late-nineteenth century and that the only parts of Britain not affected by these early counter-urbanizing trends were some of the remoter districts of Wales, Scotland and Northern England.

The reasons for migration and the characteristics of migrants within each region can also be examined. However, the data for each part of Britain closely mirrors the national picture and the details are not reported here. Two principal trends are obvious from this regional analysis. The first, reinforcing the conclusions of regional analysis in Chapter 3, is that despite obvious social, economic and cultural differences between regions, the pattern and process of migration in each large area was remarkably similar. There are some differences between the most urbanized or industrialized regions and the remoter rural regions with respect to the relative importance of movement between settlements within the same size category and, particularly, the significance of counter-urbanization in the period after 1880, but in all other respects the same trends were occurring in all parts of Britain.

Second, it is clear that although in all regions before 1880 more people moved to larger settlements than small, and thus migration was part of a general urbanizing trend, the difference between movement up and down the urban hierarchy was usually small. This again confirms the work of Welton, Cairncross and others which emphasizes the role of natural increase in urban growth in the majority of settlements. It also places the process of rural to urban migration in a broader perspective. Although migration did transfer people from countryside to town in all regions before 1880, this was a very small part of the total migration process and for most people movement within a settlement, to another settlement of a similar size, or to a smaller place was a much more common migration experience. This point is emphasized if we consider the longitudinal nature of the migration process, emphasized by the life-time residential histories used in this analysis. During a life-time an individual may have typically made several moves between and within small places, but only one move from a small place to a large. Though at the aggregate level this move contributed to the process of urbanization, for the individual it was just one of many moves in a life-time of migration. Some of these individual implications are explored in more detail below.

112

The experience of rural to urban migration

Although rural to urban moves (or more accurately moves from small settlements to large settlements) formed a relatively small part of the total migration experience, even in the period before 1880 when rates of urbanization were relatively high, it can be suggested that such moves did have particular significance for the migrant involved. Almost inevitably movement up the urban hierarchy, especially to a large city of over 100,000 population, necessitated more adjustment to a new way of life than a move within the same town or from one small place to an adjacent settlement of a similar size. It can be suggested that it is because of both the visibility of new migrants in large cities, and the problems which they had assimilating to urban life, that contemporary writers and commentators placed such emphasis on rural to urban movement. The alienating effects of large cities both for newcomers and for urban or suburban residents who found themselves in an unfamiliar part of town have been captured by many contemporary writers of fiction and social comment. Thus Dickens captures the bewilderment encountered by Florence Dombey in a strange part of London:

> Where to go? Still somewhere, anywhere! still going on; but where! She thought of the only other time she had been lost in the wide wilderness of London – though not as lost as now – and went that way. (Dickens, 1982, p. 667)

This section examines in more detail the pattern and process of movement from small to larger places using the full data set of life-time residential histories, and assesses some of the impacts of such moves using individual case histories and material drawn from diaries and life histories.

It has already been demonstrated that the majority of those who left small places of under 5,000 population and moved up the urban hierarchy, either went to towns which were only slightly larger or, and especially after 1839, went directly to very large cities of over 100,000 people. Figures 4.1 and 4.2 map these movements for the periods before and after 1880. Interpretation of the maps must, of course, bear in mind the fact that the size distribution of Britain's urban hierarchy was changing radically, and that there were relatively few very large towns in the first time period. In both time periods the larger provincial towns drew most of their migrants from the regions in which they were situated, but with a small number of longer distance migrants drawn from most parts of the country. The maps emphasize the strong regional effect of movement to larger settlements, with most people moving a relatively short distance to their local urban centre. Many migrants to London also came from relatively close by, but the metropolis

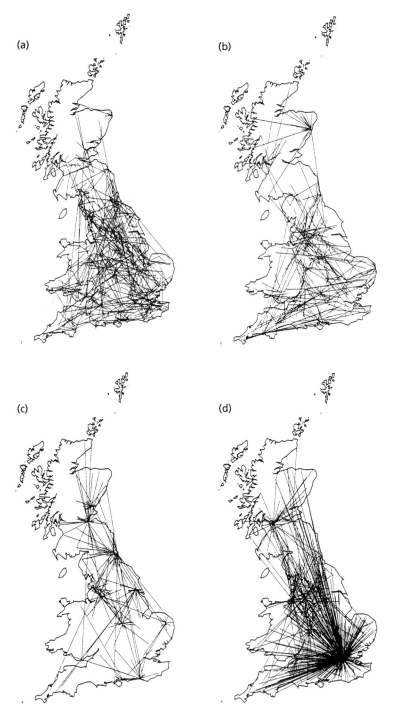

Figure 4.1 All moves 1750–1879 from places <5,000 population to places, (a) 5,000–19,999; (b) 20,000–59,999; (c) 60,000–99,999; (d) ≥100,000 population

114

Figure 4.2 All moves 1880–1994 from places <5,000 population to places, (a) 5,000–19,999; (b) 20,000–59,999; (c) 60,000–99,999; (d) ≥100,000 population

Table 4.9 Type of move to London and ten other principal cities by year of migration (%)

City	Within city and its suburbs	From other settlement where population ≥100,000	From other settlement with population between 60,000–99,999	From other settlement with population between 20,000–59,999	From other settlement with population between 5,000–19,999	From settlement with population <5,000	Proportion of moves from other settlements where settlement <5,000	Total number
1750–1879								
London	69.1	1.0	1.5	3.6	6.2	18.6	60.3	3,292
Manchester	44.6	8.3	8.0	7.6	6.9	24.6	44.4	289
Glasgow	67.6	1.6	1.3	4.9	5.8	18.8	58.0	308
Birmingham	54.0	4.3	0.5	7.2	14.4	19.6	42.7	209
Liverpool	72.0	3.8	2.1	4.0	2.7	15.4	55.1	526
Leeds	59.2	5.2	1.9	3.8	8.9	21.0	51.6	157
Edinburgh	59.6	8.1	0.9	2.7	7.6	21.1	52.2	223
Sheffield	60.8	4.5	2.0	5.5	9.0	18.2	46.2	199
Bristol	63.0	2.6	2.2	2.6	9.0	20.6	55.7	189
Bradford	50.0	7.4	5.3	9.6	9.6	18.1	36.2	94
Nottingham	53.4	7.4	3.7	2.4	7.9	25.2	54.0	163
1880–1994								
London	67.8	5.0	1.9	5.1	5.5	14.7	45.7	3,425
Manchester	55.0	20.3	2.0	4.3	9.2	9.2	20.4	305
Glasgow	66.8	6.8	1.7	3.7	8.3	12.7	38.2	458
Birmingham	67.6	6.4	2.4	7.5	3.9	12.2	37.7	466
Liverpool	75.1	6.0	2.7	3.5	3.5	9.2	36.9	490
Leeds	59.2	8.5	3.8	5.3	9.7	13.5	33.0	319
Edinburgh	69.4	7.3	0.7	2.9	6.7	13.0	42.8	451
Sheffield	72.4	6.8	2.9	5.1	2.9	9.9	36.0	453
Bristol	72.3	4.5	0.9	4.5	5.5	12.3	44.3	220
Bradford	71.4	6.1	3.6	4.1	8.2	6.6	23.2	196
Nottingham	60.8	13.5	3.6	2.9	7.6	11.6	29.9	171

Settlement size calculated from 1851 census for period 1750–1879 and from 1891 census for period 1880–1994.
Table should be read as follows: 67.6% of all moves where Glasgow was the destination in the period 1750–1879 were moves from another residence within the city or its suburbs; 1.6% were moves from another settlement with a population ≥100,000; 18.8% were moves from a settlement with a population under <5,000; 58.0% of moves into the city were moves from a settlement with a population <5,000.
Data source: 16,091 life histories provided by family historians.

also attracted a larger number of migrants from small places which were scattered much more widely around the country.

The migration patterns of migrants from small places to towns in the adjacent size categories are spatially more complex, and consist of a large number of short distance moves between the many small towns which existed (many of these moves taking place within a regional urban system), but also some longer distance moves, especially after 1880. As shown above, people who moved up the urban hierarchy on average moved further than those moving between settlements of the same size. The frequency distribution of such moves shows that the majority were over short distances, but the mean is inflated by a small number of very long distance migrations.

Because it seems probable that migration to big cities was likely to be rather more disruptive for rural to urban migrants than moving from a village to a nearby market town, which would probably have been visited on at least a weekly basis in any case, the remainder of this section will focus on migration to London and to the ten largest provincial cities in 1891. Each of these cities contained over 200,000 people in 1891; seven of the ten (Manchester, Liverpool, Edinburgh, Glasgow, Birmingham, Bristol and Leeds) had populations of over 100,000 in 1831 and in excess of 50,000 in 1801. If we take all moves with a destination in each of these cities, the overall pattern is very close to that found in the analysis of large census regions (Table 4.9). In both time periods approximately one half to three-quarters of all moves had their origin and destination within the same city. Of moves into the cities between 1750 and 1879, between 40 and 60 per cent had their origins in small places of under 5,000 people, reflecting direct movement from very small towns to large cities. After 1880 such moves remained the largest single category for most towns, but were proportionately less important, accounting for 20 to 45 per cent of all moves into the cities. The next most important categories were moves from towns with between 5,000 and 19,999 population – which would still have represented a substantial leap up the urban hierarchy in the second time period – and from other large towns with over 100,000 people. After 1880, Manchester received a particularly large proportion of its migrants from other towns with over 100,000 people, posssibly reflecting the conurbation's dominance of the northern urban system and its proximity to other large northern cities.

In 1891 London was more than five times larger than its nearest British rival (Manchester) and in 1801 it was over ten times larger than the second city. The pre-eminence of London within the British urban system merits some closer analysis of movement into the metropolis. The mean distances moved to London were rather greater than those to the northern industrial towns, but not significantly greater than those moved to the Scottish capital (Edinburgh), or to Scotland's largest city in 1891 (Glasgow). In comparison with migration to smaller places London attracted a much larger proportion of its migrants from over 100 kms away, but it was not out of line with the

117

experience of most large provincial cities. It is not surprising that this is the case as much of the long distance movement to and from London consisted of migration between the capital and large provincial towns. In other respects migration to London, and the other large cities, reflected the characteristics of the total migration system. Most people moved for work related reasons, a majority migrated as married couples or in a nuclear family, and young adults in the age range 20 to 39 years comprised the largest single age category. Not surprisingly, as a large proportion of movement to the big cities was over relatively long distances, this migration was dominated by those in skilled and professional occupations (Table 4.10a–b).

Some of these themes can be illustrated through individual case studies, which also demonstrate the way in which rural to urban moves fitted in to an individual's life-time migration experience. John H. was born in the small North Pennine (County Durham) settlement of Middleton-in-Teesdale (population 796 in 1801) in 1788. His father was a lead miner and innkeeper in a community where a large proportion of the population depended on lead mining for their living (Raistrick and Jennings, 1965; Turnbull, 1975). However, John did not seek employment in the mines, but instead at the age of 13 followed his two elder brothers and was apprenticed to the Plumbers' Livery Company in London. As many of the Middleton-in-Teesdale lead mines were owned by the Quaker-run London Lead Company, this connection may explain why this group of three brothers were apprenticed as plumbers in London. Whilst apprenticed John lived with his employers and thus his move from rural Teesdale was simplified by the provision of a job, a home and companions (his brothers) from Teesdale. Although this was possibly his first experience of a large city, his move certainly did not cast him into the metropolis without appropriate support mechanisms.

In 1807 at the age of 19 he broke his seven year apprenticeship after only five years had elapsed because he wished to marry. Apprentices would not have been allowed to marry and he needed to provide both a home and income for his wife and future family. Fortunately, with five year's training he gained a job as a plumber's assistant and he and his wife took rented rooms in the Finsbury district of London. Between 1808 and 1814 he moved on at least two occasions to rented property in Southwark and Clerkenwell Green, still working as a plumber's assistant, before in 1815 and with a family of nine children he started his own plumbing and glazing business in Carter Lane near St Pauls. However, the family was plagued with ill-health, two of his children died in 1816, his wife died in 1818 and his own health was clearly not good (possibly caused by his work with lead pipes). In 1821, mainly due to his ill-health, he moved with his seven remaining children to Southwark (South of the Thames) but died there later the same year at the age of 33. In some ways John H. is a classic rural to urban migrant in the early-nineteenth century. He moved from an industrialized rural area directly to an apprenticeship in London with the arrangements for migration clearly taken care of by his family. He then remained in London, undertaking several intra-urban moves but, like many London residents, succumbed to high urban mortality at a relatively young age.

A large proportion of migrants to large cities were skilled workers (like John H.) or professionals. Thomas W., born in the small town of Amlwch (Anglesey) in 1844 is an example of the latter category. Thomas's father was an apothecary in Amlwch (population 5,813 in 1851) who, soon after his son's birth, moved his business along the North Welsh coast to the town of Conway (population 1,528 in 1851). In 1857, when Thomas was 13 his father died, necessitating his mother and five siblings to move house in Conway, but at that time Thomas was sent away to boarding school in London during term time. In 1860, at the age of 16 he went to Manchester to study medicine for five years, lodging in hospital accommodation and, after qualifiying, he gained the post of Assistant Medical Officer at the Bridge Street Workhouse in Salford, living in furnished rooms in the workhouse. His education and career thus took him from small communities in Wales directly to London and Manchester, but like John H. he moved in very controlled circumstances with his accommodation provided through his education and work. Although in the mid-nineteenth century medical work provided status, it was not a wealthy profession, and the position of workhouse medical officer cannot have been highly sought after (Woodward, 1977; Pickstone, 1985; Loudon, 1986). Moreover, living in the workhouse would not have been compatible with family life. Thus in 1872, when Thomas married, he returned to Wales to set up in General Practice in Merthyr Tydfil (Glamorgan, population 51,949 in 1871), where he also supplemented his income with the post of Medical Officer at a steelworks. Thomas remained in the same house in Merthyr until his death in 1916. Like many migrants Thomas's migration to a large city was not permanent, and he moved down the urban hierarchy to a much smaller Welsh town at the age of 28 years.

There is some evidence that large cities attracted a disproportionate number of single migrants. This theme is exemplified by the movements of Evelyn M. in the early-twentieth century. Evelyn was born in the village of Stobo, Peeblesshire (Scotland) in 1879. Her father was an Estate Factor who lived in a tied house on the estate and, at the age of ten when her father retired, she moved with her parents and two siblings to the nearby village of Eddleston. At the age of 19, and possibly stimulated by the labour demands of the First World War, Evelyn left home and moved alone to Edinburgh to take work as a typist and clerk with the W.A.A.C. (Women's Auxiliary Army Corps). Edinburgh was less than 30 kms from her home village in Peeblesshire and it is unlikely that she had never been to the city before. Moreover, during the war accommodation was provided in a W.A.A.C. hostel, so she would have been living in a controlled environment, surrounded by girls in a similar position to herself. When the war ended and

119

Table 4.10a Summary characteristics of migrants to London and ten large cities, 1750–1879 (%)

Migrant characteristics	London	Manchester	Glasgow	Birmingham	Liverpool	Leeds	Edinburgh	Sheffield	Bristol	Bradford	Nottingham
Gender											
female	25.7	28.7	31.0	26.1	27.2	20.3	18.9	33.3	31.4	40.4	26.6
Mean distance (km)	149.4	83.4	105.3	66.5	114.5	94.4	157.6	77.6	107.7	63.3	74.9
Distance band											
<5	2.5	10.0	3.0	5.2	2.7	4.7	2.2	1.3	4.3	2.1	2.6
5–9.9	2.9	4.4	2.0	12.5	4.8	6.3	7.8	23.0	5.7	21.3	13.2
10–19.9	6.6	14.4	17.0	15.6	2.0	20.3	3.3	11.5	10.0	21.3	10.5
20–49.9	14.0	23.8	33.0	30.2	23.8	23.4	13.3	23.1	20.0	23.4	22.4
50–99.9	20.4	14.4	18.0	14.6	19.0	17.2	21.2	23.1	22.9	14.9	19.8
100–199.9	28.9	18.0	8.0	14.6	28.6	6.2	28.9	2.6	21.4	0.0	26.3
200+	24.7	15.0	19.0	7.3	19.1	21.9	23.3	15.4	15.7	17.0	5.2
Reason											
work	54.6	66.9	52.9	57.5	59.0	73.1	50.0	49.2	39.6	64.1	60.3
marriage	16.1	13.6	12.6	15.1	18.3	17.3	10.8	26.2	20.8	12.8	13.2
housing	1.0	9.8	4.6	2.7	0.9	0.0	4.1	1.7	3.8	5.1	0.0
family	2.6	9.7	1.1	0.0	1.8	0.0	0.0	0.0	1.9	2.6	1.5
other	25.7	0.0	28.8	24.7	20.0	9.6	35.1	22.9	33.9	15.4	25.0
Age											
<20	29.0	23.8	34.8	40.0	27.2	31.2	34.4	23.1	24.3	31.9	25.0
20–39	57.1	56.2	54.5	44.2	59.2	50.0	58.9	61.5	57.1	48.9	50.0
40–59	11.5	16.9	8.0	14.7	11.6	18.8	5.6	11.5	15.7	17.0	23.7
60+	2.4	3.1	2.7	1.1	2.0	0.0	1.1	3.9	2.9	2.2	1.3
Marital status											
single	50.3	40.7	45.4	53.2	41.8	45.5	62.5	27.3	38.2	39.0	39.7
married	46.3	56.0	54.6	45.7	51.5	52.7	37.5	67.5	55.9	56.1	54.4
other	3.4	3.3	0.0	1.1	6.7	1.8	0.0	5.2	5.9	4.9	5.9

Companions											
alone	28.5	25.3	20.0	14.6	18.3	16.4	33.3	13.0	20.3	12.5	17.2
couple	19.8	20.3	15.0	7.4	23.2	16.4	9.5	22.1	26.6	10.4	22.4
nuclear family	43.7	46.2	54.0	67.4	54.3	62.3	45.2	53.2	42.2	72.9	53.9
extended family	2.5	1.9	3.0	5.3	2.1	3.3	1.2	6.5	4.7	4.2	2.6
other	5.5	6.3	8.0	5.3	2.1	1.6	10.8	5.2	6.2	0.0	3.9
Position in family											
child	16.1	12.9	20.4	31.6	21.7	21.3	17.7	13.0	15.6	31.9	23.7
male head	60.8	61.5	52.0	44.2	59.4	59.0	58.8	51.9	54.7	42.6	50.0
wife	12.7	15.4	18.4	11.6	13.9	14.8	12.9	23.4	23.4	17.0	19.8
female head	4.8	3.8	4.1	2.1	2.8	1.6	4.7	5.2	4.7	6.4	2.6
other	5.6	6.4	5.1	10.5	2.2	3.3	5.9	6.5	1.6	2.1	3.9
Occupational group											
professional	12.7	7.5	6.8	10.0	5.2	21.1	19.4	7.3	22.2	4.0	19.5
farmer	0.0	0.0	2.3	0.0	0.0	5.3	0.0	7.3	0.0	0.0	0.0
intermediate	13.0	14.2	18.2	12.0	19.8	7.9	8.4	4.9	8.3	28.0	12.2
skilled non-manual	14.8	14.2	13.6	10.0	16.7	7.9	16.6	0.0	13.9	20.0	7.3
skilled manual	29.1	43.4	36.4	44.0	33.3	28.9	50.0	46.3	36.1	20.0	36.6
skilled/semi skilled agricultural	4.5	0.9	4.5	4.0	5.2	0.0	2.8	0.0	2.8	0.0	4.9
semi-skilled manual	11.9	9.4	13.6	16.0	6.3	23.7	0.0	19.5	0.0	20.0	12.2
unskilled manual	6.1	6.6	2.3	2.0	11.4	2.6	2.8	4.9	5.6	4.0	7.3
agricultural labourer	0.5	0.0	0.0	0.0	0.0	0.0	0.0	4.9	2.8	4.0	0.0
domestic servant	7.4	3.8	2.3	2.0	2.1	2.6	0.0	4.9	8.3	0.0	0.0
Total number	1,017	160	102	103	147	64	90	78	73	47	76

Data source: 16,091 life histories provided by family historians.

Table 4.10b Summary characteristics of migrants to London and ten large cities, 1880–1994 (%)

Migrant characteristics	London	Manchester	Glasgow	Birmingham	Liverpool	Leeds	Edinburgh	Sheffield	Bristol	Bradford	Nottingham
Gender											
female	42.6	33.6	40.4	41.7	32.8	39.2	45.7	37.6	37.7	46.4	35.8
Mean distance	135.6	102.9	140.0	104.8	125.9	100.3	160.3	116.2	127.1	66.2	93.1
Distance band (km)											
<5	4.7	9.5	5.3	3.3	2.5	4.6	1.4	0.8	1.6	12.5	8.9
5–9.9	7.4	8.8	2.7	13.3	14.0	3.8	4.4	7.2	1.6	28.6	7.5
10–19.9	11.8	8.0	7.3	5.3	4.1	7.7	9.4	11.2	6.6	8.9	4.5
20–49.9	12.0	16.8	20.6	16.8	15.7	22.3	15.9	16.0	26.2	16.1	16.4
50–99.9	19.6	13.1	23.3	14.7	18.2	30.8	20.3	21.6	16.4	10.7	19.4
100–199.9	19.9	19.0	10.0	34.0	12.4	16.9	17.4	16.8	24.6	12.5	37.3
200+	24.6	24.8	30.8	12.6	33.1	13.9	31.2	26.4	23.0	10.7	6.0
Reason											
work	47.2	57.9	54.8	56.9	53.3	57.0	64.9	56.2	46.2	58.5	64.9
marriage	8.8	8.7	9.8	2.8	9.8	5.8	7.0	10.7	5.6	5.7	8.8
housing	7.2	4.8	9.0	8.3	2.5	4.1	0.9	4.2	7.4	13.2	5.3
family	4.0	5.6	6.8	1.4	4.9	12.4	5.3	1.6	5.6	0.0	10.5
other	32.8	0.0	19.6	30.6	29.5	20.7	21.9	27.3	35.2	22.6	10.5
Age											
<20	25.6	22.6	27.2	20.0	22.1	19.3	27.6	21.6	18.0	25.0	20.9
20–39	48.5	57.7	46.3	52.0	59.8	46.9	38.4	54.4	42.6	46.4	59.7
40–59	17.3	14.6	19.1	22.0	11.5	19.2	18.1	17.6	19.7	23.2	9.0
60+	8.6	5.1	7.4	6.0	6.6	14.6	15.9	6.4	19.7	5.4	10.4
Marital status											
single	47.2	38.9	43.4	40.9	36.7	36.6	46.8	39.0	31.7	37.5	44.6
married	46.5	54.8	54.6	53.7	55.8	53.6	47.4	55.9	56.7	55.4	49.2
other	6.3	6.3	2.0	5.4	7.5	9.8	5.8	5.1	11.6	7.1	6.2

Companions											
alone	33.9	25.2	22.3	23.3	28.3	27.8	35.6	28.0	32.8	12.5	41.6
couple	14.2	16.3	17.9	10.1	14.2	16.9	14.8	17.6	24.6	8.9	13.8
nuclear family	42.8	50.4	49.7	58.0	51.6	52.3	41.5	46.4	36.0	67.9	32.3
extended family	4.6	3.7	5.8	3.3	4.2	1.5	5.1	4.0	6.6	3.6	0.0
other	4.5	4.4	4.3	5.3	1.7	1.5	3.0	4.0	0.0	7.1	12.3
Position in family											
child	15.1	18.6	19.3	20.3	18.3	16.9	12.6	19.2	8.2	23.2	10.8
male head	44.9	49.6	47.3	43.2	53.3	47.7	42.2	50.4	52.5	41.1	60.0
wife	16.6	18.5	18.0	16.2	18.3	20.8	20.7	16.0	16.4	23.2	12.3
female head	17.6	7.4	10.0	15.5	5.8	12.3	19.3	10.4	18.0	7.1	12.3
other	5.8	5.9	5.4	4.8	4.3	2.3	5.2	4.0	4.9	5.4	4.6
Occupational group											
professional	19.1	27.7	12.0	23.3	30.2	24.6	27.0	33.8	26.6	23.4	22.4
farmer	0.0	0.0	0.0	0.0	0.0	0.0	0.0	0.0	0.0	0.0	0.0
intermediate	7.9	10.8	18.1	13.4	13.3	11.6	9.5	10.8	8.8	23.3	10.2
skilled non-manual	29.1	19.3	21.8	17.8	22.4	18.8	14.3	16.2	29.4	20.0	28.6
skilled manual	19.1	20.6	27.7	23.3	19.7	18.8	27.0	21.6	23.5	13.3	22.4
skilled/semi skilled agricultural	1.6	0.0	4.8	1.1	5.3	2.9	0.0	1.4	0.0	0.0	0.0
semi-skilled manual	4.3	10.8	6.0	17.8	3.9	14.6	3.2	10.8	2.9	13.3	8.2
unskilled manual	3.4	1.2	2.4	2.2	2.6	5.8	3.2	1.4	2.9	0.0	0.0
agricultural labourer	0.4	0.0	0.0	0.0	0.0	0.0	0.0	0.0	0.0	0.0	0.0
domestic servant	15.1	9.6	7.2	1.1	2.6	2.9	15.8	4.0	5.9	6.7	8.2
Total number	1,103	137	152	164	122	130	138	125	66	60	88

Data source: 16,091 life histories provided by family historians.

the W.A.A.C. was disbanded in 1919, Evelyn took work as a shorthand typist in the Peebles Hydro Hotel, Peebles (population 5,539 in 1921). Although living in staff accommodation in the hotel she was only about 6 kms from her home village and working in what would have been her local town for much of her childhood.

Two years later Evelyn returned to Edinburgh, now living in lodgings and working as a shorthand typist with the Inland Revenue. Civil Service clerical work would have been seen as a good and respectable career for a single woman in the early twentieth century, and the prospects which this offered were probably the main reason for returning to Edinburgh. She worked with the Inland Revenue until her retirement in 1964, but the remainder of her life was not uneventful. Following the death of her widowed mother and the sale of the family home in 1933, Evelyn's sister joined her in Edinburgh which necessitated a move from lodgings into a rented house. She progressed to a Clerical Officer grade in the Inland Revenue, but her career was interrupted in 1936 when, at the age of 37, she took a period of leave and moved to stay with friends in Sunderland because she was expecting an illegitimate child. Following the birth of her daughter she got a transfer to Inland Revenue offices in Birmingham and from 1936–9 she lived with her daughter and sister in Solihull. In 1939, she moved to a Senior Clerical Officer post with the Inland Revenue in Glasgow, living in Uddingston on the edge of the city. In 1946 she moved about 45 kms to the village of Crossford in Fife due to her transfer and promotion to the post of Assistant Tax Collector in Dunfermline. All these moves were undertaken with her daughter and her sister.

Evelyn M. remained in Crossford until 1966 when, due to ill-health and at the age of 69, she moved to her daughter's home in Thurso (Caithness, population 8,037 in 1961). She died later the same year. This complicated life history provides a good example of the way in which rural to urban migration, in this case over a relatively short distance to a nearby city, was part of a longer life-course which also included moves between large urban areas and from cities back to smaller settlements. As with many single women, Evelyn's moves were stimulated by a combination of employment and family reasons, and it is reasonable to assume that her move from Edinburgh to Birmingham reflected a desire to move away from the area in which she was known at a time when having an illegitimate child would still have been frowned upon. As with the other case studies, her early life in a large city was aided by the provision of accommodation, and there is no evidence that any of these migrants found assimilation to urban life particularly difficult. Whilst novelists and some observers were at pains to stress what was strange about big cities, it seems that for many migrants movement up the urban hierarchy was a relatively straightforward event surrounded by a well developed set of support networks. The life histories also illustrate the way in which, for many people, rural to urban migration was part of a much more complicated pattern of movement. Some of these other types of move will be examined more briefly in subsequent sections.

The experience of moving between towns

From around 1850 more than half of the British population lived in towns. Thus it is not surprising that movement between urban areas became an increasingly important part of the migration system. As shown above, movement to the largest cities of Britain came increasingly from other large places in the late-nineteenth and twentieth centuries. However, it would be misleading to suggest that movement between urban areas was all the same. It can be suggested that migration between the largest cities, by definition mostly over long distances, was a rather different experience to migration between adjacent market towns. The extent to which movement between towns of different sizes had distinctive characteristics, and the degree to which it differed from other sorts of migration, can be explored using the full data set drawn from family historians. Figures 4.3 to 4.7 plot movement

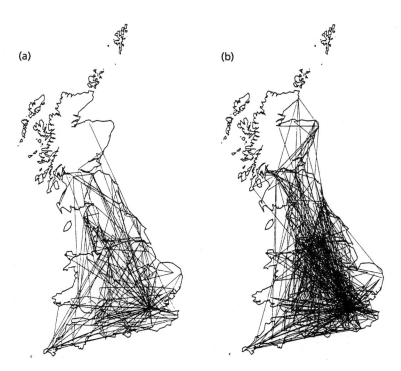

Figure 4.3 All moves from places 5,000–19,999 population to other places 5,000–99,999 population, (a) 1750–1879; (b) 1880–1994

125

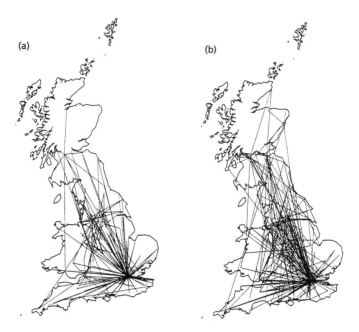

Figure 4.4 All moves from places 5,000–19,999 population to places ⩾100,000 population, (a) 1750–1879; (b) 1880–1994

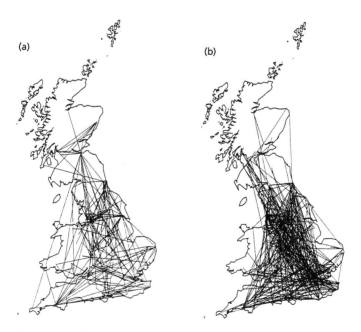

Figure 4.5 All moves from places 20,000–99,999 population to other places 5,000–99,999 population, (a) 1750–1879; (b) 1880–1994

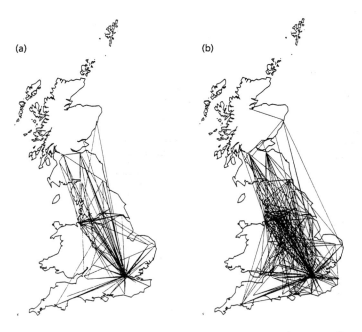

Figure 4.6 All moves from places 20,000–99,999 population to places ≥100,000 population, (a) 1750–1879; (b) 1880–1994

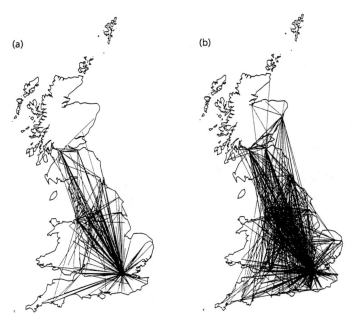

Figure 4.7 All moves from places ≥100,000 population to all other places >5,000 population, (a) 1750–1879; (b) 1880–1994

127

Table 4.11 Summary characteristics of migrants between towns with populations ⩾5,000 and <100,000 by year of migration (%)

Migrant characteristics	1750–1879	1880–1994
Gender		
female	25.6	39.8
Mean distance (km)	116.7	109.8
Distance band (km)		
<5	6.9	12.5
5–9.9	5.5	7.6
10–19.9	10.2	9.6
20–49.9	18.0	15.8
50–99.9	18.7	16.0
100–199.9	21.6	18.6
200+	19.1	19.9
Reason		
work	60.1	54.2
marriage	11.5	8.6
housing	1.2	6.4
family	6.0	5.3
other	21.2	25.5
Age		
<20	26.3	20.4
20–39	57.6	50.9
40–59	13.6	18.4
60+	2.5	10.3
Marital status		
single	42.5	36.7
married	52.6	56.0
other	4.9	7.3
Companions		
alone	19.2	22.7
couple	17.1	17.0
nuclear family	55.0	51.7
extended family	3.0	4.4
other	5.7	4.2
Position in family		
child	21.2	16.6
male head	56.5	48.3
wife	14.2	19.2
female head	3.2	10.8
other	4.9	5.1
Occupational group		
professional	18.8	21.8
farmer	0.2	0.6
intermediate	11.6	13.3
skilled non-manual	10.2	23.0
skilled manual	39.1	22.8
skilled/semi skilled agricultural	2.8	1.4
semi-skilled manual	11.9	8.1
unskilled manual	2.6	2.8
agricultural labourer	0.2	0.3
domestic servant	2.6	5.9
Total number	1,121	2,552

Data source: 16,091 life histories provided by family historians.

128

between towns in three size bands over two time periods. What is clear from these maps, and from a more detailed examination of migration patterns broken down into a larger number of sub-groups, is that inter-urban movement was much more likely to be between regions than rural to urban migration. Although London and other cities over 100,000 population did draw migrants from a wider range of destinations and over longer distances, moves between smaller towns did not display the same regional structure as moves from places under 5,000 population to larger towns. Thus the characterics of migration between towns were very stable regardless of town size or regional location. This is borne out by a detailed analysis of the characteristics of inter-urban movement (Table 4.11).

Most urban to urban moves were stimulated by employment change, initiated by the migrant, as the result of promotion or an enforced move determined by an employer. Both Evelyn M. and Thomas W. (above) illustrate such moves as part of their life-time migration experience, and some urban to urban moves occurred in the life history of a large number of individuals. Henrietta D. provides a good example of a woman whose migration experience was tied up with that of her family, and who embraced an urban to urban move within a much wider migration experience. Henrietta was born in a small village in Kirkcudbrightshire in 1838, the daughter of an agricultural worker. However at the age of eight she was orphaned and lodged (with friends or relatives) in a nearby village whilst attending school. At the age of 15 in 1853 she went into service in the town of Lanark (population 5,304 in 1851). A rural orphan would have had few alternative employment opportunities in the mid-nineteenth century, and she moved around 60 kms to a nearby town.

In 1861 she married a man who was in the army and moved with him to Birkenhead (Cheshire, population 65,971 in 1861) where his regiment was stationed. This was a substantial move of around 250 kms to a completely new environment in a much larger town, but it is likely that contacts within the regiment would have provided a social and support network for the young wife. Her husband was discharged in 1863 and they then moved back to Scotland when her husband obtained a job as a house factor in Glasgow. This move between two large towns took them back to a locality with which they were more familiar, and the flat they lived in went with her husband's job. They lived in the same flat in Glasgow for 23 years, but following her husband's death in 1886 she moved elsewhere in Glasgow to her daughter's house. She lived with her daughter and her family for the rest of her life, moving several times within Glasgow and the surrounding suburbs until her death in 1923.

Henrietta D.'s inter-urban moves were, in part at least, stimulated by her husband's enlistment and discharge from the army, and later moves were made with her daughter's family. As such, she had little direct control over the migration decision, a situation common to many women in the

129

past. However, the subject of the second case study had a more direct input to his own migration pattern. Henry H. was born in the small Yorkshire village of Thorne in 1849, the son of a carpenter. His early employment is not known, and he may have worked with his father as a carpenter, but by 1875 he was married with one son, living in Leeds (population 259,212 in 1871) and working as a warehouseman. Three years later he sought a change of trade and moved some 50 kms to the smaller town of Ashton-under-Lyne (population 37,040 in 1881) to work as a draper's assistant. In 1889 he moved towns again, migrating to the nearby but larger town of Oldham (population 111,343 in 1881) to take up a new position in the drapery trade. This pattern of inter-urban movement, linked to his career progression continued as in 1890 he moved into Manchester (population 505,368 in 1891) to what seemed like a better position, and in 1897 migrated with his wife and six children back across the Pennines to Bradford (population 235,436 in 1901) to take up new employment as a commercial traveller in the wholesale clothing trade. He moved once more within Bradford in 1900, but following his wife's death in 1911 emigrated to Canada to join one of his sons. He died in Krugersdorf, Ontario in 1925. Henry H. thus lived with his family in five different towns in Northern England, moving both up and down the urban hierarchy, and with all his inter-urban moves stimulated mainly by employment change and attempts to advance his career. Such a pattern was typical of many inter-urban migrants in the nineteenth century.

There is little evidence that movement either between nearby towns within an urban region (as in the case of Henry H.'s moves between Leeds, Ashton, Oldham, Manchester and Bradford), or longer distance moves from one city to another (as with Henrietta D.'s moves between Glasgow and Merseyside) was especially difficult or traumatic. Most such migrants moved with their families, their migration decisions were usually linked to employment change and career advancement, and there were often support networks to aid assimilation and help find accommodation. The main difference between shorter distance moves within an urban region and those that took the migrant to a new urban environment was in the extent to which migrants could keep in touch with relatives and friends. The fact that migrants often returned to their home area and frequently lived with relatives in times of ill health or personal crisis suggests that family networks were able to survive long distance movement between towns. Some of these themes are explored in more detail in Chapter 6.

The experience of rural circulation

As shown in Chapter 3, movement between small places was a very common migration experience in all time periods. Even after 1850, many moves took the form of rural circulation and it can be suggested that movement between villages and small towns within one locality enabled migrants to adjust to employment and family changes without losing touch with either their kin or the place associations of their local area. Such migration must have been easier, cheaper and less disruptive than movement between towns. For these reasons it was a taken-for-granted part of the British rural landscape, rarely commented on by contemporaries, though novelist such as Elliot and Hardy make some reference to the rural circulation of agricultural workers.

Some local history studies have also focused on the importance of rural circulation within a limited area (Holderness, 1970; Escott, 1988). It was not only agricultural workers who moved from place to place within the countryside. Skilled craftsmen and industrial workers also moved frequently, and it can be suggested that in the late-eighteenth century sites of rural industry were particularly important foci of rural migration. The data set collected from family historians allows us to test the validity of these generalizations, and to compare movement between small places with other forms of migration. Figure 4.8 maps all moves between places of under 5,000 people in four time periods. The first three maps are essentially similar, confirming the suggestion that most such moves took place within one locality and rarely involved longer distance moves to another part of the country. Within each region there was a complex network of moves between villages and to expanding small market towns. However, the picture after 1920 is rather different. Although some short-distance local circulation remained, the overall impression is of a much more national migration system, with many rural to rural migrants moving quite long distances from one part of the country to another. The characteristics of migrants between small places are summarized in Table 4.12. In comparison with movement to and between towns the mean distances moved between small places were very short with over 60 per cent of moves under 10 kms in both time periods (before and after 1880). However, after 1880 there was a significantly larger proportion of moves over 100 kms, increasing the mean distances moved to 36kms and confirming the impression given in Figure 4.8.

In most other respects the characteristics of migrants between small places were much the same as those of other sorts of migrants, although in comparison with those who moved to and between towns there were a larger proportion of married and slightly older migrants and, not surprisingly, before 1880 the migration flows were dominated by those associated with agricultural work. Most moves were for employment reasons and in both time periods the movement of skilled craft and industrial workers was particularly important. The migration patterns of agricultural and industrial workers were rather different, with those with craft and industrial skills moving over longer distances than those employed in agriculture. These and other issues relating to migration and employment are explored in Chapter 5. It might be assumed that, if movement over longer distances to

131

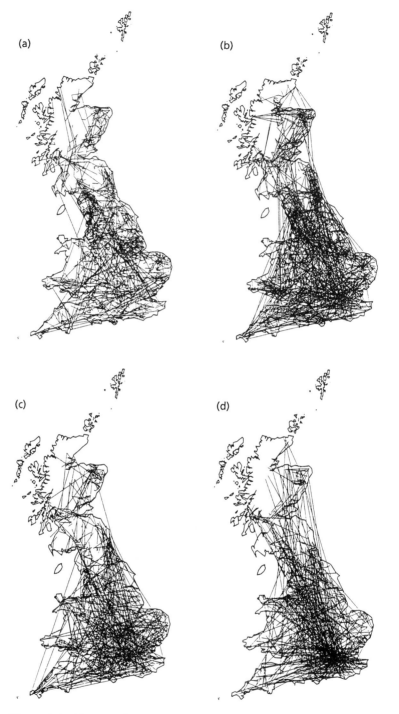

Figure 4.8 All moves from places <5,000 population to other places <5,000 population, (a) 1750–1839; (b) 1840–79; (c) 1880–1919; (d) 1920–94

Table 4.12 Summary characteristics of migrants between places <5,000 population by year of migration (%)

Migrant characteristics	1750–1879	1880–1994
Gender		
female	28.4	42.6
Mean distance (km)	23.3	36.5
Distance band (km)		
<5	47.8	51.0
5–9.9	15.2	9.0
10–19.9	11.6	9.2
20–49.9	13.5	11.7
50–99.9	6.4	7.7
100–199.9	3.5	6.4
200+	2.0	5.0
Reason		
work	53.9	33.8
marriage	23.8	11.4
housing	6.5	22.8
family	2.4	4.5
other	13.4	27.5
Age		
<20	25.2	17.3
20–39	52.9	38.6
40–59	15.7	24.3
60+	6.2	19.8
Marital status		
single	33.1	24.5
married	62.4	65.1
other	4.5	10.4
Companions		
alone	12.1	13.2
couple	21.1	21.3
nuclear family	58.2	55.6
extended family	5.0	6.7
other	3.6	3.2
Position in family		
child	19.5	15.8
male head	55.7	47.0
wife	17.2	23.7
female head	4.4	9.2
other	3.2	4.3
Occupational group		
professional	4.3	13.2
farmer	13.8	6.6
intermediate	8.7	14.3
skilled non-manual	3.1	14.5
skilled manual	25.2	21.2
skilled/semi skilled agricultural	8.8	7.7
semi-skilled manual	6.4	7.0
unskilled manual	5.6	3.3
agricultural labourer	20.3	5.7
domestic servant	3.8	6.5
Total number	7,829	5,965

Data source: 16,091 life histories provided by family historians.

towns or between towns was mostly undertaken with family support and was not as disruptive as some contemporary commentators suggested, then movement between small settlements within one region might have been accomplished without any difficulty. For some this was undoubtedly the case, with many such migrants moving to adjacent villages to work (for instance) as farm servants for a relative or friend. However, in other cases moves between small places could create difficulties. Two case histories illustrate these themes.

Henry D. was born in the Sussex village of Plumpton in 1753. His father was a shepherd and, from 1762 (aged nine) he was employed as a shepherd's boy on a neighbouring farm. Seven moves between small villages in the same area of Sussex are recorded as Henry progressed from shepherd's boy to shepherd, and it is very likely that other moves also occurred but were not revealed by available sources. Throughout this mobility, Henry remained in the same occupation, close to family and friends and in a neighbourhood he knew well. It is likely that jobs were gained through local knowledge and contacts and such frequent short-distance rural mobility was a normal and relatively undisruptive part of late-eighteenth century life.

The Shaw family (mentioned in previous chapters) also moved frequently between small settlements, but their experience of migration was not entirely happy. In this case, the existence of a detailed life history allows the personal impact of migration to be examined in more detail. During his life-time (1748–1823) Joseph Shaw moved 10 times in North Lancashire, Yorkshire and South Cumbria. Although most moves (in and around the village of Dent) were unproblematic, a decision in 1791 to move from Dent to the industrial village of Dolphinholme proved disastrous for the family. Joseph had combined handloom weaving and clock repairing whilst living in Dent, and the main attraction of a move to Dolphinholme was the offer of more regular employment in a newly established rural textile mill. However, despite moving to Dolphinholme with other families from his home village, Shaw did not settle to factory work and quickly quarrelled with his employer. The Shaws lived unhappily for two years in Dolphinholme, during which time there was much sickness and three of their children died. In 1793 Joseph and most of his family left Dolphinholme and returned to handloom weaving in a village near his previous home of Dent. In this case migration over approximately 50 kms from one rural community to another proved disastrous for Shaw's employment and family life, and led after less than two years to return migration which also entailed removal from factory back to domestic employment. Thus migration between rural communities could be at least as disruptive and traumatic as migration from the countryside to towns, and unsuccessful migration could lead to return movement which also entailed (albeit temporarily) transfer from a more modern to a less modern economic sector.

When the experiences of these migrants, and the characteristics of all movers between small places, are compared with other types of migrants, it is the similarities rather than the differences which are most striking. Although moves to and between towns tended to be over longer distances, and more commonly involved single migrants, families moving over relatively short distances were a very large component of all types of move. The individual case studies also illustrate the ways in which long distance moves could be accomplished without difficulty, whereas some short distance migration could prove quite traumatic. The experiences of migrants were thus highly varied, and may have depended more on particular individual and family circumstances, on individual personalities (for instance, how well a person adapted to a new environment) and simply on luck, rather than on the nature of the move itself.

The experience of moving within a single settlement

In the late-twentieth century the most common residential move is a short distance relocation within the same urban area (Stillwell, Boden and Rees, 1992). It is likely that in the past such moves were even more common, due to the ease with which people could move between rented property. However, very frequent short distance moves which left few traces in the documentary records which have survived to the present are the most difficult to trace for individual life histories. Although, as shown in Chapter 3, many of the individuals for whom we have information moved very frequently, it seems likely that intra-urban movement was under-recorded, especially in the eighteenth and early-nineteenth century. Overall, moves with an origin and destination in the same settlement formed 38 per cent of all migration and, after 1839, it was clearly the most important component of migration.

It can be suggested that the nature and impact of intra-urban migration changed over time and varied between different types and sizes of town. Although many such moves were literally between adjacent streets, thus retaining migrants in the same local community, increasingly as towns grew intra-urban moves took the form of suburbanization, removing those families who could afford to move to the expanding suburbs. Although this trend was especially obvious in London, the same processes would have been operating in all large towns (Jackson, 1973; Thompson, 1982). For at least some such migrants removal to the suburbs could have been as disruptive as migration to another nearby town. Some of the characteristics and effects of such movement can again be explored both through the full data set and a series of individual case studies.

The basic characteristics of intra-urban migrants vary little (Table 4.13). In comparison with other types of moves, migration for work was rather

Table 4.13 Summary characteristics of migrants moving within one settlement by settlement size and year of migration (%)

Migrant characteristics	<5,000 population		≥5,000 population	
	1750–1879	1880–1994	1750–1879	1880–1994
Gender				
female	28.0	40.5	29.2	40.5
Reason				
work	34.9	18.4	24.7	12.6
marriage	33.2	15.9	31.3	19.4
housing	15.9	41.3	24.9	43.4
family	2.1	2.2	1.9	3.4
other	13.9	22.2	17.2	21.2
Age				
<20	22.0	17.1	24.6	18.7
20–39	48.6	38.8	50.0	44.5
40–59	19.9	25.3	19.4	23.0
60+	9.5	18.8	6.0	13.8
Marital status				
single	25.7	22.4	29.5	25.1
married	68.2	67.9	65.7	66.4
other	6.1	9.7	4.8	8.5
Companions				
alone	7.1	8.5	5.5	8.0
couple	24.2	22.2	17.6	19.9
nuclear family	56.8	59.0	68.6	63.0
extended family	7.7	7.3	5.0	6.5
other	4.2	3.0	3.3	2.6
Position in family				
child	18.4	17.5	24.0	20.4
male head	56.4	48.0	53.3	46.4
wife	17.3	23.4	17.1	23.2
female head	4.3	7.3	3.2	6.4
other	3.6	3.8	2.4	3.6
Occupational group				
professional	2.8	10.0	4.9	9.5
farmer	11.6	4.7	0.7	0.4
intermediate	12.2	16.5	18.8	15.5
skilled non-manual	2.4	13.1	7.7	19.0
skilled manual	26.1	26.6	43.0	31.9
skilled/semi skilled agricultural	7.9	6.2	3.9	2.2
semi-skilled manual	7.6	9.4	11.5	12.4
unskilled manual	6.6	4.5	6.0	5.1
agricultural labourer	20.0	4.6	1.1	0.2
domestic servant	2.8	4.4	2.4	3.8
Total number	3,864	3,910	6,039	11,525

Data source: 16,091 life histories provided by family historians.

(a)

(b)

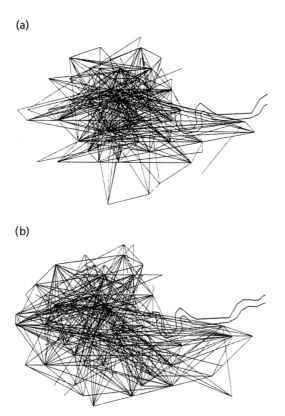

Figure 4.9 Moves within the county of London, (a) 1750–1879; (b) 1880–1994

less important, and movement for housing reasons was much more significant, accounting for over 40 per cent of such moves in the period after 1880. In the period before 1880 work-related moves still accounted for up to a third of all intra-urban moves. This reflects both the need to ensure a short journey to work in the past, especially for those working in the large casual labour market, and the fact that many people lived and worked in the same place and thus residential moves were by definition linked to employment change. This aspect is explored in more detail below. In comparison with moves to and between larger towns, intra-urban migrants were rather older and were more likely to move as a family group. Such migrants were thus adjusting their housing needs at crucial stages of their life-cycle in a way which has been demonstrated in a number of present-day studies (Champion and Fielding, 1992).

Due to its size, London had a much more complex pattern of intra-urban migration than most other British cities. This included both local movement within and between local communities and suburbanization to the edges of the built-up area, especially in the period after 1880 (Figure 4.9).

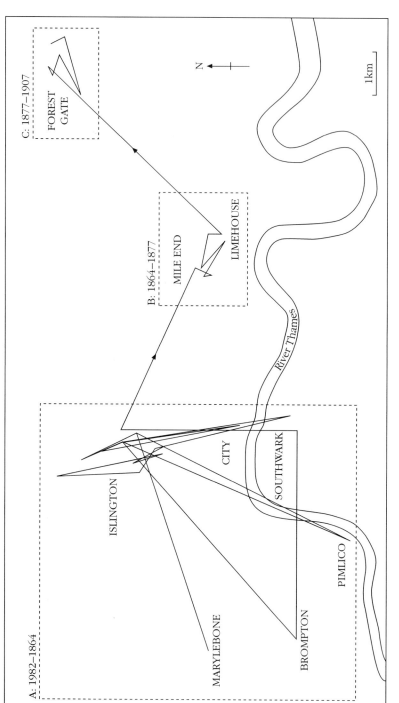

Figure 4.10 Residential moves of Henry Jaques in London, 1842–1907

However, although the area over which people moved increased in the late-nineteenth century, the maps of moves before and after 1880 show remarkably similar patterns for both time periods. Although covering a smaller area, similar patterns are seen in the larger provincial cities of Britain. Many of the case studies already cited include examples of intra-urban migration within their life histories. In this section we will concentrate on one detailed example drawn from London, and one from the Scottish capital of Edinburgh.

Henry Jaques (mentioned briefly in earlier chapters) was born in London in 1842 and undertook some 25 residential moves within the metropolis before his death in 1907. These included local moves within the same community, suburbanization to the outskirts of London and movement for both housing and work related reasons (Figure 4.10). The complex residential history of Henry Jaques can best be understood in relation to his life stages. For the first 16 years of his life Henry lived at home, attending school until the age of 12 and then working, first, as an errand boy, and then apprenticed to his father's trade of watchmaking. During this period he moved four times with his parents and siblings, all short distance adjustments for housing reasons in and around the Islington area of London. Henry did not enjoy working with his father as a watchmaker and, at the age of 16, in 1858 he left home for the first time to take up an apprenticeship with a draper in Southwark. As with all apprentices at this time he lived in his master's house and although only some 5 kms from his parental home the move took him to a new part of London, south of the river Thames.

Henry did not complete his apprenticeship (he was dismissed under suspicion of stealing) and over the next six years he worked as a draper's assistant at various levels in both the wholesale and retail trade in a further seven different locations. In each case he lived-in at his workplace, and thus all moves were stimulated by a change of work. When between employment he returned to live at home for several weeks. Although on one occasion he moved briefly to the West of London (Pimlico), all his other homes and workplaces were in east London quite close to his parental home. This was clearly important to him as he returned home every Sunday and remained in close touch with his parents and his local community.

In 1864 Henry was intending to marry and this required him to reappraise both his home and his workplace. Employment as a living-in draper's assistant provided neither sufficient income nor a suitable home for a married man. With financial help from a contact in the clothing trade Henry set up his own small retail drapers on the Mile End Road and, at the same time, taught himself the skill of shirt-making so that he was engaged in both home-based shirt manufacture and in the retail clothing trade. His move was thus stimulated by his changed circumstances on marriage, but the precise location was determined by his employment. After only three years his retail business failed and he was forced to move into two rooms with his wife and growing family. For the first time he now had a separate home and workplace, seeking employment where he could, mostly in the wholesale clothing trade. However, he also continued to work at home at his shirt manufacture, which at some periods provided the bulk of his income. This home-working, together with a growing family, was the major stimulus for the next series of residential moves as he sought cheap but larger accommodation in East London to provide space for his family and home working. Over a period of ten years from leaving the shop he moved a total of six times to rented property in East London, close to the Mile End Road. This location was important to him as it enabled him to exploit his business contacts in the East End clothing trade and retain links with the area in which he had lived for most of his life.

By the 1870s Henry's life was moving into a more prosperous phase. Although he continued with his home working, he now had a more secure job as manager and buyer for a City shirt manufacturer. Mainly because he was worried about the health of his family, and with some trepidation, in 1877 he moved to the London suburb of Forest Gate. Although only about 6 kms from his previous homes around the Mile End Road (and 10 kms from his work in the City), Forest Gate was clearly seen as a new community. He now had a longer journey to work, and took some time to adjust to life in Forest Gate, often travelling back to his old neighbourhood to visit friends, and attend church. However, like many other London residents, he eventually settled in the London suburbs and remained there for the rest of his life, moving a further five times within Forest Gate for housing reasons. The life of Henry Jaques thus illustrates a wide variety of different types of intra-urban moves, demonstrates the ways in which such moves could be stimulated by employment as well as housing reasons, and shows that although most people who moved to the suburbs eventually settled there, the initial step could be almost as daunting as a move to a completely new settlement.

A similar pattern of frequent short distance migration linked, in some cases, with eventual suburbanization, can be discerned in many of the life histories returned by family historians. Duncan M., born in the East Lothian village of Athelstaneford in 1824, moved to Edinburgh with his parents at the age of two. Here he moved eight times in and around the city until his death in 1899. The first three moves in Edinburgh were all undertaken with his parents as over the space of ten years his father changed his business several times. Initially he had opened a coffee house, then ran two different spirit dealers before opening a grocer's shop in 1835. All these moves were thus stimulated by his father's work and Duncan lived at home and was employed in his father's grocery business until the age of 30. In 1854 Duncan M. left home for the first time and took employment as a grocer's warehouseman in Leith, an industrial and dockside part of the Edinburgh built-up area. He lived in lodgings and presumably gained the work through contacts he already had in the grocery trade.

Two years later in 1856 Duncan married and this forced him to seek a larger home and a higher income. He took another post as a grocer's warehouseman in the small town of Musselburgh, renting a house in what was rapidly becoming a residential suburb of Edinburgh. Three year's later he moved back to Edinburgh to work at another grocer's and rent a house in Cannonmills, a fashionable area on the edge of Edinburgh's new town. In 1872 he took over the running of his father's grocery business in nearby Clarence Street and, with increasing income, invested in two houses in Glenogle Terrace, living in one and renting the other. This property was on the edge of the built-up area, less than half a kilometre from his place of work. His father died in 1877 and Duncan inherited his father's business. However, he remained in Glenogle Terrace until in 1891 at the age of 66 he moved into a house previously occupied by his sister and conveniently located opposite his grocery business. He remained there until his death eight years later. Most of Duncan M.'s moves were thus stimulated by his employment, though with some desire for housing improvement as well. For most of his life he lived in a fashionable district on the edge of Edinburgh, but convenience to his workplace was clearly important.

Both the aggregate analysis of the full data set, and the individual case histories cited above, bear out evidence from the limited number of other studies of intra-urban residential mobility that exist (Pritchard, 1976; Dennis, 1977a; Pooley, 1979). Short distance movement within an urban area was very common, the need for a short journey to work meant that even within quite compact towns such movement was often stimulated by employment change, but that housing improvement was also an important motive. For those that could afford it, a move to a residential suburb was a common experience in the second half of the nineteenth century; a trend which became increasingly commonplace after the First World War.

The experience of urban to rural migration

Migrating from larger settlements to small towns and rural districts occurred for a variety of reasons. For some it was an extension of suburbanization, moving a short distance to a small settlement on the edge of a larger city in the same way that Duncan M. (above) moved less than 10 kms from Edinburgh to Musselburgh, before moving back to suburban Edinburgh. However, other migrants who moved down the urban hierarchy travelled over much longer distances. For some this may have been a return migration to a place they had lived in earlier in their life, but for others the move necessitated adjustment to a new community. This process of counter-urbanization has been well documented in the late-twentieth century (Fielding, 1982; Champion, 1989) but, as outlined above, there is considerable evidence to

Table 4.14 Summary characteristics of migrants from places ⩾5,000 to places <5,000 by year of migration (%)

Migrant characteristics	1750–1879	1880–1994
Gender		
female	28.9	42.2
Mean distance (km)	62.9	73.9
Distance band (km)		
<5	16.0	14.4
5–9.9	13.8	14.3
10–19.9	15.3	14.7
20–49.9	21.7	16.8
50–99.9	14.4	14.8
100–199.9	10.6	14.1
200+	8.2	10.9
Reason		
work	54.5	47.4
marriage	15.9	13.4
housing	4.4	4.5
family	4.3	3.6
other	20.9	31.1
Age		
<20	23.8	16.7
20–39	57.4	42.8
40–59	14.5	22.8
60+	4.3	17.7
Marital status		
single	33.0	28.5
married	62.5	62.6
other	4.5	8.9
Companions		
alone	12.5	17.8
couple	18.8	22.2
nuclear family	60.4	50.3
extended family	4.8	5.4
other	3.5	4.3
Position in family		
child	19.9	14.1
male head	55.8	47.1
wife	17.3	22.5
female head	3.5	10.6
other	3.5	5.7
Occupational group		
professional	12.3	20.8
farmer	5.5	2.0
intermediate	14.4	15.1
skilled non-manual	8.6	21.8
skilled manual	31.4	20.4
skilled/semi skilled agricultural	4.6	3.3
semi-skilled manual	9.3	6.4
unskilled manual	4.4	2.7
agricultural labourer	5.8	1.5
domestic servant	3.7	6.0
Total number	2,387	4,923

Data source: 16,091 life histories provided by family historians.

suggest that movement from large to small places, if not the dominant form of migration, was an important part of many life-time migration experiences over a much longer period of time. This section examines in more detail the pattern and characteristics of such moves and illustrates trends through a series of case histories.

The fact that such movement was more than suburbanization is emphasized by the distances moved (Table 4.14) with migrants from places over 5,000 population to those under 5,000 people moving on average at least 62 kms, and with over one third of all such moves being over 50 kms in each time period. Although almost 30 per cent of such moves were under ten kms, and thus could arguably be classed as suburbanization, many people were moving over substantial distances down the urban hierarchy. As with all moves the main reasons for moving to smaller places were for employment, although retirement migration became increasingly important in the twentieth century. People moving to small places from larger towns were, on average, rather older than all migrants, and were most likely to move in family groupings. The spatial pattern of such moves (Figure 4.11), shows that for most moves down the urban hierarchy there remained a strong regional effect, but with some moves from one part of the country to another.

Many individual migrants incorporated some movement down the urban hierarchy in their life histories. Two case studies illustrate the way in which movement both to and from large settlements was a routine part of many migration histories. John B. was born in the small town of Bishop Auckland (population 1,961 in 1801), County Durham, in 1767. His father was a potter, and he lived at home following his father's trade (from the age of 13) until he married in 1796 at the age of 29. He then established his home and his own pottery business in the village of Newbottle (population 970 in 1801) near Houghton-le-Spring, some 30 kms from his parental home. However, within a year of his marriage his wife died (probably in childbirth), and this stimulated a move to a new house in Newbottle with his baby son. In 1806, at the age of 39, he married again and moved some 120 kms to his wife's home city of Leeds (population 35,951 in 1811), establishing himself as a potter and earthenware dealer. However, he clearly did not settle in a larger city, and a year later he returned to Houghton-le-Spring with his wife and two sons, working as an itinerant pottery salesman. He moved again in 1811 to the small town of Hexham (population 3,427 in 1811) in Northumberland, and in 1817 to Richmond (population 3,546 in 1821) in Yorkshire, both moves related to his work as a potter and travelling pottery salesman. However, in 1830 at the age of 63 he moved some 60 kms with his wife, three sons and three daughters, back to Newbottle. Although he continued to work as a potter, the main reason for this move was to secure better employment opportunities for his sons in the Durham coal industry. John B. died in Newbottle in 1855. The life-time migration experience of John B. thus included movement up the urban hierarchy to the

143

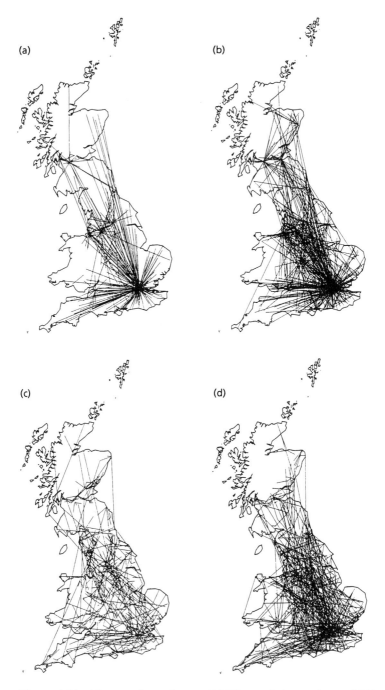

Figure 4.11 All moves from places ≥100,000 population to places <5,000 population, (a) 1750–1879; (b) 1880–1994. Moves from places 5,000–99,999 population to places <5,000 population, (c) 1750–1879; (d) 1880–1994

eighth largest city in Britain (in 1801), but for most of his life he moved between smaller settlements and returned in old age to the area in which he had spent his early married life. His moves were mainly stimulated by employment reasons (for himself or his children), though marriage and the effect of the death of his wife were also important, and movement between small settlements, and down the urban hierarchy, was a more frequent occurrence than movement towards larger settlements.

Archibald M. was born in 1856 in the small Scottish town of Forres (population 3,339 in 1851), Elginshire. His father was a farm labourer and his mother worked as a seamstress, but in 1858 (when Archibald was only two) his father died and the family of four children moved with their mother to a different house in Forres. The family moved once more in Forres, where Archibald was apprenticed as a painter from the age of 14, until in 1875 he moved with his two brothers to London where he continued to work as a painter. In 1883 he moved from London to Glasgow (no specific reason is stated), but by 1885 he was again living in Forres with his mother and two sisters, where he had established himself as a master painter and decorator. In 1887, at the age of 31 he married and moved to a new house in Forres, where he remained until 1909 when he entered a nursing home following a stroke. He died in Forres three years later. The migration history of Archibald M. thus includes movement to the largest cities of England and Scotland, probably in both cases for employment reasons, but he was drawn back to his family in the small town of Forres when he established himself in business on his own account. Such movement directly from a small place to a large city and then back to a small settlement was typical of the experience of many migrants in the nineteenth century.

Conclusion

In conclusion, we can return to the questions posed at the beginning of this chapter, and summarize the main findings of this analysis of movement between places of different sizes. Contrary to the impression given by many contemporaries, rural to urban movement only outstripped movement down the urban hierarchy by a relatively small margin, and from the 1880s movement to small places was at least as common as migration to larger places. As with the previous analysis of migration patterns, all regions experienced more-or-less the same trends, suggesting a high degree of uniformity in the British migration experience. The analysis emphasizes the importance of counterflows with, overall, movement down the urban hierarchy matching movement from small to large places. There is strong evidence that significant counter-urbanization was occurring from at least the 1880s.

Most migration was clearly undertaken in family groups, and often involved the exploitation of kinship networks, although movement from small places to large cities was most likely to be undertaken by single migrants moving alone. However, such migrants often moved into a controlled environment in which their employment and housing was taken care of prior to their move. Moving over long distances from a village to a large urban area was not necessarily more traumatic than a short distance move. Much depended on local and individual circumstances. Finally, the analysis emphasizes the extent to which most people encountered a wide variety of migration experiences during their life-course. Subsequent chapters explore in more detail the ways in which the migration experience linked with key life-course events, including changes in employment, family life-cycle stages, personal crises, and external factors such as war, which could disrupt individual lives.

CHAPTER 5

Migration, employment and the labour market

Introduction: industrialization and the changing structure of regional labour markets

Britain has been characterized as the 'first industrial nation', but progress towards a predominantly industrial society has not been straightforward (Wrigley, 1972; Thompson, 1973; Mathias, 1983; Mathias and Davis, 1989; Hudson, 1992). The experience of economic modernization, and associated technological, social, cultural and political transformations were not spread evenly, but varied between regions and economic sectors. This chapter begins by outlining the main dimensions of change and hypothesizing the ways in which the processes of industrialization and labour market change were related to population migration. Subsequent sections then examine in detail the extent to which mobility was stimulated by economic factors, variations in mobility by region and economic sector, the links between migration and social mobility and the changing relationship between home and workplace.

Economic historians have identified a series of national economic cycles which affected Britain from the mid-eighteenth century (Schumpeter, 1939; Deane and Cole, 1967; Lee, 1986; Crafts, Leyburn and Mill, 1989). It can be suggested that these trends had wide-ranging influences, not just on industrial and commercial development, but also on social and cultural life (Thompson, 1968; Perkin, 1969; Royle, 1987; Thompson, 1990), including the pattern and process of population migration. The first long cycle initially identified by Kondratieff and elaborated by Schumpeter and others, incorporating periods of economic prosperity, recession, depression and revival, ran from the 1780s to the 1840s and parallels Rostow's (1960) stage of economic 'take off' to sustained growth in the early phase of industrialization and urbanization. This period is characterized by high investment and high returns concentrated on leading economic sectors such as cotton textiles and iron manufacture. The second long wave, from the 1840s to the 1890s, coincides with Rostow's 'drive to maturity' and saw significant investment across a wider range of economic activity and the widespread adoption of new technology across most economic sectors. New transport technology, particularly the railways, transformed the distribution of raw materials and finished products. The third long cycle from the 1890s to the Second World War was characterized by new industries based on chemicals, motor manufacture and the production of household goods, and the adoption of new forms of power, especially electricity and the petrol engine. Rostow identified the period from the 1930s as the 'age of mass consumption' and this broadly coincides with a possible fourth cycle from the 1940s to the 1990s, focused on the expansion of micro-electronic technology and new communications systems as Britain's economic fortunes became increasingly bound up in a global economy over which it had little direct control (Wallerstein, 1979; Lash and Urry, 1987; Harvey, 1989).

Some would argue that the crucial period, which transformed Britain's economic structure and the lives of its inhabitants, was in the eighteenth century when evolution from a pre-industrial to an industrial society was most marked. However, the transformation from an 'organic' economy based on the land and handicraft industry and using wood as the major fuel and construction material to an 'inorganic' society based on coal, steam power, iron and machines was much more drawn out than this, with repercussions that spanned the seventeenth to the nineteenth centuries (Wrigley, 1988). In many parts of Britain early industrial change, or 'proto-industrialization' was crucial to later developments as domestic handicraft production became increasingly capitalistic, adopted new technology and utilized surplus labour both within and outside the family to create the economic and social environment from which entrepreneurs could develop larger scale factory industry (Mendels, 1972; Kreidte, Medick and Schulmbohm, 1981; Berg, Hudson and Soneuscher, 1983). These changes had far-reaching effects on people's lives, including patterns of residential change.

There have been marked variations in the impact of economic and technological change between different industrial sectors: the type of work undertaken fundamentally affected experiences of economic change and, in turn, influenced patterns of migration. New technology disproportionately affected a small number of industrial sectors, particularly coal, iron manufacture, engineering and textiles. In these industries output expanded rapidly from the mid-eighteenth century and the nature of work for many employed in such sectors altered dramatically (von Tunzelmann, 1978; Gregory, 1982; Berg, 1985; Hudson, 1986; Church, 1994; Jenkins, 1994; Pollard, 1994). Thus British coal output increased from some 4.4 million tons per annum in the 1750s to 133.3 million tons in 1871 and a peak of 287.4 million tons in 1913 (Pollard, 1980; Flinn, 1984; Church, 1986), and the number of cotton operatives in Britain increased from 254,800 in 1838 to 525,200 in 1898/9 (Chapman, 1972; Farnie, 1979; Rose, 1996). However, not all sectors changed rapidly and new technology was only selectively applied to much of British industry. Large occupational sectors such as agriculture, domestic service, much office and commercial employment, the manufacture of clothing, shoemaking, brewing, and most food processing, together employing the greater proportion of the British population, remained relatively unchanged until the late-nineteenth century (Lee, 1984; Armstrong, 1988; Gourvish and Wilson, 1994; Hoppit and Wrigley, 1994; Michie, 1994).

The distinctive geography of industrial production in Britain also meant that experiences of the impact of economic change varied considerably from region to region and between urban and rural areas. Spatial patterns of migration were likely to have been closely associated with this economic geography. Britain was characterized by a relatively small number of rapidly modernizing regions, in which the leading industrial sectors expanded rapidly, adopted new technology, and changed work practices; and a spatially much greater number of areas which, to varying degrees, missed many of the implications of industrial change (Hudson, 1989; Gregory, 1990). Modernizing regions were principally in the North of England, Central Scotland, parts of the West Midlands and South Wales, based mainly on the availability of coal and the development of textile, iron and steel and engineering industries. In large urban areas such as London, Manchester, Liverpool and Glasgow, although total employment expanded rapidly with the urban population, much of this was in relatively un-modernized sectors. Thus the textile industry of the East End of London remained concentrated in small workshops and domestic premises well into the twentieth century (Stedman Jones, 1971; Schmiechen, 1984; Fishman, 1988; Green, 1996) and, although office employment and the retail trade expanded rapidly in the nineteenth century, they remained both male dominated and traditional in their practices until the twentieth century (Jefferys, 1954; Anderson, 1976; Winstanley, 1983). The census captures some of these spatial variations in regional economic structure, but there were many local variations which nuanced this broader picture. Other areas, including East Anglia, much of Southern England and South-West England experienced de-industrialization during the nineteenth century. Unable to compete with the new technologies and economies of scale achieved in coalfield locations, domestic, workshop and small factory industry in these areas either diversified into more specialist markets or gradually declined (Hudson, 1989). Such economic experiences could be expected to have fundamental effects on inter-regional population flows as labour moved from de-industrializing to expanding regions.

There has been considerable debate about the impact of the industrial revolution in Britain. The traditional view is that change was indeed revolutionary, that levels of investment, economic growth and technological innovation were significant, and that these changes fundamentally affected people's lives (Deane and Cole, 1967). More recently there have been revisionist arguments suggesting variously that change in the key economic

149

indicators was limited, or that such change that did occur was highly concentrated into particular sectors and regions (Crafts, 1985; Gregory, 1990). A more balanced view has been taken by Hudson (1992), who argues that, although the economic impact of industrial change may have been overstated in the past, this is a somewhat narrow view of the impact of the processes that were operating in Britain from the mid-eighteenth century. Both industrializing and de-industrializing regions were affected by change, though in different ways, and if the impacts of the industrial revolution are extended beyond the narrowly economic to include social, cultural and political change and, most significantly, the effects on the everyday lives of individuals and families, Hudson argues that there remain grounds for viewing the changes as truly revolutionary. In the context of population migration, and the ways in which people made decisions about their residential location, it is such wide ranging ramifications that are of most significance.

It is possible to develop a series of more specific hypotheses about the links between economic change and population migration. These will be examined during the course of this chapter. First, it can be suggested that the volume, direction and characteristics of inter-regional migration flows were related to the economic fortunes of specific regions. Thus de-industrializing regions should lose population and industrializing regions gain through in-migration. Second, there were similar sectoral movements of labour from declining industries to those that were expanding. This could take the form of inter-regional flows, but it may also be reflected in shorter distance intra-regional movement as labour shifted from one sector to another within a geographical area. Third, the characteristics of labour migrants were directly related to the nature of labour demand at any particular time. Thus it can be suggested that labour migration flows for men and women might be very different, that migrants with skills that were relevant to the modernizing sectors behaved differently from those with skills that were no longer needed, and that young labour migrants had very different opportunities to those who were older. Fourth, in some cases, the pattern and process of migration was directly related to the characteristics of an occupation. Some jobs demanded mobility, and increases in the level of mobility required from some employees are likely to have been especially important in the twentieth century. Fifth, the experience of labour migration was related to whether associated employment change was viewed as a success or failure. Labour migration associated with upward social mobility may be viewed very differently from that which was forced by redundancy. Lastly, it can be suggested that, as transport opportunities changed in the twentieth century, there were quite fundamental alterations in the relationship between home and workplace which affected both decisions to change home and work. For instance, a job change which would have necessitated a change of address in the nineteenth century may have been within commuting distance in the mid-twentieth century.

However, the hypothesized relationships between migration and economic change are not always as clear-cut as might initially seem to be the case. There are a number of complicating factors that must be taken into account in the ensuing discussion. First, modernizing industrial sectors and regions were usually associated with technological change and capital investment which could lead to less labour intensive industry and demand for workers with new skills (Feinstein and Pollard, 1988). Both processes could influence patterns of labour migration, and the extent to which labour moved from economically declining to expanding areas. Second, even if labour in declining regions had appropriate skills, transport constraints may have restricted the extent to which inter-regional labour migration took place. The cost of long distance travel remained relatively high during the nineteenth century (Dyos and Aldcroft, 1969) and expanding industries were most likely to recruit from nearby labour surpluses. Third, it can be suggested that in periods of economic recession there may be strong psychological factors encouraging labour immobility (Salt, 1985; Green, 1985). Most people would prefer to be unemployed in a familiar area where they had support networks of family and friends. Perceptions of widespread recession may have reduced the likelihood of long-distance labour mobility. Fourth, patterns of labour migration were affected by the structure of recruitment in different industries and by the flow of information about job vacancies. Even in the twentieth century, whilst some employment is advertised nationally, most lower-skilled employment depends on local and regional labour markets (Saunders, 1985). Lack of information and variations in recruitment patterns will have influenced patterns of labour migration between sectors and regions.

The structure of labour migration has also been influenced by government and other agencies. In the eighteenth and nineteenth centuries operation of the Laws of Settlement could remove those without means of support to their parish of settlement, thus restricting labour migration and deterring some from leaving their home area (Marshall, 1985; Rose, 1986; Taylor, 1989). In the twentieth century the government intervened directly in labour migration during the inter-war recession through the development of Labour Transfer Schemes, designed to move workers from depressed regions to the South East and West Midlands. Unpopular with most workers, such schemes had only limited impact (Pilgrim Trust, 1938). Less directly, government has also influenced labour migration through the development of regional policy which has directed investment and new industry to selected development areas since the 1930s (Jones, 1984). Such policies may have both retained labour in declining regions and encouraged migration to areas in which investment was taking place. Lastly, since the 1840s, emigration has been viewed, at least in part, as a safety valve which has removed labour surpluses. Encouraged by government and promoted by private emigration companies, high levels of emigration to North America, South

Africa and Australasia have reduced pressure in many regions with declining economies (Baines, 1985; Constantine, 1990; Erickson, 1994).

It can thus be suggested that the relationship between labour migration and economic and industrial change is not straightforward. Whilst, a priori, it might be expected that there was a clear relationship between economic prosperity and migration, with distinct regional and sectoral differences, in practice there were many complicating factors which may have reduced such differences and influenced migration patterns. This chapter uses data provided by family historians to examine the changing characteristics of migration for work. Some of the methodological problems involved with these data were outlined in Chapter 2 but, in the context of this chapter, the difficulties of comparing occupational data over time must be re-emphasized. The 200 years since 1750 have seen massive changes in the structure and organization of most occupations. Thus comparing occupational status over such a long time period is fraught with difficulty. In this research we have adopted a standard occupational and industrial classification, based on OPCS definitions, but the limitations of such data must be borne in mind in subsequent analyses.

Labour migration in Britain

Although moving for work-related reasons was the single most important cause of migration, accounting for 37.8 per cent of all moves, it should be stressed that many moves were undertaken for a multiplicity of reasons. Although work may have been a significant reason and, from documentary evidence was recorded by family historians as the main reason for moving; family, housing and other factors may also have played a part. Moreover, some moves which were recorded as being stimulated for non-work reasons subsequently necessitated a change of employment. The interaction of a variety of reasons for migration is best illustrated from diaries and accounts which begin to reveal the complexities of migration decisions (see below). Analysis in this section focuses, first, on those moves where the main reason for migration was recorded as employment, but possible interaction with other factors must be borne in mind. Variations in the pattern and characteristics of work-related moves are examined for different occupational groups. Second, the characteristics of all moves are examined in relation to changes in work status following migration. In conclusion, this evidence is related to patterns of sectoral and spatial change in British labour markets.

Given their volume, it is not surprising that the spatial pattern of work-related moves reflects the national pattern of all migration (separate maps are not reproduced here). Essentially, work-related moves mirror the total data set, but without the element of local circulation which was so important in the national migration system. The pattern of work-related moves was also similar for those in different occupational categories (Figures 5.1–5.3), but those in higher-status occupations undertook more inter-regional moves, and local migration for employment was more important for those in low-skilled work (as before, local moves cannot be shown on the maps). The spatial pattern of work related moves for agricultural workers is most distinctive, with a large amount of local circulation between neighbouring employers, but domestic servants are one low-status occupation where longer distance migration was common, as many servants moved with their employers' households.

These details are confirmed in Table 5.1a–b which summarizes the characteristics of all work related moves for ten occupational groups. Those in professional and skilled non-manual occupations consistently moved over the longest distances (over 95 kms on average in each time period), whilst farmers and unskilled agricultural labourers moved over much the shortest distances (under 25 kms on average). For all other groups the average distance moved was in the range of 40–70 kms, with distances increasing in the time period after 1880. Perhaps the most surprising feature of these migration distances, and the accompanying maps, is the similarity of the spatial characteristics of work related moves for different groups. Those with unskilled occupations moved over similar mean distances for employment reasons as those in skilled manual and intermediate occupations, although those with few skills undertook fewer very long distance moves. Whilst a high income (such as amongst professional workers) obviously made long distance mobility easier, much more important for work related moves was the structure of the labour market. Professional and skilled non-manual workers were mainly urban based and operated within a national labour market for their skills. Most other workers tended to move within regional labour markets, restricted by more localized demand for their skills and more restricted information flows about work opportunities. Thus a cotton spinner was unlikely to move outside Lancashire in the nineteenth century, although there was some long distance work-related movement of those with specialist skills (Jackson, 1982; Pooley and Doherty, 1991). Farmers obviously had a strong attachment to a local area and thus, despite relatively high status, moved infrequently and over short distances. Unskilled agricultural workers too remained mainly within local labour markets, constrained by both low income, lack of relevant skills and restricted knowledge of alternative opportunities. If being hired as a farm-hand depended on building a reputation as a reliable worker, there were clear advantages in remaining in an area where you were known.

In most respects the other characteristics of work related moves for different occupational groups were similar to each other and reflect trends in the whole data set. There are, however, some differences which require comment. Those in professional and intermediate occupations, together

153

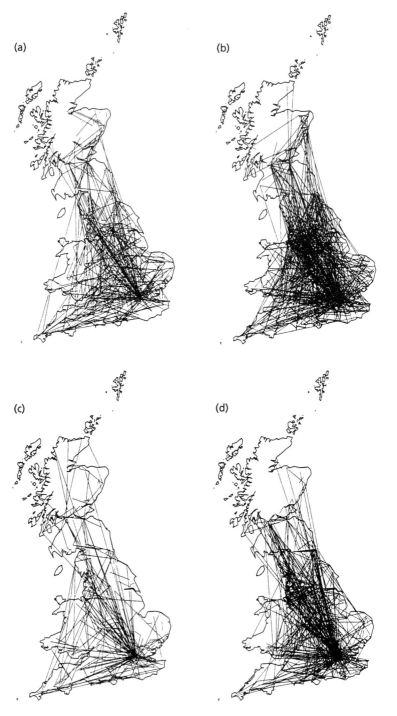

Figure 5.1 All work-related moves by professionals, (a) 1750–1879; (b) 1880–1994; and by skilled non-manual workers, (c) 1750–1879; (d) 1880–1994

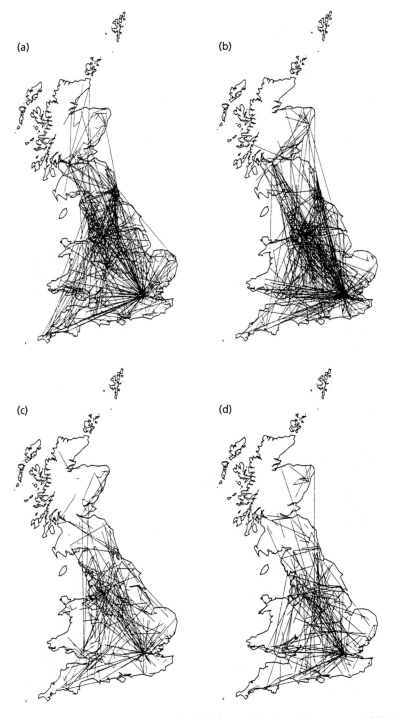

Figure 5.2 All work-related moves by skilled manual workers, (a) 1750–1879; (b) 1880–1994; and by semi-skilled and unskilled manual workers, (c) 1750–1879; (d) 1880–1994

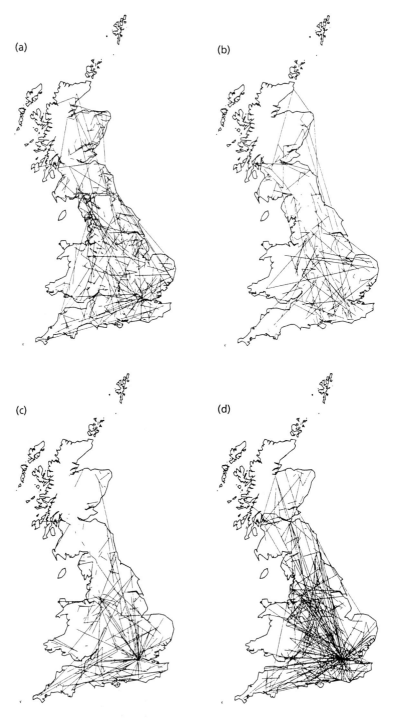

Figure 5.3 All work-related moves by agricultural workers, (a) 1750–1879; (b) 1880–1994; and by domestic servants, (c) 1750–1879; (d) 1880–1994

Table 5.1a Summary characteristics of all work-related moves by occupational group, 1750–1879 (%)

Migrant characteristics	Professional	Farmer	Intermediate	Skilled non-manual	Skilled manual (industrial)	Skilled/ semi-skilled agricultural	Semi-skilled industrial	Unskilled industrial	Agricultural labourer	Domestic servant
Gender										
female	4.6	0.9	13.6	3.6	3.4	1.7	4.2	0.4	3.0	72.8
Mean distance (km)	109.6	19.4	39.3	95.1	57.5	45.2	57.3	40.0	20.1	53.7
Distance band (km)										
<5	12.7	40.9	47.0	16.3	24.6	29.0	23.9	31.2	39.1	25.8
5–9.9	4.7	17.8	9.2	7.9	11.0	10.3	10.1	12.2	20.1	14.4
10–19.9	7.9	15.9	8.3	8.6	13.3	15.9	13.5	15.6	16.1	13.8
20–49.9	15.3	14.9	13.8	15.6	20.3	22.3	21.3	16.0	14.8	16.3
50–99.9	17.2	7.2	8.9	20.3	13.6	10.8	14.9	11.8	6.8	11.9
100–199.9	25.1	2.6	7.4	16.1	9.8	6.4	8.3	9.8	1.9	12.9
200+	17.1	0.7	5.4	15.2	7.4	5.3	8.0	3.4	1.2	4.9
Age										
<20	3.7	2.7	2.2	14.6	12.0	19.5	17.5	15.9	26.8	59.9
20–39	71.8	59.9	61.9	66.2	70.4	62.3	64.8	67.3	55.8	35.6
40–59	22.9	31.7	32.0	18.4	16.8	15.2	16.6	14.4	15.0	4.2
60+	1.6	5.7	3.9	0.8	0.8	3.0	1.1	2.4	2.4	0.3
Marital status										
single	42.2	17.1	16.3	36.0	31.3	40.3	37.4	35.7	42.5	83.3
married	53.9	81.2	80.8	61.0	66.3	58.1	61.2	62.9	56.1	14.4
other	3.9	1.7	2.9	3.0	2.4	1.6	1.4	1.4	1.4	2.3
Companions										
alone	36.8	9.1	12.3	27.9	19.0	25.7	19.6	20.0	29.1	66.8
couple	13.0	11.3	11.4	10.8	11.7	8.0	10.3	11.0	9.5	4.2
nuclear family	44.7	67.3	69.0	55.6	63.3	58.6	61.6	62.1	54.2	16.9
extended family	2.0	5.9	3.7	1.9	2.4	3.3	3.8	3.7	2.5	0.6
other	3.5	6.4	3.6	3.8	3.6	4.4	4.7	3.2	4.7	11.5
Position in family										
child	1.1	3.4	1.5	5.2	8.1	11.0	12.1	11.4	9.8	6.0
male head	93.2	92.9	93.3	89.1	86.0	84.1	80.7	83.7	83.6	22.9
wife	1.6	0.8	2.2	2.1	1.6	0.5	1.6	0.4	0.2	4.0
female head	2.3	0.2	1.0	0.4	0.7	0.8	0.9	0.0	2.2	55.6
other	1.8	2.7	2.0	3.2	3.6	3.6	4.7	4.5	4.2	11.5
Total number	735	557	832	498	1,541	405	450	255	704	379

Data source: 16,091 life histories provided by family historians.

Table 5.1b Summary characteristics of all work-related moves by occupational group, 1880–1994 (%)

Migrant characteristics	Professional	Farmer	Intermediate	Skilled non-manual	Skilled manual (industrial)	Skilled/ semi-skilled agricultural	Semi-skilled industrial	Unskilled industrial	Agricultural labourer	Domestic servant
Gender										
female	25.4	1.3	12.6	16.7	5.8	4.1	8.7	0.0	4.0	80.8
Mean distance (km)	125.7	24.2	62.9	98.7	77.2	49.3	68.3	65.5	24.9	72.2
Distance band (km)										
<5	9.6	39.9	35.9	16.7	23.2	24.7	24.9	27.8	36.8	26.3
5–9.9	4.2	8.2	7.9	6.5	8.1	15.4	9.9	6.0	15.4	9.7
10–19.9	6.9	18.3	8.8	11.3	11.8	10.8	9.7	6.6	18.4	10.5
20–49.9	15.0	20.2	14.5	15.8	17.1	21.3	19.3	20.6	19.9	15.9
50–99.9	17.4	8.2	11.9	16.4	13.0	12.4	11.8	19.2	4.4	14.3
100–199.9	23.8	4.8	11.2	17.1	14.0	10.5	13.4	11.9	2.2	11.1
200+	23.1	0.4	9.8	16.2	12.8	4.9	11.0	7.9	2.9	12.2
Age										
<20	2.6	1.8	1.0	13.9	8.1	20.0	9.8	22.6	22.7	41.8
20–39	67.6	40.4	46.8	61.2	68.2	50.8	68.4	54.8	52.7	41.8
40–59	26.3	44.9	45.4	23.7	21.8	24.4	19.2	17.6	18.0	14.2
60+	3.5	12.9	6.8	1.2	1.9	4.8	2.6	5.0	6.6	2.2
Marital status										
single	42.9	12.9	11.5	45.3	34.1	37.3	37.6	46.8	39.0	72.6
married	55.9	82.9	84.4	53.3	64.6	61.0	61.2	51.6	60.3	22.0
other	1.2	4.2	4.1	1.4	1.3	1.7	1.2	1.6	0.7	5.4
Companions										
alone	37.4	6.2	12.7	35.2	30.8	29.8	31.8	41.2	26.0	61.2
couple	13.5	13.3	14.2	10.3	7.4	9.3	8.9	6.2	7.5	2.6
nuclear family	42.3	67.1	64.4	47.3	55.9	54.8	53.2	44.8	58.4	23.6
extended family	1.5	5.8	5.0	2.4	3.3	2.9	2.9	3.2	5.5	1.9
other	5.3	7.6	3.7	4.8	2.6	3.2	3.2	4.6	2.6	10.7
Position in family										
child	3.0	3.1	2.0	11.4	8.4	9.4	8.2	9.3	13.0	5.7
male head	71.5	89.8	84.4	73.1	85.4	84.9	81.6	87.1	82.9	19.0
wife	7.0	0.0	8.8	4.7	1.3	0.6	1.8	0.0	0.0	6.7
female head	13.6	1.3	2.4	5.9	1.3	1.9	4.0	0.0	2.7	57.4
other	4.9	5.8	2.4	4.9	3.6	3.2	4.4	3.6	1.4	11.2
Total number	1,589	225	1,261	1,411	1,454	315	460	199	150	693

Data source: 16,091 life histories provided by family historians.

with farmers, were less likely to move for work-related reasons when under the age of 20 (reflecting longer periods of training and delayed entry into the relevant occupation) whereas, not surprisingly, domestic servants were most likely to move for work related reasons when young. The recruitment of domestic servants was dominated by young women leaving home for the first time and 76 per cent of domestic servants who moved for work-related reasons were single. For all other occupations most moves were undertaken by nuclear families and married couples, although a relatively high proportion of those in professional occupations were also unmarried and moved alone. This probably reflects movement for career advancement in a national labour market in the early stages of a career. However, despite these differences, the overwhelming picture which emerges is that work-related migration had remarkably similar characteristics in all occupational groups, and that the differences which emerge can be easily explained by obvious variations in the career and labour market structure of different occupations.

Some of these features can be further illustrated from diary evidence in which the complexity of migration decisions for work can be more clearly determined. Amos Kniveton was born in the small village of Astley Green, some 10 kms to the west of Manchester in 1835. He was the second child of a large family of 14 children, and his father worked as a boot and shoemaker in the village. Amos began work as a half-time worker in a local cotton mill at the age of seven, but two years later he left the mill and unbeknown to his parents hired himself as a ploughboy to a nearby farmer. Here he received board and lodgings: his main motivation for leaving home at the age of nine was to relieve pressure on the overcrowded family home. However, he maintained close ties with his family and at the age of 14 returned to the family home (some of his siblings having themselves left) to learn the trade of boot and shoemaking. All these early moves were thus stimulated by a combination of employment and family reasons. Amos remained at home until 1857 when, aged 22, he married a local girl and rented a house in Astley Green. Initially he continued to work for his father, but within two years he had established his own boot and shoemaking business in the village.

Amos was successful at his trade, but this created problems as he was taking trade from his father. In 1863 he decided to look for new employment away from Astley Green primarily because the village could not support two shoemakers and he did not wish to compete with his father. Amos answered an advertisement in the *Methodist Recorder* for a 'bootmaker and manager' for a business in Morecambe (Lancs) owned by John Gorrill of Lancaster. The advertisement stated that a teetotaller and local preacher would be preferred, and it is clear that Amos' contacts in the Methodist church helped him to gain the post some 70 kms from his family home. Again the move was stimulated by a combination of family and business reasons, but it is clear that Amos viewed his new situation as an opportunity to broaden his experience and better himself. He was able to achieve this two years later when he applied for the post of manager of a shop and foreman over 15 men in a shoemaker's business in Blackburn. This job gave him both a higher income and more responsibility and was clearly made for employment reasons, though the fact that it brought him closer to his family home also seemed to be important to him. However, after 18 months he moved again for work reasons as his employer asked him to manage a new shop in Bolton. Although even nearer his parental home, Amos was reluctant to move because the post offered less responsibility. However, he stuck with his employer until 1867 when he was in a position to set up his own boot and shoemaker's business in Leigh, just 4 kms from his home village where many of his family still lived.

Amos and his family lived above the shop in Leigh for 27 years, but at the age of 59 a combination of circumstances led to the collapse of his business and forced a move. First, Amos contracted smallpox and customers avoided the shop for several months; second, the premises had to be closed for rebuilding leading to a loss of trade and a subsequent increase in rent; third, Amos lost a substantial sum of money when his brother's business (which Amos had invested in) failed; and, fourth, he suffered heavy losses when a series of wholesale customers went bankrupt. The combined effect of these circumstances, mostly outside Amos' control, led to the sale of the business in 1894 and a move to a rented house, also in Leigh. Through business contacts Amos gained work as an assurance salesman, though this was a definite step down in status, and he continued to live in the same house until his death in 1927 at the age of 92. Like many with limited education and mainly manual skills, Amos' moves were confined within a single region. Although work played a part in the motivation for most of his movement, this interacted with his family commitments and loyalties, and also with his sense of attraction to his home area. It is difficult to record such complex migration reasons from any sources other than diaries, but Amos's moves for work and other reasons within Lancashire seem typical of many migrants drawn from a wide spectrum of the labour market.

The longer distance movement of those in professional occupations is illustrated by the case of Annie Madin who was born in 1893 in Leek (Staffordshire). She was born into a reasonably affluent household (her father was a surveyor with the local authority) and she received a good education, graduating as a doctor from Edinburgh University Medical School in December 1916. She had experienced frequent mobility as a child, her parents moved seven times before she left home, including a long distance removal from Staffordshire to Northamptonshire when her father took a new job, and this pattern was initially repeated in her own life. In 1917 Annie moved from Edinburgh to Guildford (Surrey) to take her first post as a junior hospital doctor, and the requirements of further training and promotion led her to move four more times in as many years to Winchester, back to Edinburgh, to Sheffield and Birmingham. Whilst in Birmingham she was an Assistant Medical Officer of Health in Smethwick, but after her marriage in 1922 she worked in General Practice with her husband in Cannock (Staffordshire). She moved four more times before her death in 1975, but all within the same part of the West Midlands. The characteristic pattern of long distance mobility alone, in the early stages of the life-cycle, was thus produced by the career requirements of Annie's profession. Once she married and settled to a career, subsequent migration was short distance and contained within one region. In this respect it was similar to the migration experiences of those in many other occupations from across the social spectrum.

It is also possible to examine the extent to which migration was related to a change in work status, irrespective of the reason for migration. This helps to place work-related migration in a broader perspective, and emphasizes that not all migration led to employment change and that job changes could occur without residential migration. As Tables 5.2a–b demonstrate, overall, just under one third of moves did not coincide with a change in employment; in around one quarter of moves the same type of work was undertaken in a different place; in one fifth of moves different work was undertaken in a new location; and in one fifth of moves the migrant either entered or left the labour market. There were some changes over time, with a smaller proportion of moves which did not entail a change of work in the period before 1880. However, this could relate to the source material, as it is likely that very short distance moves were under-reported for the eighteenth century. Likewise, variations in the proportion entering and leaving the labour force are an artefact of the sample which focused on people born 1750–1930. Thus, the first time period will have a surfeit of young people and the last time period a surplus of the elderly. The most notable feature is the stability of trends over time.

Examination of the characteristics of migrants with different work status following migration again shows a high degree of stability. Variations relate mainly to the characteristics of the full data set, or to obvious factors. Thus it is not surprising that the young were most likely to enter the labour market, and the old most likely to leave! Of more interest is the fact that those entering the labour market moved on average over a longer distance than those leaving. The differences are only some 15 kms, but it suggests that those undertaking retirement migration may have had slightly stronger ties to a home area. The distance moved was also much the same, irrespective of whether there was a change in the type of work undertaken. There is a clear gender difference reflecting male and female experiences of the labour market. For women, a residential move was most likely to be associated with either entering or leaving the labour market, whereas for men this was a minority experience. This clearly demonstrates the more precarious position of women in the labour market and suggests that throughout the

161

Table 5.2a Changes in work status following migration by migrant characteristics, 1750–1879 (%)

Migrant characteristics	Same work in same place of work	Same work in different place of work	Different work in same place of work	Different work in different place of work	Entered labour market	Left labour market
Gender						
female	4.0	4.3	2.9	2.9	25.2	45.8
Mean distance (km)	7.0	46.0	6.1	44.6	50.7	37.4
Distance band (km)						
<5	84.1	33.5	88.8	39.5	35.4	48.0
5–9.9	6.0	13.3	4.1	10.7	12.5	12.6
10–19.9	2.8	12.3	1.2	11.3	10.9	9.6
20–49.9	3.7	16.5	2.2	14.7	14.8	11.8
50–99.9	1.9	10.3	2.2	9.9	9.9	6.2
100–199.9	0.8	8.4	1.5	7.8	9.5	6.6
200+	0.7	5.7	0.0	6.1	7.0	5.2
Reason						
Work	18.3	64.9	41.4	62.2	59.5	8.2
Marriage	44.0	14.6	22.1	12.5	8.4	34.5
Housing	23.9	3.6	21.0	3.1	2.0	1.5
Family	1.6	1.5	1.1	1.6	1.9	6.4
Crisis	4.6	2.6	5.0	4.7	6.4	12.9
Other	7.6	12.8	9.4	15.9	21.8	36.5

Age						
<20	7.9	6.8	5.5	7.5	59.1	8.1
20–39	65.6	68.2	63.2	65.7	35.1	47.6
40–59	21.4	21.3	25.8	22.7	4.7	16.0
60+	5.1	3.7	5.5	4.1	1.1	28.3
Marital status						
single	13.8	20.4	19.3	24.8	76.5	16.1
married	82.3	75.7	77.7	71.5	19.3	67.2
other	3.9	3.9	3.0	3.7	4.2	16.7
Companions						
alone	5.4	13.5	4.4	19.2	46.6	24.9
couple	27.0	18.5	19.7	18.7	10.0	37.0
nuclear family	58.7	59.4	64.6	52.8	33.2	26.5
extended family	5.4	3.2	5.5	4.4	3.5	8.3
other	3.5	5.4	5.8	4.9	6.7	3.3
Position in family						
child	8.0	4.3	6.6	3.5	23.2	1.8
male head	87.1	87.9	88.2	89.9	51.2	49.4
wife	1.8	1.7	1.5	1.0	4.8	33.8
female head	0.7	1.2	0.4	1.0	13.9	9.7
other	2.4	4.9	3.3	4.6	6.9	5.3
Total number	4,702	6,246	277	3,792	2,090	1,036
% of total	25.9	34.4	1.5	20.9	11.6	5.7

Data source: 16,091 life histories provided by family historians.

Table 5.2b Changes in work status following migration by migrant characteristics, 1880–1994 (%)

Migrant characteristics	Same work in same place of work	Same work in different place of work	Different work in same place of work	Different work in different place of work	Entered labour market	Left labour market
Gender						
female	11.5	16.7	4.1	10.5	44.9	49.8
Mean distance (km)	7.2	79.4	11.1	76.3	71.6	53.4
Distance band (km)						
<5	79.3	35.8	74.6	31.2	33.8	43.3
5–9.9	7.8	9.2	9.0	7.9	7.1	8.8
10–19.9	5.5	10.7	6.0	8.6	7.8	7.7
20–49.9	3.7	15.1	5.3	13.9	13.7	10.2
50–99.9	2.2	13.0	2.3	12.6	12.6	11.4
100–199.9	0.9	13.0	1.8	12.8	12.9	10.6
200+	0.6	13.2	1.0	13.0	12.1	8.0
Reason						
Work	10.2	59.0	28.4	54.4	53.4	9.8
Marriage	26.4	6.9	18.8	5.3	3.9	29.0
Housing	43.2	7.7	32.2	4.4	8.2	6.5
Family	2.6	1.8	1.5	1.6	2.1	4.3
Crisis	5.4	4.2	3.7	4.2	7.0	9.4
Other	12.2	20.4	15.4	30.1	25.4	41.0

Age						
<20	5.3	5.6	3.9	10.5	45.4	2.8
20–39	59.2	59.1	55.9	62.4	42.4	44.6
40–59	29.3	30.1	35.0	22.7	10.1	15.2
60+	6.2	5.2	5.2	4.4	2.1	37.4
Marital status						
single	15.2	25.8	13.2	37.0	71.7	14.5
married	80.7	70.3	84.1	59.7	24.5	74.9
other	4.1	3.9	2.7	3.3	3.8	10.6
Companions						
alone	7.1	20.1	7.1	34.5	42.3	18.8
couple	25.8	15.8	22.4	13.0	6.8	45.3
nuclear family	58.9	52.6	62.7	39.1	38.6	26.7
extended family	5.6	3.3	4.4	3.6	4.0	6.8
other	2.6	8.2	3.4	9.8	8.3	2.4
Position in family						
child	8.8	4.5	5.9	4.5	23.3	2.1
male head	80.8	75.2	87.8	77.0	34.2	47.0
wife	4.5	5.7	2.0	3.0	11.7	35.6
female head	2.6	6.7	1.2	5.3	21.6	10.1
other	3.3	7.9	3.1	10.2	9.2	5.2
Total number	8,604	5,724	412	6,151	2,364	3,235
% of total	32.5	21.6	1.6	23.2	8.9	12.2

Data source: 16,091 life histories provided by family historians.

Table 5.3 All labour migration* by type of move by region and year of migration (%)

Region	1750–1879				1880–1994			
	Within region moves	Into region moves	Out of region moves	Total number	Within region moves	Into region moves	Out of region moves	Total number
London	31.2	40.8	28.0	984	23.3	39.2	37.5	1,163
South East	53.9	22.8	23.3	1,008	37.9	32.7	29.4	1,812
South Midland	45.1	23.8	31.1	676	26.0	37.8	36.2	1,230
Eastern	55.9	16.7	27.4	592	37.3	33.6	29.1	888
South West	58.4	15.7	25.9	849	41.2	25.9	32.9	851
West Midland	41.6	37.8	20.6	1,165	28.4	48.8	22.8	1,690
North Midland	61.2	17.6	21.2	754	43.5	27.3	29.2	984
North West	62.9	23.5	13.6	1,410	48.9	27.2	23.9	1,802
Yorkshire	54.9	22.6	22.5	960	47.2	26.6	26.2	1,311
Northern	70.1	15.8	14.1	1,112	50.2	20.7	29.1	840
Monmouth	40.4	34.8	24.8	141	31.4	38.4	30.2	188
North Wales	60.6	17.8	21.6	180	43.4	22.1	34.5	249
South Wales	54.2	12.3	33.5	203	25.9	36.2	37.9	116
Glamorgan	43.8	41.6	14.6	230	49.7	26.6	23.7	429
Highland	58.9	16.4	24.7	146	29.7	32.8	37.5	64
NE Scotland	77.1	11.7	11.2	330	64.7	13.8	21.5	312
Scottish Midland	65.8	12.4	21.8	450	42.9	24.8	32.3	348
SW Scotland	49.8	27.5	22.7	308	26.0	38.7	35.3	331
SE Scotland	36.0	30.3	33.7	211	21.5	38.2	40.3	228
Southern Scotland	53.6	19.5	26.9	220	63.5	8.7	27.8	115

* Migration for work-related reason.

Data source: 16,091 life histories provided by family historians.

period under study residential moves were most likely to be undertaken with male careers in mind.

National patterns of work related migration and changes in employment status following migration thus reveal few real surprises. What is consistently surprising is the lack of variation in the data set. With a small number of easily explained exceptions, labour migration had much the same characteristics irrespective of the type of work undertaken. This suggests that the factors influencing migration patterns in Britain go beyond employment factors and that life-course and regional influences were similar irrespective of the nature of employment. Because work-related moves were numerically important, their characteristics are similar to those of the whole data set.

The significance of regional labour markets

Having examined the overall structure of labour migration in Britain, we can now assess the extent to which there were differences related to the characteristics and fortunes of specific regional labour markets. As demonstrated in Chapter 3, one of the more surprising results of the analysis of migration characteristics was the fact that, in general, all British regions were experiencing very similar trends. Certainly, the similarities between regions were much greater than the differences. However, within this overall structure, it may be that work related moves had specific regional characteristics, and it is reasonable to hypothesize that expanding and contracting regions would have had rather different patterns of in- and out-migration. However, one problem in pursuing this analysis is the scale at which regional economies are examined. Although it is possible to ascribe broad economic characteristics to regions, and relate these to migration patterns, in reality there would have been a large amount of local variation. For many individual migrants local economic conditions in particular communities may have been more important than regional economic indicators. Unfortunately, such characteristics can only be discerned through detailed local research. In this section we, first, relate labour migration patterns to broad regional characteristics, second, focus on case studies of selected regions and, third, illustrate these through some individual case studies which once again show the complexities of the labour migration process.

In the period 1750–1879, during which regionally differentiated processes of industrialization and de-industrialization were most marked, there were only small variations in the extent to which labour migration flows were contained within regions or were between regions (Table 5.3). In all regions apart from London local redistribution within a region was a more important component of labour migration than movement between regions. London and the West Midlands were the two areas in which movement into the region was relatively most important: both regions had a high demand

Figure 5.4 All work-related migration to London, (a) 1750–1879; (b) 1880–1994; and to the West Midlands, (c) 1750–1879; (d) 1880–1994

for labour in expanding industrial and urban areas and attracted migrants mainly from surrounding counties, though London had a significantly larger sphere of influence (Figure 5.4). Predictably, regions which were undergoing industrial decline and restructuring experienced the least in-migration and the highest proportions of out-migration. Thus, eastern England, south-western England and much of Wales and Scotland had relatively high rates of labour migration out of the regions (Figure 5.5). However, not all industrializing and de-industrializing areas experienced high rates of inter-regional labour migration. Northern and north-western England, together with north-eastern and central Scotland, all experienced particularly high rates of within-region migration, with regional redistribution more important than inter-regional flows.

Similar trends continued in the period after 1880, though within-region movements were less important in all areas. Most marked was the fact that large urban areas, most notably London and the conurbations of Central Scotland, had the highest rates of both in- and out-migration. Both the West Midlands and South Wales had high rates of labour migration into the regions and some mainly rural areas continued to lose labour migrants, though at a slower rate than in the earlier time period. In both time periods there were very few differences by gender, with male and female labour migrants behaving in very similar ways, and only small variations by age (Table 5.4). There was a slight tendency for labour migrants over the age of 60 to make more within-region moves, representing short distance career adjustments in a relatively late stage of the life-course, and the West Midlands seemed to attract more young adult migrants than any other region, but otherwise all regions were very similar. Overall, the differences between regions were small and clearly related to levels of industrialization and urbanization at particular times in the past. However, the most persistent picture is of essential similarities in the pattern and process of labour migration in different regions, with most moves for employment (as for other reasons) being relatively short distance and contained within one regional economy. Although inter-regional migration related to the relative economic prosperity of different areas obviously did occur, for most people and in most places it was less important than shorter distance labour migration within a region.

The influence of occupation on migration

Analysis has so far concentrated on the significance of employment change in stimulating movement, and on the ways in which the characteristics of such moves varied over time and space. We can now focus on the related concept of the way in which particular occupations influenced residential mobility. By concentrating on those in paid employment, this is inevitably a male-dominated picture of migration – female participation in the labour

Figure 5.5 All work-related migration out of South-West England, (a) 1750–1879; (b) 1880–1994; and out of Northern Scotland, (c) 1750–1879; (d) 1880–1994

Table 5.4 Labour migration by gender, age, type of move and region (%)

Gender/Age	London	South East	South Midland	Eastern	South West	West Midland	North Midland	North West	Yorkshire	Northern	North Wales	*South Wales	**Northern Scotland	***Southern Scotland
Males														
Within region	27.4	43.0	33.1	46.1	50.3	33.8	51.3	55.5	50.2	61.5	54.7	42.0	62.5	41.1
Into region	40.5	29.0	32.0	26.0	20.3	44.2	22.7	25.7	25.1	18.4	18.8	31.6	15.9	27.6
Out of region	32.1	28.0	34.9	27.9	29.4	22.0	26.0	18.8	24.7	20.1	26.5	26.4	21.6	31.3
Total number	1,536	1,861	1,284	1,047	1,184	1,941	1,229	2,221	1,563	1,366	268	817	1,104	971
Females														
Within region	25.6	44.7	32.1	41.3	48.8	33.9	50.9	54.1	51.1	61.6	44.1	46.3	58.2	35.0
Into region	38.3	29.5	34.6	28.9	21.9	44.4	24.1	25.3	24.4	16.9	23.0	28.6	17.1	33.2
Out of region	36.1	25.8	33.3	29.8	29.3	21.7	25.0	20.6	24.5	21.5	32.9	25.1	24.7	31.8
Total number	616	954	630	433	516	914	511	995	714	593	161	490	543	444
Age <20 years														
Within region	26.2	46.7	35.3	48.9	50.9	41.3	56.8	59.1	55.9	67.6	53.5	52.8	66.8	44.6
Into region	45.1	25.0	34.3	21.6	19.1	38.4	17.5	25.3	21.6	13.9	13.9	27.9	13.3	27.9
Out of region	28.7	28.3	30.4	29.5	30.0	20.3	25.7	15.6	22.5	18.5	32.6	19.3	19.9	27.5
Total number	572	651	464	403	408	683	451	783	555	545	101	362	452	419
Age 20–39 years														
Within region	25.3	42.6	30.4	43.5	46.9	29.6	48.7	50.1	46.4	58.7	45.7	35.0	63.3	35.5
Into region	39.5	29.6	33.4	27.8	21.8	47.0	25.6	28.8	27.2	19.5	22.6	34.1	21.2	30.9
Out of region	35.2	27.8	36.2	28.7	31.3	23.4	25.7	21.1	26.4	21.8	31.7	30.9	15.5	33.6
Total number	1,171	1,446	1,012	719	855	1,502	906	1,626	1,192	998	208	677	712	719
Age 40–59 years														
Within region	32.7	41.8	33.2	45.2	53.1	34.2	48.2	62.2	55.2	60.8	59.8	52.4	64.6	38.6
Into region	34.9	33.3	30.8	28.4	20.2	45.0	24.4	18.4	23.4	19.3	17.6	26.0	15.4	28.2
Out of region	32.4	24.9	36.0	26.4	26.7	20.8	27.4	19.4	21.4	19.9	22.6	21.6	20.0	33.2
Total number	358	610	383	292	377	582	336	656	462	352	102	227	319	241
Age 60+ years														
Within region	31.1	48.1	51.0	32.2	67.9	49.3	64.9	48.2	48.3	58.8	38.9	59.0	79.3	54.3
Into region	22.2	25.9	26.5	40.4	19.6	36.6	21.6	25.9	20.7	15.7	44.4	15.4	11.3	22.9
Out of region	46.7	26.0	22.5	27.4	12.5	14.1	13.5	25.9	31.0	25.5	16.7	25.6	9.4	22.8
Total number	45	104	49	62	56	71	37	108	58	51	18	39	53	35

* Includes Monmouth and Glamorgan.

** Includes Highland, North East Scotland and Scottish Midland census regions.

*** Includes South West Scotland, South East Scotland and Southern Scotland census regions.

Data source: 16,091 life histories provided by family historians.

force was quite stable at around 35 per cent from the 1870s to the 1950s (Mitchell and Deane, 1962) with women heavily concentrated in a small number of occupations – although many women would have had a direct input to household migration decisions. It is also likely that female labour force participation was under-represented in most sources (Roberts, 1988).

Employment could influence the migration decisions of individuals and households in a number of ways. Most obviously, and as demonstrated above, those in higher status occupations had more resources and knowledge and were likely to be able to travel further regardless of the reason for migration. Other factors of importance included the extent to which particular occupations were geographically concentrated or dispersed and patterns of advertisement and recruitment in different areas of employment. Clearly, if a post was only advertised locally then long distance migration to take up that appointment would have been limited and workers with skills required in only a few areas were less likely to be highly mobile than those with more transferable skills. Even the diaries analyzed give only limited information on how prospective migrants gained information about new employment, and this is an area ripe for further research. Employment which gave easier access to transport (such as in the case of railway workers) may have eased migration for some, and it can be suggested that some employment depended particularly heavily on the development of contacts and networks within the local community, providing a further brake on mobility. Thus casual dockworkers needed to be known by local foremen to ensure work, and shopkeepers might be tied to a particular neighbourhood by business contacts. These themes can be illustrated by examining five selected occupational categories, designed to represent a range of broadly working-class experiences and focusing on the aggregate analysis of mobility characteristics (Table 5.5, Figures 5.6–5.10).

Those employed in textile industries included a relatively high proportion of female workers in comparison with other occupations, but their age structure (at migration) was much like that of other groups. They were however distinctive in that they tended to move over very short distances with 74 per cent of moves in the period after 1880 being under 5 kms. This is a much higher rate of short distance mobility than any other group (including agricultural workers), and the predominance of short distance movement actually increased in the twentieth century. It can be suggested that this pattern was directly related to the nature of employment in the textile industries. From the mid-nineteenth century different branches of textile employment were increasingly concentrated in a limited number of areas with considerable differentiation of function and skill within these areas. Thus, in the Lancashire cotton industry, weaving was dominant in the northern part of the county and spinning to the south, although a number of large combined mills also developed in south Lancashire (Laxton, 1986; Rose, 1996). Similarly in the East Midlands the knitting manufactures of Leicestershire, Derbyshire and

Table 5.5 Characteristics of all migrants in selected occupational groups by year of migration (%)

Migrant characteristics	Textile workers	Miners	Shopkeepers	Clerical workers	Engineers
1750–1879					
Gender					
Female	20.1	0.4	7.9	3.5	0.0
Age					
<20 years	15.9	13.5	2.8	14.4	8.1
20–39 years	61.7	67.5	61.0	63.9	76.7
40–59 years	18.5	16.8	29.3	19.0	14.0
60+ years	3.9	2.2	6.9	2.7	1.2
Mean distance (km)	22.1	32.6	26.5	54.6	30.9
Distance band (km)					
<5	63.2	39.3	59.3	43.4	59.9
5–9.9	7.6	12.5	8.3	9.0	9.5
10–19.9	7.4	11.8	7.3	6.4	5.3
20–49.9	11.4	17.6	10.0	11.2	10.2
50–99.9	4.8	10.3	6.7	12.0	4.9
100–199.9	2.4	5.1	4.8	9.4	3.9
200+	3.2	3.4	3.6	8.6	6.3
Total number	868	834	1,170	1,124	323
1880–1994					
Gender					
Female	28.4	0.5	18.8	20.9	0.3
Age					
<20 years	15.7	8.0	1.0	11.4	2.1
20–39 years	59.3	60.2	46.1	62.4	66.9
40–59 years	20.3	27.0	42.3	23.0	26.4
60+ years	4.7	4.8	10.6	3.2	4.6
Mean distance (km)	16.8	24.5	29.2	45.4	32.3
Distance band (km)					
<5	74.0	59.8	59.6	50.2	72.3
5–9.9	6.0	9.3	6.5	8.8	4.9
10–19.9	6.9	10.8	7.7	9.1	3.6
20–49.9	5.5	8.6	9.3	9.4	3.3
50–99.9	2.2	4.4	7.5	8.5	5.2
100–199.9	2.7	4.4	5.9	6.9	4.8
200+	2.7	2.7	3.5	7.1	5.9
Total number	598	729	1,587	4,171	330

Data source: 16,091 life histories provided by family historians.

Nottinghamshire were based on different raw materials and used different skills (Smith, 1963). Thus textile workers with specific skills were tied to those areas in which the relevant processes were carried out. If they changed employment it was likely to be to a nearby mill, and if they moved for other

Figure 5.6 All moves undertaken by textile workers, (a) 1750–1879; (b) 1880–1994

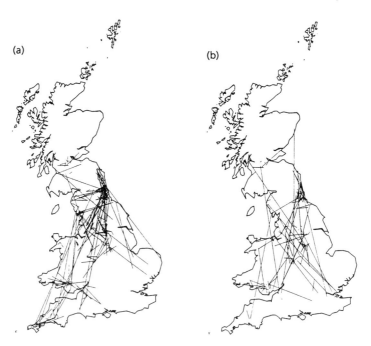

Figure 5.7 All moves undertaken by miners, (a) 1750–1879; (b) 1880–1994

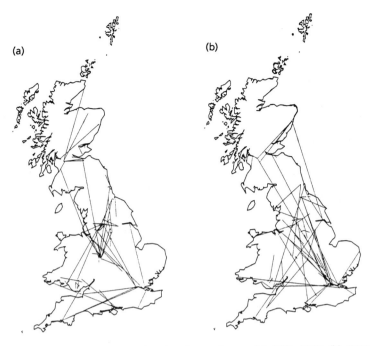

Figure 5.8 All moves undertaken by engineers, (a) 1750–1879; (b) 1880–1994

Figure 5.9 All moves undertaken by clerks, (a) 1750–1879; (b) 1880–1994

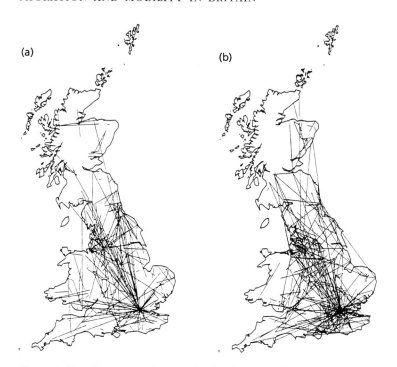

Figure 5.10   All moves undertaken by shopkeepers, (a) 1750–1879; (b) 1880–1994

reasons there would be strong constraints not to move outside the local area.

In contrast to textile workers, miners were predominantly male and, although most moves were short distance, they undertook more longer distance moves than textile workers, especially before 1880. Thus in the period 1750–1879 36.4 per cent of moves by miners were over 20 kms and 39.3 per cent were under 5 kms. Like textile workers miners were obviously tied to particular geographical areas, but as some mines contracted and others expanded they would be required to move between collieries. Thus, although miners who remained in the employment of one colliery for a long period of time would move infrequently and/or over short distances, others who were forced to seek work elsewhere because a pit was at the end of its working life, would be required to move to a new pit village, but probably on the same coalfield. There were strong community and family bonds among coal miners, who traditionally worked as family groups, and these factors would have discouraged long distance migration away from the local community (Church, 1994; Langton, 1998).

Engineers employed in heavy industry were also predominantly male but had an age structure which was rather older than most other groups. More migration later in life probably reflected both the length of time taken to serve an engineering apprenticeship, and the fact that skilled engineers

could expect to improve their income and status over the life-course. However, they too moved mainly over very short distances and in their mobility
characteristics were more similar to textile workers than to miners. They
probably experienced similar mobility constraints to textile workers. The
engineering industry was quite highly differentiated and, although some
skills were transferable, most workers would specialize in a particular branch
of the industry. Their movement would thus be largely constrained to the
areas in which they had trained and, if moving between establishments,
recommendations from employers known in the area would be of great
assistance (Southall, 1991a; Pollard, 1994).

In terms of income there was often little difference between skilled
manual workers such as engineers and the increasing number of clerical
workers who were employed in Victorian offices. However, it is often suggested that the lower middle class of clerical workers and similar occupations had higher aspirations which more secure employment often allowed
them to fulfil (Crossick, 1977). We can investigate the extent to which such
trends were reflected in mobility patterns. Although mainly male in the nineteenth century, clerical workers included an increasing number of female
employees after 1880, but their age structure and mobility was similar to that
of most other groups. Most moves were over short distances (50.2 per cent
under 5 kms in the period after 1880), but they did also have a relatively
high proportion of longer moves (30 per cent over 50 kms in the period
1750–1879). Thus although very short distance movement was the most
common experience for clerical workers, they did undertake more longer
distance migration than groups such as skilled engineers. This bi-modal
distribution may relate to the fact that most clerical employment was found
in larger urban centres and thus office workers either moved within one
urban area or moved some distance to a different town.

Although often living on very limited and precarious incomes, small
shopkeepers were a rung further up the social ladder as they had a degree
of independence (through self-employment or managing a shop); though
this was often obtained at the cost of very long working hours and poor
living conditions (Winstanley, 1983). Like clerical workers the proportion of
females increased markedly after 1880, and shopkeepers undertook a higher
proportion of moves later in life. This could relate either to increased opportunities for advancement or to the precarious nature of many businesses
which could force a move (see the examples of Amos Kniveton and Henry
Jaques discussed elsewhere). Unlike most other groups, loss of employment also usually meant loss of a home as most shopkeepers lived on the
premises. Like other groups most moves were over very short distances
with 59 per cent of moves under 5 kms in both time periods. It can be
suggested that shopkeepers were as tied to a particular neighbourhood as
textile workers or engineers. They would depend upon a network of business contacts, they may have had local creditors to chase, and they would

want to retain the loyalty of local customers. Although some shopkeepers did move over longer distances, as Amos Kniveton moved from Astley Green to Morecambe, Blackburn, Bolton and Leigh, there were also strong incentives to stay within the same area.

Migration and social mobility

It has been demonstrated that much migration was stimulated by the desire or need to change employment, and that many jobs themselves imposed constraints or offered particular opportunities for migration. However, we have not yet examined the extent to which migration was associated with changes in social status. Data collected from family historians on changes in both residence and workplace allow some assessment of the extent to which migration was associated with social mobility. There are many difficulties associated with the analysis of social mobility only some of which can be considered here. Most fundamental is the difficulty of measuring social position within society. In many ways this is a subjective concept related not only to income and power, but to the amount of status a person is accorded in the community. It is perfectly possible, for instance, to have a high social position in terms of the esteem in which a person is held but a relatively low income. In practice, most studies ascribe social position from occupational descriptions based upon standard occupational classifications. As explained in Chapter 2 such an approach has been used in this study, and assumptions of upward or downward mobility are based solely on a simple socio-economic classification of occupational descriptions. This process is made even more difficult because the nature and status of some occupations changed fundamentally over the 200 years from 1750 (though less so within the working life of any one individual), and because many women employed in part-time or casual occupations fit uneasily into socio-economic classifications based mainly on male occupational categories. These caveats must be taken into account in the ensuing discussion (Kaelble, 1985; Goldthorpe, 1987; Breiger, 1990).

Analysis presented in this section is confined to the relationship between migration and occupational change. Other studies have examined social mobility over the life-course and between generations (Prandy, 1997), but there have been few previous attempts to link migration and social mobility in the past. It can be hypothesized that opportunities for upward social mobility following migration may have been greater for some groups than for others, depending on age, gender, occupation and stage in the life-course. It can also be suggested that opportunities for upward mobility may have been greater in some regions than in others, and that upward mobility was most likely to have been linked to longer distance migration. The relatively unusual event of a long distance move was most likely to have

178

Table 5.6 Changes in socio-economic status following migration by year of migration (%)

Status change	1750–1839	1840–79	1880–1919	1920–94
Same socio-economic position	43.3	42.8	38.8	38.7
Higher socio-economic position	6.3	7.5	5.0	3.5
Lower socio-economic position	2.0	3.2	2.3	2.1
Into labour force	9.5	7.3	7.2	4.9
Out of labour force	2.6	4.5	6.8	10.2
Remained out of labour force	36.3	34.7	39.9	40.6
Total number	7,444	18,858	20,591	17,986

Data source: 16,091 life histories provided by family historians.

been undertaken if it led to significant material improvement. These propositions will be tested against data on residence and work change provided by family historians. Not all moves resulted in upward mobility, and the links between downward social mobility and migration will also be explored. This links to themes explored in Chapter 6 (on the family) and Chapter 8 (on migration due to crises). For instance, it can be suggested that in any household a migration experience may have led to improved chances for some but reduced opportunities for others: in some cases migration may have deliberately produced upward mobility for one generation but downward mobility for another.

As with much of the analysis presented in this volume, one of the most surprising characteristics of data on the links between migration and social mobility is the high degree of stability over both time and space (Tables 5.6, 5.7). In all time periods and most regions the proportion of people who experienced upward or downward mobility following migration was very similar. Overall, around 80 per cent of those who moved experienced no change in their social position (they either remained in the same class of employment or remained out of the labour market altogether), about five per cent experienced upward mobility within their employment, and half that proportion downward mobility. There was rather more upward mobility following migration in the period before 1920 (and especially 1840–79), suggesting that opportunities for social improvement were greater and more readily grasped during Victorian periods of rapid economic and social change. Variations between regions are small and hard to interpret, but there does seem to be some link between both upward and downward mobility and marginality. Thus there were relatively high rates of upward mobility in parts of Scotland in both time periods. It can be suggested that this feature was related to the relatively high rates of longer distance migration into and out of such regions (see Chapter 3), and this is confirmed by the association between social mobility and distance moved (Table 5.8). There were consistent and predictable links between social mobility following migration

Table 5.7 Changes in socio-economic status following migration by region and year of migration (%)

Region	Same socio-economic position	Higher socio-economic position	Lower socio-economic position	Into workforce	Out of workforce	Remained out of workforce	Total number
1750–1879							
London	45.5	6.6	2.8	7.6	3.0	34.5	3,491
South East	42.0	7.0	3.3	8.6	5.0	34.1	2,595
South Midland	41.8	7.7	2.9	9.8	3.1	34.7	1,788
Eastern	43.4	8.3	3.0	8.6	3.5	33.2	1,646
South West	42.3	7.5	3.2	10.0	3.4	33.6	2,070
West Midland	40.1	6.5	2.0	10.3	4.0	37.1	2,700
North Midland	39.5	8.5	2.8	10.0	4.8	34.4	1,877
North West	45.2	7.9	4.1	6.6	4.1	32.1	3,785
Yorkshire	43.9	7.2	3.2	6.8	2.9	36.0	2,583
Northern	43.7	6.9	2.8	6.5	2.7	37.4	2,215
Monmouth	31.7	7.3	3.3	7.7	3.3	46.7	246
North Wales	37.6	6.3	2.4	8.6	3.2	41.9	370
South Wales	35.7	7.8	3.6	14.1	3.4	35.4	412
Glamorgan	33.4	10.6	4.5	7.9	4.7	38.9	470
Highland	34.3	9.3	2.5	8.5	4.2	41.2	236
NE Scotland	38.8	8.7	3.0	7.2	4.9	37.4	770
Scottish Midland	39.0	6.6	2.4	6.6	4.4	41.0	860
SW Scotland	43.1	5.4	2.0	7.7	4.8	37.0	771
SE Scotland	32.9	6.2	2.3	8.4	5.7	44.5	488
Southern Scotland	45.0	9.8	2.1	8.4	3.9	30.8	380

1880–1994							
London	41.6	4.2	2.0	7.1	7.5	37.6	4,516
South East	34.6	3.9	1.9	6.0	9.5	44.1	5,581
South Midland	36.9	3.6	2.2	6.2	8.6	42.5	4,224
Eastern	37.8	4.9	2.2	5.4	8.5	41.2	2,834
South West	33.8	4.3	2.8	6.8	9.0	43.3	2,237
West Midland	37.2	4.9	2.1	7.0	8.1	40.7	4,285
North Midland	37.4	5.5	2.7	6.7	8.3	39.4	2,818
North West	39.7	4.6	2.1	5.6	8.3	39.7	6,073
Yorkshire	38.5	4.5	2.1	5.6	8.4	40.9	4,053
Northern	38.0	4.7	2.2	5.2	6.9	43.0	2,577
Monmouth	33.3	4.4	1.7	5.0	7.2	48.4	517
North Wales	29.1	4.7	1.1	8.7	8.9	47.5	621
South Wales	35.4	3.8	1.9	9.1	8.7	41.1	263
Glamorgan	33.2	5.4	2.7	6.4	7.4	44.9	978
Highland	32.5	5.5	3.1	9.2	13.5	36.2	163
NE Scotland	37.3	6.2	3.2	7.2	8.9	37.2	843
Scottish Midland	36.1	5.7	2.4	6.4	8.2	41.2	889
SW Scotland	41.6	4.2	1.6	5.6	7.5	39.5	1,054
SE Scotland	35.8	2.4	2.0	8.9	7.3	43.6	749
Southern Scotland	33.1	6.6	3.5	7.8	10.5	38.5	257

Data source: 16,091 life histories provided by family historians.

Table 5.8 Changes in socio-economic status following migration by migrant characteristics, distance and year of migration (%)

Migrant characteristics	1750–1879						1880–1994					
	Same socio-economic position	Higher socio-economic position	Lower socio-economic position	Into workforce	Out of workforce	Remained out of workforce	Same socio-economic position	Higher socio-economic position	Lower socio-economic position	Into workforce	Out of workforce	Remained out of workforce
Percentage by column												
Gender												
Female	4.2	1.4	1.2	25.2	45.8	62.3	13.7	5.7	6.0	44.9	59.0	70.6
Average distance (km)	29.1	35.3	35.4	50.7	37.4	35.0	35.7	50.8	54.9	71.6	53.4	45.8
Distance band (km)												
<5	55.7	45.1	45.7	35.4	48.0	49.4	58.6	42.4	43.4	33.8	43.3	51.1
5–9.9	10.0	11.1	9.3	12.5	12.6	10.7	8.3	8.5	9.1	7.1	8.8	7.9
10–19.9	8.2	9.6	13.0	10.9	9.6	9.7	7.3	9.1	8.3	7.8	7.7	7.5
20–49.9	11.0	14.1	12.5	14.8	11.7	12.2	8.1	12.6	12.3	13.7	10.2	10.0
50–99.9	6.5	9.6	8.1	9.9	6.3	7.8	6.4	10.6	8.1	12.6	11.4	8.1
100+	8.6	10.5	11.4	16.5	11.8	10.2	11.3	16.8	18.8	25.0	18.6	15.4
Age												
<20	7.1	4.5	3.5	59.1	8.1	51.3	5.3	4.3	4.4	45.4	2.8	33.1
20–39	66.9	69.5	57.7	35.1	47.7	34.8	59.0	59.7	50.6	42.4	44.5	27.6
40–59	21.7	21.6	30.9	4.7	16.0	9.1	30.0	30.5	35.3	10.1	15.2	15.8
60+	4.3	4.4	7.9	1.1	28.2	4.8	5.7	5.5	9.7	2.1	37.5	23.5

Companions											
alone	10.0	13.6	11.4	46.6	24.8	6.0	12.6	13.7	16.7	42.3	18.8	13.3
couple	21.9	20.7	15.7	10.0	37.2	13.2	21.7	19.5	15.8	6.8	45.3	14.9
nuclear family	59.3	57.2	63.2	33.2	26.4	72.0	56.2	58.8	57.8	38.6	26.7	61.5
extended family	4.1	4.7	6.6	3.5	8.3	5.2	4.6	4.6	5.0	4.0	6.8	7.2
other	4.7	3.8	3.1	6.7	3.3	3.6	4.9	3.4	4.7	8.3	2.4	3.1
Position in family												
child	5.7	3.7	2.2	23.2	1.9	46.8	6.7	3.7	3.8	23.3	2.2	30.4
male head	87.7	92.1	93.6	51.2	49.4	5.9	78.7	88.4	86.2	34.2	47.0	11.5
wife	1.7	0.9	0.5	4.8	33.8	38.8	5.0	3.0	3.1	11.7	35.6	40.7
female head	1.1	0.2	0.5	13.9	9.7	5.3	4.5	1.5	1.9	21.6	10.1	12.3
other	3.8	3.1	3.2	6.9	5.2	3.2	5.1	3.4	5.0	9.2	5.1	5.1
Total number	11,291	1,870	771	2,090	1,036	9,238	14,931	1,670	855	2,364	3,235	15,522
Percentage by row												
Occupational group												
Professionals	86.8	0.0	5.9	N/A	7.3	N/A	79.5	0.0	3.8	N/A	16.7	N/A
Agricultural workers	67.9	20.1	6.6	N/A	5.4	N/A	56.3	19.5	9.2	N/A	15.0	N/A
Textile workers	73.7	11.7	6.6	N/A	8.0	N/A	72.6	8.7	3.3	N/A	15.4	N/A
Miners	82.1	10.6	5.2	N/A	2.1	N/A	79.9	8.9	4.6	N/A	6.6	N/A
Domestic servants	57.8	38.8	3.4	N/A	0.0	N/A	75.7	19.3	3.5	N/A	1.5	N/A
Shopkeepers	64.8	7.7	26.8	N/A	0.7	N/A	63.2	12.7	23.2	N/A	0.9	N/A
Clerical workers	75.3	14.2	4.3	N/A	6.2	N/A	74.3	9.0	3.4	N/A	13.3	N/A
Engineers	85.1	9.6	2.9	N/A	2.4	N/A	80.8	8.0	2.9	N/A	8.3	N/A

Data source: 16,091 life histories provided by family historians.

Table 5.9 Changes in socio-economic status by sequence of moves over the life-course by birth cohort (%)

Socio-economic status	Move 1	Move 2	Move 3	Move 4	Move 5	Move 6	Move 7	Move 8	Move 9	Move 10+
1750–1849										
Same socio-economic position	33.7	47.7	50.6	52.8	51.3	51.9	49.6	51.2	50.3	51.9
Higher socio-economic position	5.9	9.1	9.0	7.6	7.5	5.9	6.8	6.1	4.5	2.4
Lower socio-economic position	1.8	3.3	4.4	4.1	4.2	4.6	3.0	3.4	3.4	2.7
Into workforce	14.2	6.6	4.4	2.9	2.6	1.4	1.9	2.3	2.0	1.3
Out of workforce	2.4	5.5	5.5	7.0	7.6	8.1	7.6	7.5	9.2	8.7
Remained out of workforce	42.0	27.8	26.1	25.6	26.8	28.1	31.1	29.5	30.6	33.0
Total number	6,720	5,814	4,651	3,382	2,297	1,463	962	604	382	670
1850–1930										
Same socio-economic position	15.6	31.8	39.2	43.4	44.3	44.7	42.9	42.8	42.9	42.6
Higher socio-economic position	2.3	4.7	4.9	4.8	4.6	4.0	4.8	3.7	4.3	2.9
Lower socio-economic position	0.9	1.5	2.2	2.2	2.1	2.5	2.6	2.4	2.1	2.0
Into workforce	10.8	11.2	8.2	6.1	5.2	4.0	3.2	3.8	2.3	3.4
Out of workforce	2.8	7.2	8.8	8.6	8.1	7.6	9.1	8.2	9.7	8.8
Remained out of workforce	67.6	43.6	36.7	34.9	35.7	37.2	37.4	39.1	38.7	40.3
Total number	6,456	5,890	5,305	4,629	3,880	3,080	2,392	1,796	1,309	3,189

The table should be read as follows: for those born 1750–1849, 33.7% remained in the same socio-economic position after their first move, 47.7% after their second move, etc.

Data source: 16,091 life histories provided by family historians.

and the life-course, with upward social mobility most common among young adults (age 20–39) in the career building stage of their life-cycle. Conversely, downward mobility occurred most frequently for older adults (age 40–59) who were finding it increasingly difficult to compete in the labour market (Table 5.9). These trends are quite stable over time, though they are most marked prior to 1880. Both upward and downward social mobility was most likely to be experienced by men (women tended to remain in the same position and entered or left the labour market more frequently than men), but otherwise there were no significant variations with position in the household.

For individual families the process of upward or downward social mobility, and its connections with migration decisions could often be complex. These nuances can only be brought out through the use of individual case studies. The extent to which migration may have had varied effects on different family members is demonstrated by one event in the life history of Benjamin Shaw (already mentioned in previous chapters). In 1791 the Shaw family were persuaded to move from their home in Dent (Yorkshire) to the mill village of Dolphinholme in Lancashire. The mill manager was recruiting labour from families in the Dent area in the knowledge that times were hard for handloom weavers and that factory employment might seem attractive. It is clear from the life history that the Shaw family's move was calculated to maximize economic benefits for all the family rather than suit individual members. Joseph Shaw had been a self-employed weaver and clock repairer and, although he gained employment in the machine shop of the mill, this would almost certainly have been seen as a downward step in status. However, the main attraction of moving to Dolphinholme was the opportunity for apprenticeships and what promised to be secure employment for Joseph's children. In the event things did not turn out as anticipated, but it is clear that the decision to migrate was taken in the expectation that job opportunities would be improved for the children but probably worsened for their father.

Changing residence and workplace

The changing nature of the journey to work is a neglected area of study, due mainly to the paucity of data which deals directly with the relationship between home and workplace. Some aggregate analysis is possible from twentieth century censuses (Lawton, 1963, 1968b; Warnes, 1972), and a variety of sources have been used to try to infer journey to work patterns in the nineteenth century (Warnes, 1970; Dennis, 1984a, pp. 138–40; Green, 1988; Barke, 1991). Data collected from family historians, giving information on change of residence and change of occupation and workplace, provide a unique insight into changes in the journey to work over a period of 200 years. It is widely assumed that journey to work distances have increased in the twentieth century, but there is little evidence of by how much or of regional and socio-economic variations in such changes. Almost nothing is known of temporal and spatial variations in the journey to work prior to 1900. Data collected from family historians allows some of these questions to be answered.

Moreover, journey to work considerations are fundamentally related to the process of labour migration. For most people the distance over which it was possible and convenient to travel to work acted as a constraint on both residential and job choice and, crucially, on the relationship between the two. The need for a short journey to work, due to transport constraints, cost or time factors made it more likely that a change of employment would lead to a change of residence, due to the need to live close to the workplace. Such considerations were obviously more important in the past, when transport constraints were greater, but journey to work considerations are also more significant for some jobs than for others. Work which entails long or irregular hours, where someone is on call day and night, or where employment is unpredictable, means that a short journey to work is likely to be desirable. Thus in the nineteenth century when dock workers were employed by the half day depending on whether work was available, it was essential for labourers to live near the docks so that they knew when to present themselves for work. In the second half of the twentieth century a General Practitioner who can be on-call at night will need to live not only reasonably close to the surgery but also in a reasonably central location with respect to the population that is served. For many people in the past and the present home and workplace were the same: living above the shop or working as an apprentice in a master's house. Thus although the journey to work was not a problem, unemployment could also lead to homelessness and life-course changes may require finding a new home away from the workplace. The journey to work is thus an important consideration affecting the residential choice of labour migrants.

The data used in this analysis are not perfect and they do need to be interpreted with some caution. The original data collection forms were not specifically designed to collect information on journeys to work – the possibility of doing this and its significance was only fully appreciated during later analysis – and thus place of employment was not always specified precisely. Due to this, although it is possible to identify and measure most journeys to work over about 2 kms, any separation of home and workplace less than this has to be aggregated into a single category and an arbitrary distance assigned. Although this affects micro-scale changes, it should not influence the broader trends that are identified in this section. Apart from this, the quality of information on place of work seems as good as that for place of residence and offers some unique opportunities for analysis.

At the aggregate level journey to work distances changed little in the eighteenth and nineteenth centuries. Between 1750 and 1879 over 92 per cent of all journeys to work were less than 5 kms, and even in the period 1880 to 1919, almost 88 per cent were less than 5 kms. Although after 1920 almost three quarters of journeys to work were under 5 kms, there was a marked increase in the mean journey to work with almost a quarter of all journeys to work between five and 50 kms (Table 5.10). The distance between home and workplace varied relatively little with personal characteristics, but there was a tendency for women to have slightly shorter journeys to work than men, possibly reflecting the fact that many women worked part-time and had to juggle work and family commitments. They were thus unable to spend a long period of time travelling to work each day. The journey to work also increased slightly with age, presumably reflecting more mature and affluent families making residential moves to improve their housing and accepting a longer journey to work, and those in professional and skilled non-manual occupations had longer than average journeys to work. Although a relatively short journey to work remained the norm for all commuters, it is notable that skilled white collar workers extended their journey to work more than any other group (including professionals), presumably reflecting a desire to suburbanize to good quality but relatively low cost accommodation. In contrast, agricultural workers and domestic servants lived especially close to their workplaces (for many, of course, their home was their workplace). These trends are consistent throughout the time period studied, but with an overall increase in commuting distances in the twentieth century.

Regional variations in commuting distances were quite small, especially in the eighteenth and nineteenth centuries (Table 5.11). Before 1880 those living in London were slightly more likely to have a longer journey to work (reflecting the spatial extent of the metropolis), though this had little effect on the mean distances travelled. After 1880 there was a clearer north–south split, with people living in London, South East, South Midland and Eastern England all experiencing longer than average commuting distances. This clearly reflects the expanding labour market area of London, with workers commuting from throughout the outer South East. Elsewhere, there were few differences in the distances that people travelled to work. Variations in commuting distances by settlement size of residence also reflect similar trends (Table 5.12). In the period before 1880 variations were small with most people constrained to a short journey to work wherever they lived. There was a slight tendency for commuting distances to be shorter in medium sized towns (40,000–100,000), but cities over 100,000 population had slightly longer commuting distances. In the period after 1880 there is a much clearer pattern. Commuting distances were markedly longer for workers living in small settlements, reflecting the centralization of employment opportunities in larger centres and an increasing trend towards counter-urbanization. Residents of cities, on average, had a shorter journey to work than those who lived in villages and small towns.

Table 5.10 Journey to work distances by migrant characteristics and year of migration (%)

Migrant characteristics	1750–1879						1880–1994					
	<5 km	5–9.9 km	10–19.9 km	20–49.9 km	50–99.9 km	100 km+	<5 km	5–9.9 km	10–19.9 km	20–49.9 km	50–99.9 km	100 km+
Gender												
male	92.5	3.1	1.4	1.8	0.8	0.4	80.7	8.1	5.4	3.6	1.5	0.7
female	96.3	1.4	0.7	1.0	0.5	0.1	84.7	6.2	4.1	2.7	1.7	0.6
Age												
<20	94.1	2.1	1.1	1.3	1.1	0.3	87.2	5.3	3.3	2.3	1.0	0.7
20–39	92.5	3.3	1.5	1.7	0.7	0.3	81.0	8.0	5.6	3.3	1.5	0.6
40–59	92.6	2.7	1.4	1.9	0.8	0.6	79.7	8.3	5.4	4.2	1.6	0.8
60+	93.3	2.4	1.4	1.8	0.7	0.4	83.1	6.0	4.2	3.5	2.4	0.8
Position in family												
child	92.5	3.0	1.6	1.4	1.2	0.3	81.3	8.9	4.8	2.5	1.7	0.8
male head	92.6	3.0	1.4	1.8	0.8	0.4	80.6	8.1	5.4	3.7	1.5	0.7
wife	94.6	1.1	0.7	2.5	0.7	0.4	83.9	5.4	4.2	3.6	2.4	0.5
female head	98.2	1.4	0.0	0.0	0.0	0.4	88.7	4.2	3.7	1.4	1.2	0.8
other	93.8	1.9	1.9	2.0	0.2	0.2	82.9	7.4	5.4	2.5	1.2	0.6
Occupational group												
professional	91.7	4.2	2.0	1.2	0.2	0.7	75.4	9.1	6.5	5.5	2.5	1.0
farmer	97.2	1.0	0.5	0.7	0.5	0.1	98.2	0.6	0.3	0.9	0.0	0.0
intermediate	92.9	2.6	1.3	1.6	1.0	0.6	83.2	6.7	5.2	3.0	1.3	0.6
skilled non-manual	87.1	7.3	2.2	1.6	1.2	0.6	68.5	11.6	10.7	6.1	2.3	0.8
skilled manual	93.0	3.2	1.2	1.8	0.7	0.1	84.3	8.1	4.0	2.3	1.0	0.3
skilled/semi-skilled agricultural	91.5	2.0	1.6	4.0	0.9	0.0	89.9	4.5	1.6	2.3	1.2	0.5
semi-skilled manual	90.1	3.5	2.6	2.3	1.0	0.5	87.5	7.0	2.1	2.1	1.1	0.2
unskilled manual	91.4	2.6	2.7	2.2	0.9	0.2	88.0	5.3	2.5	1.9	1.9	0.4
agricultural labourer	96.4	0.9	0.9	0.6	1.0	0.2	93.0	3.4	2.3	1.3	0.0	0.0
domestic service	97.3	1.6	0.2	0.7	0.0	0.2	95.3	2.3	0.7	1.0	0.3	0.4
Total number	13,048	415	196	240	112	55	14,827	1,412	952	632	278	127

Data source: 16,091 life histories provided by family historians.

Table 5.11 Journey to work distances (km) by region of residence and year of migration

Region	1750–1879	1880–1994
London	3.3	5.4
South East	2.3	7.7
South Midland	1.8	8.3
Eastern	2.3	7.1
South West	4.6	6.4
West Midland	2.8	3.8
North Midland	2.8	4.8
North West	2.8	5.3
Yorkshire	3.6	4.1
Northern	3.5	4.1
Monmouth	3.2	3.6
North Wales	1.1	4.4
South Wales	4.5	6.1
Glamorgan	3.9	4.7
Highland	0.3	0.2
North East Scotland	1.7	5.1
Scottish Midland	3.4	8.9
South West Scotland	2.2	6.8
South East Scotland	1.7	3.9
Southern Scotland	1.9	0.5

Data source: 16,091 life histories provided by family historians.

Table 5.12 Journey to work distances by settlement size of residence and year of migration (%)

Settlement size	<5km	5–9.9km	10–19.9km	20–49.9km	50–99.9km	100km+
1750–1879						
<5,000	92.1	2.6	1.9	2.3	0.8	0.3
5,000–9,999	93.0	1.2	2.3	2.3	0.7	0.5
10,000–19,999	93.9	1.6	0.9	2.1	0.9	0.6
20,000–39,999	94.6	1.0	0.8	1.8	1.4	0.4
40,000–59,999	96.3	0.2	0.5	0.8	1.1	1.1
60,000–79,999	98.1	0.2	0.8	0.3	0.3	0.3
80,000–99,999	98.0	0.5	1.0	0.5	0.0	0.0
100,000+	91.5	6.4	0.7	0.4	0.5	0.5
Total number	13,048	415	196	240	112	55
1880–1994						
<5,000	73.1	9.4	7.7	6.8	2.2	0.8
5,000–9,999	80.0	6.3	6.1	4.1	2.9	0.6
10,000–19,999	81.3	5.7	7.7	3.3	1.4	0.6
20,000–39,999	84.1	3.5	7.5	2.8	1.0	1.1
40,000–59,999	90.8	2.5	2.6	0.9	2.7	0.5
60,000–79,999	87.8	6.5	1.2	2.0	1.3	1.2
80,000–99,999	90.0	4.8	2.8	0.9	0.6	0.9
100,000+	86.8	9.5	1.9	0.6	0.7	0.5
Total number	14,827	1,412	952	632	278	127

Data source: 16,091 life histories provided by family historians.

Table 5.13 Journey to work characteristics (%) for six large labour markets (by location of workplace)

Migration characteristics	London	Birmingham	Glasgow	Manchester	Liverpool	Leeds
1750–1879						
Mean distance (km)	4.1	2.5	4.1	3.4	1.1	0.7
Distance band (km)						
5–9.9	83.8	93.0	92.5	91.5	93.5	94.4
10–19.9	11.4	4.6	1.7	5.2	5.2	2.2
20–49.9	2.4	0.8	2.3	1.3	0.3	2.2
50–99.9	1.3	0.0	1.2	0.8	0.5	1.2
100+	1.1	1.6	2.3	1.2	0.5	0.0
Total number	1,841	129	173	236	369	90
1880–1994						
Mean distance (km)	11.6	3.7	5.2	9.9	5.3	4.5
Distance band (km)						
5–9.9	50.1	75.5	86.7	58.4	71.1	83.7
10–19.9	21.2	15.6	4.1	21.9	22.1	6.0
20–49.9	16.9	4.3	1.7	9.4	2.8	4.9
50–99.9	7.6	4.3	4.4	4.0	2.3	3.8
100+	4.2	0.3	3.1	6.3	1.7	1.6
Total number	2,764	326	294	397	470	184

Data source: 16,091 life histories provided by family historians.

These trends can be examined in more detail if we focus on selected labour market areas. When focusing on labour markets it is obviously more useful to examine commuting distances by place of work rather than place of residence. Table 5.13 summarizes these data for six large British cities. In the period before 1880 all cities drew most of their workforce from a very small area. London and Glasgow had the largest mean commuting distances (4.1 kms), and London had the smallest proportion of moves under 5 kms (83.8) per cent. Leeds and Liverpool had particularly low commuting distances with most people travelling less than one km to work. In the period after 1880 mean commuting distances increased most markedly in London (11.6 kms) with only half of all those working in London travelling less than five kms. Workers in Manchester also experienced longer than average commuting distances (9.9 kms), but in all other cities the mean journey to work remained low and over 70 per cent of workers commuted less than 5 kms. These trends can be explained in terms of the range of employment opportunities in the cities, the availability of housing close to places of work, the convenience of public transport into cities, and the nature of local labour markets. London not only offered more employment than any other city, but it had by far the best-developed suburban transport links.

Manchester too had good radial bus, tram and rail connections to surrounding areas and drew labour from many adjacent towns. In Liverpool the continued concentration of job opportunities in dockside areas, meant that many workers lived quite close to their workplace, and relative economic depression on Merseyside meant that the city was less likely to attract workers over long distances than either Manchester or London.

Conclusion

Analysis presented in this chapter has suggested that the relationship between migration and employment is, at different levels, both straightforward and complex. Whilst much migration was to some degree stimulated by the need or desire to change employment, only a relatively small proportion of such moves were clearly related to the economic and labour market characteristics of particular areas. Whereas London did consistently attract labour migrants over long distances, in most other areas labour migration patterns were remarkably similar over both time and space, though specialized labour markets could attract workers over long distances from other areas with similar economic structures. Thus although work could be an important factor in stimulating migration, for most people that stimulus was as likely to come from individual, family life-course and local circumstances. It is very difficult to sustain a convincing argument about consistent links between migration and the relative economic prosperity of particular regions. Employment was a factor which consistently affected migration, but this influence was exercised in a variety of complex ways.

For many people the ways in which particular occupations either restricted or provided opportunities for migration were rather more important. Thus in some areas of employment there were strong benefits to be gained from remaining in the same neighbourhood, building up local networks, and exploiting these in the process of career advancement. In other jobs the nature of career progression required mobility and it was necessary to move to a new location to gain relevant experience and career advancement. In this sense the career path that an individual chose also affected their subsequent migration path. Employment could also affect residential decisions in other ways. Opportunities for upward social mobility may have allowed migrants to improve their housing and to move to a better residential area, whilst other jobs required employees to live close to their workplace and thus constrained the journey to work and residential location.

In most households migration decisions related to employment opportunities were taken to benefit the career of a male household head, or possibly to provide opportunities for male children. Female household members often had to fit their careers around such moves and hence women were much more likely to enter or leave the labour market following migration. However, it is clear from the individual case studies in particular that migration for employment reasons was often closely intertwined with movement stimulated by other factors. Although job-related movement was the single largest category, migration for housing improvement, marriage, family reasons and as a response to crises were all important and may have disproportionately affected women and children. Subsequent chapters will examine the pattern and process of migration for this wider range of reasons in much more detail.

CHAPTER 6

Migration, family structures and the life-course

Introduction

Although employment opportunities were the single most important factor stimulating migration, there were also strong links between migration decisions, family obligations and the life-course. In addition, many moves could alter family responsibilities and separate kin, leading to changes in family structures and obligations. Over the past two decades there has been a large volume of research on the changing nature of the family in the past, but this literature had been only loosely related to the pattern and process of migration (Flandrin, 1979; Anderson, 1980, 1985, 1990; Davidoff, 1990; Drake, 1994). In addition, contemporary migration studies have focused on the ways in which migration decisions vary over the life-course, related not only to age but, especially, to family position and responsibility (Champion and Fielding, 1992). This chapter draws on such studies and explores the ways in which migration in Britain from the eighteenth to the twentieth centuries was related to changing family structures and the nature of the life-course.

A large number of connections between family structures, the life-course and migration can be hypothesized. The constraints of living in an overcrowded household, leading to lack of privacy and independence, may have been significant push factors for young single migrants leaving the parental home for the first time. Conversely, for others, close family ties and a sense of obligation to parents or other relatives, may have delayed movement from the family home or constrained migration away from the locality in which various family members were settled. It has already been demonstrated that movement as a family group was a common experience in the past and, for at least some members of such groups input into a migration decision may have been limited. The experience of moving as a dependent (child, wife, elderly relative) with little control where and when a move took place must have raised some conflicts between family obligations and personal preferences. In some instances the process of migration itself could be directly influenced by family ties. Chain migration, where relatives who had moved either within Britain or overseas, sent back information and encouraged other members of the family to follow, must have been a common occurrence, though the ways in which such family motives for migration interacted with employment and other factors are little explored.

The effects of migration could also vary significantly for different family members. This could have been related to specific intentions, with a move planned to enhance the employment opportunities of some family members in the knowledge that other (perhaps older) members of the family would find it harder to gain work after a move, or the effects could be unintended. Thus movement from a rural domestic economy to an urban factory economy could not only alter employment prospects, but also may have changed the nature of relationships and responsibilities within the family. A household in which family members worked together in the domestic economy must have had very different relations to one in which different family members dispersed each morning to work in various locations, only meeting again in the evening. Superimposed on all these factors were broader changes in the nature of families, and especially the role of women in society, which must in turn have influenced migration decisions. These questions interact with issues raised elsewhere in this volume: this chapter explores some of these themes stressing the links between family circumstances, migration and the life-course.

It is first necessary to briefly explore some of the changes in family structures which occurred in Britain from the mid-eighteenth century and which, in turn, affected both the nature of the life-course and individual migration decisions. Most studies have emphasized the relative stability of patterns over time and the increasing homogeneity of experiences between different regions and groups of the population, especially from the mid-nineteenth century (Anderson, 1990; Drake, 1994). Mean household size in Britain was typically small (under five) and remarkably constant during the eighteenth and nineteenth century. Thus mean household size in England for the period 1750–1821 was 4.81, for Britain in 1851 it was around 4.75, and in 1911, 4.4. In the twentieth century mean household size fell to 3.18 by 1951 (Wall, 1983, p. 497; Anderson, 1990, p. 56). However, mean figures for household size can be misleading. Of greater importance is the nature of the distribution around the mean and the way in which household size varied over the life-course. Thus children were much more likely to live in large households than older people, and the nature of the size distribution has meant that most people have always lived in households larger than the mean (Laslett, 1972; Anderson, 1980, 1990). The relatively large size of young households, coupled with severely restricted housing opportunities (see Chapter 7), thus created overcrowding which was a potential push factor in migration.

The apparent stability in household size masks significant variations in the composition of households over time. In all periods the number of household members who did not belong to the nuclear family were small, but four groups were significant: lineally-related kin (parents, grandparents etc.); other kin (siblings, nieces, nephews etc.); live-in servants; and lodgers. The proportion of married couples living with parents or parents-in-law appeared to increase from the mid-eighteenth to the mid-twentieth century, reflecting both changes in the age of marriage and the relative availability of housing. Over half of couples marrying in the early 1950s initially shared accommodation with kin. Such trends could also reflect migration experiences. Whilst most migration was highly localized, within particular communities children could retain contact with and care for ageing parents without taking them into their own home. As families became more scattered this was more difficult and, thus, more adult children and their parents shared accommodation. The largest group of co-resident relatives were 'parentless kin', siblings, nieces, nephews etc. living apart from their parents. These formed three quarters of co-resident kin in 1851 (Wall, 1983; Anderson, 1990 pp. 60–1). It can be suggested that many of these household patterns were a direct result of migration. In some cases a family may have moved away leaving one or more child with a relative, in others a teenage child may have left home for the first time to gain work and end up lodging with a relative. Such processes are hard to prove from aggregate census evidence, but can be explored through the life-time residential histories.

Both the number and composition of live-in servants changed from the eighteenth to the twentieth centuries. In the period 1750 to 1821 servants comprised 10.7 per cent of the population and were as likely to consist of male farm servants as female domestics. During the nineteenth century farm and other trade servants were largely replaced by day labourers, but the number of female domestics expanded peaking at over 1.5 million in 1891. In 1851, 27 per cent of single women aged 18–27 were employed as domestic servants, but domestic service declined during the twentieth century (Wall, 1983; Anderson, 1990, p. 62). As demonstrated in Chapter 5, domestic servants were likely to move over quite long distances, either initially to take up a position or, more commonly, with the family with whom they were employed. It is also clear that some servants were related to the head of the household in which they worked, whilst others married (or had illegitimate children by) male members of the household in which they were employed. Servants were thus an important component of many households, interacting with family members, and their migration experiences were often structured by their relationship with their employing family. Lodgers are particularly hard to define as a group though they undoubtedly formed an important component of many households. In 1851 lodgers were present in 12 per cent of all households, and the typical lodger was male, single and aged under 35. In the twentieth century the incidence of lodging declined rapidly (Anderson, 1990, pp. 64–5). Moving into lodgings was clearly related to migration, with the status of lodger frequently following the first move from the parental home. How frequently such young people lodged with distant relatives or family friends is hard to ascertain.

Summarizing changes in family and household composition over time, Anderson (1990) stresses that although very large and complex households were relatively rare at all times, the simple nuclear family (consisting of parents and children only) was much less common in the mid-nineteenth century and earlier than in the the mid-twentieth century. In 1851, 36 per cent of households consisted only of a married couple and at least one child, and 44 per cent of households contained at least one extra person (lodger, servant, visitor or relative). Such household patterns obviously influenced decisions about migration as larger and more complex household structures meant that more voices could contribute to migration decisions and the effects of a move would impact upon a larger number of people. For instance, a lodger might be forced to seek new lodgings because his landlord moved away, a servant might be torn between moving with an employer or remaining in a neighbourhood close to kin, and co-resident adult children may have been forced to make choices between parents, friends and employment opportunities following migration decisions taken by the head of the household in which they lived.

A number of other factors also interacted to affect the structure of families and influence migration. These included the changing nature of employment and the household economy from the late-eighteenth century, the decline in fertility from the 1880s, changes in the position of women in society, the development of state-sponsored welfare, and changes in attitudes towards family obligations and responsibilities. There is a large literature on what can be broadly termed the household economics approach to the family (Anderson, 1980, pp. 65–84). Recent research has tended to stress that, although the period of industrial change in Britain from the eighteenth century transferred many people from domestic to factory employment and, in particular, provided new individual opportunities freed from the constraints of families, there was also a high degree of continuity. Rather than destroying traditional patterns, it has been argued that much proto-industrial labour required working for long periods away from home, and many aspects of the family economy were transferred to factories as employers took on whole families to fulfil different roles within the workforce. Moreover, many migration decisions were based on family needs and the optimization of household rather than individual incomes (Medick, 1976; Levine, 1977; Hareven, 1978). Data on life-time residential histories can demonstrate the way in which migration decisions fitted into the whole life-course, and examine in more detail the extent to which family structures were changed during the process of industrialization (see Chapter 5).

The decline in marital fertility which occurred in Britain from the 1870s is now well established. Women born before about 1830 had, on average, five or six children whereas those born in the 1880s had half as many.

Moreover, more births were concentrated in the early years of married life (Woods, 1987, 1992; Anderson, 1990). These changes had implications not only for household size, but also for the ways in which families organized their lives. Fewer children could be given more material possessions, a better education, more parental care and a higher standard of living. However, fewer children also meant that the burdens of caring for elderly relatives fell on a smaller number of children and, at the societal level, the elderly dependent population has increased dramatically due to both falling mortality and fertility in the twentieth century. There are a number of possible implications for migration. Smaller families with more resources are likely to be able to move more freely over longer distances, fewer kin may place fewer constraints on migration decisions, and women freed from constant childbearing are more likely to enter the labour force and move for employment reasons. Although welfare provision in the twentieth century has lessened dependence on children in old age (Thane, 1980; Fraser, 1984), a large and relatively healthy elderly population will also be more likely to move for family reasons to remain near younger relatives.

It can be suggested that there have also been significant changes in the social and cultural meaning of the family, linked to broader changes in the relative positions of men and women in society (Shorter, 1976; Flandrin, 1979; Anderson, 1980; Davidoff, 1990). Much of this work on the 'idea' of the family is concerned with the extent to which notions of privacy, domesticity and emotion within family relationships have changed over time. Evidence for or against change in the nature of relationships within the family is inconclusive (Macfarlane, 1979; Anderson, 1980, 1994; Finch 1994) but, at the very least, it can be suggested that the ways in which individual families organized their lives, their attitudes to such issues as privacy and family responsibilities, and the closeness of relationships may have affected migration decisions. It may be that such factors varied considerably from one family to another at all times, rather than changing systematically over time, but there were undoubtedly certain societal trends which have affected the meaning of the family. For instance, the increased involvement of women in the workforce in the second half of the twentieth century, linked to changing notions of responsibility for household duties and increased labour-saving devices, have changed the nature and extent of housework for many women, though it can be argued that such changes are only limited and that many women still fulfil very traditional roles within the family (Oakley, 1974; Roberts, 1994). In the context of migration, it can be suggested that at all times many women have moved as dependents and have played a subordinate role in family decision-making strategies.

The remainder of this chapter attempts to explore some of these complex issues using data on life-time residential histories collected from family historians and genealogists. The very nature of family relationships, rarely disclosed in any documents, means that many of the issues examined above

197

cannot be directly investigated. However, the life-time histories provided in this study do allow examination of the migration process at all stages of the life-course, though (as explained in Chapter 2) there is a bias in the data to those who eventually married. Subsequent sections of this chapter focus, first, on an examination of migration undertaken for family related reasons; second, on the way in which migration varied at different stages of the family life-course; and, third, on the experience of migration from the perspective of different family members. There are many links between moves for family reasons and for other motives, and the data examined in this chapter leads on to the more detailed study of movement as a result of personal and family crises discussed in Chapter 8.

Migration for family reasons

As shown in Chapter 3, overall, movement for family reasons was relatively unimportant. This section examines the aggregate characteristics of movement for family reasons in more detail and illustrates selected themes through the use of a number of case studies. Obviously the ascription of reasons for movement must be treated with some caution, and the data were checked to ensure that the sources used by family historians could reasonably have yielded the evidence cited. The interpretation of what was meant by a family related reason for moving is also problematic. The most common reasons included in this category were moves to be nearer relatives, to live with relatives, to care for relatives or to be cared for by kin; but in addition there were a wide range of other family related motives. Movement for marriage is excluded. At this aggregate level, the moves thus do not separate out those who were moving to undertake a caring role from those who moved because they required care themselves.

The spatial pattern of moves for family reasons (Figure 6.1) suggests that there was little difference from the pattern of all moves in the relevant time periods. However, detailed examination of the characteristics of these moves does reveal some significant differences (Table 6.1). Most marked is the fact that women were much more likely to move for family reasons than men, particularly in the period after 1880. Those who moved for family reasons were also much older than all migrants, with a significant over-representation of those over 60 years. This was especially true after 1880 when the over-60s were the largest single category, forming more than half of all those moving for family reasons. In relation to the whole data set, those moving for family reasons were more likely to move alone, and were much more likely to be widowed, separated or divorced (especially after 1880). Those migrating for family reasons were also much more likely than all migrants to be female household heads. This formed the second largest

Figure 6.1 All moves undertaken for family reasons, (a) 1750–1879; (b) 1880–1994

category in both time periods and female household heads accounted for over 30 per cent of family related migrants after 1880.

The typical migrant whose movement was ascribed to family reasons was thus a widowed or separated female, aged over 60 and moving alone. This trend was apparent in both time periods, but was most marked after 1880. Whilst men, nuclear families with children, and the young did sometimes move for family reasons, these categories were all under-represented in relation to the data set as a whole. These trends were also very stable between different parts of Britain (Table 6.2). Each region had broadly similar characteristics and, given the small number of family related moves in some regions, any differences were quite likely to have occurred by chance. Reasons for these characteristics are not hard to determine. Family related reasons for migration were most likely to occur at particular stages of the life-course, and at times when movement for the more common reasons of employment and marriage were less likely to occur. If we assume that most such moves incorporated a need or desire to be near relatives, it can be suggested that they were most likely to occur at a time when the migrant was vulnerable and in need of support and security. It is thus not surprising that elderly women who had been left on their own form a large part of this group of migrants. Such women would probably have

199

Table 6.1 Moves for family reasons by migrant characteristics, distance and year of migration (%)

Migrant characteristics	1750–1879	1880–1994
Gender		
female	39.8	56.5
Mean distance (km)	46.6	77.2
Distance band (km)		
<5	36.1	32.7
5–9.9	13.2	6.6
10–19.9	12.6	8.4
20–49.9	15.6	12.5
50–99.9	8.5	13.1
100–199.9	6.8	12.6
200+	7.2	14.1
Age		
<20	26.6	12.3
20–39	34.2	23.0
40–59	11.0	12.5
60+	28.2	52.2
Marital status		
single	39.0	22.2
married	37.6	41.0
other	23.4	36.8
Companions		
alone	34.1	40.4
couple	13.3	18.4
nuclear family	36.8	27.9
extended family	10.4	10.2
other	5.4	3.1
Position in family		
child	17.0	8.6
male head	44.1	35.7
wife	9.9	13.4
female head	17.6	30.9
other	11.4	11.4
Total number	495	1,583

Data source: 16,091 life histories provided by family historians.

been dependent wives for much of their lives, would find it difficult to gain employment, and may have had limited experience of making their own way in the world. A move to live near relatives would have been a perfectly rational way of providing a degree of security whilst also maintaining independence. The fact that this pattern was most pronounced after 1880 suggests the consequences of a period during which families had become increasingly scattered, but welfare provision had failed to provide adequately for those left on their own.

Table 6.2 Characteristics of all moves for family reasons by region (%)

Migrant characteristics	London	South East	South Midland	Eastern	South West	West Midland	North Midland	North West	Yorkshire	Northern	North Wales	*South Wales	**Northern Scotland	***Southern Scotland
Gender														
female	51.5	58.6	54.6	54.9	54.0	58.6	51.2	52.5	46.5	57.1	53.7	56.7	45.6	47.6
Mean distance (km)	84.1	111.8	78.7	99.4	116.8	76.8	97.2	73.9	99.4	110.6	97.3	69.5	133.5	129.6
Age														
<20	20.0	13.8	11.3	18.6	23.3	16.7	17.9	11.6	13.8	13.2	12.2	12.4	16.7	11.7
20–39	39.2	23.3	24.3	19.8	27.3	22.4	25.3	23.5	20.1	26.3	29.3	32.0	17.8	27.2
40–59	9.8	7.6	14.8	14.0	9.6	11.4	8.6	13.0	16.1	10.8	12.2	13.4	11.1	10.7
60+	31.0	55.3	49.6	47.6	39.8	49.5	48.2	51.9	50.0	49.7	46.3	42.2	54.4	50.4
Marital status														
single	34.0	20.2	25.2	28.4	32.5	24.3	27.2	18.0	24.6	24.2	30.0	23.7	32.5	22.0
married	42.2	42.6	38.8	39.6	38.6	40.6	34.4	46.1	40.2	45.2	45.0	40.0	28.9	33.0
other	23.8	37.2	36.0	32.0	28.9	35.1	38.4	35.9	35.2	30.6	25.0	36.3	38.6	45.0
Companions														
alone	36.1	37.8	41.5	39.2	30.5	40.2	40.9	39.6	41.1	41.7	34.1	30.9	49.4	48.0
couple	16.5	21.1	15.6	17.6	19.0	17.6	18.9	17.8	17.9	20.8	26.8	16.0	14.9	17.6
nuclear family	33.0	28.5	28.7	29.2	32.2	29.9	28.3	27.0	29.9	25.0	29.3	30.9	27.6	25.5
extended family	10.4	9.3	9.2	11.7	13.8	9.3	6.9	11.4	8.0	9.5	4.9	19.1	4.6	7.8
other	4.0	3.3	5.0	2.3	4.5	3.0	5.0	4.2	3.1	3.0	4.9	3.1	3.5	1.1
Position in family														
child	10.0	9.8	7.4	10.5	16.7	10.8	10.7	9.1	8.9	7.7	14.6	8.5	8.0	5.9
male head	38.7	33.7	36.9	35.7	32.2	31.8	39.6	36.8	44.2	37.5	39.0	34.0	42.5	43.1
wife	13.9	13.4	13.1	9.4	10.3	11.8	12.6	14.1	11.2	19.0	17.1	14.9	9.2	12.8
female head	26.5	32.1	30.2	33.3	27.6	33.3	27.7	28.0	25.9	28.0	19.5	24.5	31.0	30.4
other	10.9	11.0	12.4	11.1	13.2	12.3	9.4	12.0	9.8	7.8	9.8	18.1	9.3	7.8
Total number	235	370	284	173	176	210	162	364	226	168	41	97	90	103

* Includes Monmouth and Glamorgan.

** Includes Highland, North East Scotland and Scottish Midland census regions.

*** Includes South West Scotland, South East Scotland and Southern Scotland census regions.

Data source: 16,091 life histories provided by family historians.

Table 6.3 Sequential moves for different reasons by birth cohort (%)

Reason	Move 1	Move 2	Move 3	Move 4	Move 5	Move 6	Move 7	Move 8	Move 9	Move 10+
1750–1849										
Work	41.8	45.3	48.6	49.5	47.4	46.8	45.1	43.7	42.8	43.7
Marriage	34.3	25.0	14.1	7.8	5.9	5.7	4.3	3.6	3.1	1.8
Housing	3.2	7.4	11.3	12.7	14.0	15.6	13.3	13.5	15.6	13.3
Family	1.7	2.3	3.4	4.6	5.0	6.3	5.6	7.0	6.2	6.6
Crisis	4.7	5.8	6.8	8.2	9.4	8.2	9.3	10.6	10.5	15.1
Retirement	0.4	1.6	2.9	5.3	5.5	6.2	6.7	8.1	9.3	8.2
Armed forces	1.9	2.0	2.0	2.2	2.2	2.6	2.9	3.6	1.6	2.8
Other	12.0	10.6	10.9	9.7	10.6	8.6	12.8	9.9	10.9	8.5
Total number	5,755	4,573	3,331	2,343	1,577	974	656	394	257	497
1850–1930										
Work	39.1	35.3	33.8	33.7	31.7	31.8	29.1	31.7	29.4	29.6
Marriage	16.4	22.0	18.8	14.4	9.7	7.2	6.0	4.1	3.8	2.4
Housing	15.9	13.8	16.5	19.7	22.7	23.8	25.0	24.9	24.3	22.1
Family	2.3	2.2	2.9	2.9	4.7	5.0	5.8	6.4	6.8	6.6
Crisis	7.3	6.0	6.5	7.0	8.1	8.6	8.9	9.3	11.4	11.5
Retirement	1.0	1.5	2.2	3.3	3.7	4.4	5.8	6.1	6.9	7.9
War service	4.8	7.0	6.9	6.2	5.9	4.8	4.5	4.6	8.2	6.0
Other	13.2	12.2	12.4	12.8	13.5	14.4	14.9	12.8	9.2	13.9
Total number	5,635	5,444	5,195	4,513	3,801	2,993	2,347	1,732	1,269	3,110

The table should be read as follows: for those born 1750–1849, 34.3% of first moves were for marriage and 3.2% were for housing reasons.
Data source: 16,091 life histories provided by family historians.

The other main difference between all moves and those undertaken for family reasons relates to the distance moved (Table 6.1). Although, as with all moves, very short distance moves of under 5 kms were the most common category, the mean distances moved for family reasons were substantially greater than those moved by all migrants. This trend is again most marked after 1880 when over one quarter of all those moving for family reasons migrated over 100 kms within Britain. Although distances moved were shorter before 1880, they were still significantly longer than those for all migrants. Thus this group of predominantly elderly women migrating alone were prepared to travel over quite long distances to retain contact with their families. It is notable that the distances moved were comparable with some of the work related moves undertaken by young males (Chapter 5). This finding is significant because it is often assumed that long distance migration was dominated by males moving for employment reasons (Ravenstein, 1885/1889; Grigg, 1977). Data on life-time residential moves suggests that some older women could also move over quite long distances. One advantage of longitudinal data is that it allows investigation of the way in which migration for particular reasons fits into the whole life-course. This can be summarized for the whole data set and illustrated through a series of case studies (outlined below). Table 6.3 summarizes the sequential reasons for migrating at each successive move. Whereas migration for work occurred at all stages of the adult life-course, and for marriage at earlier stages, movement for family reasons (and for housing and crises) was most likely to take place at later stages in the sequence of life-time moves.

Individual case studies not only provide more intimate details about why people moved, but allow further assessment of the way in which movement for family reasons was related to other mobility. Mary B. represents many of the typical characteristics of those who moved for family reasons. Born in Toxteth Park, Liverpool in 1840, daughter of a journeyman joiner, she moved three times within Liverpool with her parents until at the age of 20 she left home to live with her fiance's sister (in Liverpool) who employed her as a dressmaker. The following year she married, gave up paid employment and, with her husband, rented a house nearby. The couple remained in Liverpool (moving once) for 14 years, but in 1875 moved with their two children to Tamworth, Staffs due to her husband's employment. They moved once more in Tamworth to improve their housing until in 1925, at the age of 84, the recently-widowed Mary moved some 40 kms to Dudley (Warwickshire) to live with her son. Mary died later the same year. Family thus played an important part in many of Mary B.'s moves. For much of her life she moved as a dependent, first as a child and then with her husband. The move away from her home city of Liverpool was entirely for her husband's employment and this must have removed her from a well established circle of family and friends. Future family also played a part in gaining her employment and housing when she worked for and lived with her fiance's sister prior to marriage. Most typically, in old age and after her husband's death, she moved some distance to a completely new community to live with her son. After 50 years living in Tamworth this move must have been something of a wrench as, although she was living with family, she was leaving a well established circle of friends and an area in which she had lived for over half a century.

Frances B. also moved to live with relatives late in life, though in her case movement was within the community in which she had always lived despite leading what was, in many ways, a highly varied life. Frances was born in Newport, Monmouthshire in 1832, the daughter of a coal shipper. She lived at home until the age of 22, moving once in Newport with her parents and working as a domestic and laundress. However, in 1854 she married and travelled with her soldier husband to the Crimea where she worked as a nurse. Her husband was killed in the Crimea and Frances returned to Newport a young widow. Initially she returned to work as a domestic, but at the age of 28 married again and moved to a new home in Newport. During her life Frances was married and widowed a total of four times but, apart from her excursion to the Crimea, remained in Newport. At the age of 67, and widowed for the fourth time, Frances moved (still in Newport) to live with her brother and his family. All her marriages had been childless and in old age she turned to her closest living relative, although she must have had many other relatives in and around Newport. She lived with her brother for ten years before her death in 1909. This example clearly shows the importance of both family and home community to a woman who had considerable experience of the world, yet remained largely dependent on her husbands.

It was not only women who moved to live with relatives in old age, as shown by the case of Ben M., born in Holmfirth, Yorkshire in 1887. Ben lived at home, working as a cloth presser in a worsted mill (and moving once in Holmfirth with his family in 1911) until called up for the war in 1915. He served in France until 1919 when he returned to his parent's house in Holmfirth and resumed the work he had left four years earlier. Five years later Ben married and established his own home in Holmfirth, but in 1935, aged 48, he moved with his wife and three children some 20 kms to the town of Halifax. The move had been stimulated by a dispute with his employer over union membership, and he was fortunate to gain employment as a cloth finisher in a Halifax woollen mill. Three years later he moved again to the nearby town of Huddersfield, the main reason being the opportunity to return to work (as a cloth finisher) in a worsted rather than a woollen mill. In Halifax Ben had lived in accommodation provided by the mill owner but on his move to Huddersfield in 1938 he moved into owner occupation. Ben M. remained in the same house for 25 years until, at the age of 76, he and his wife moved right away from the locality in which they had lived all their lives to be near their daughter in Wylam-on-Tyne, near

Newcastle, Northumberland. He then followed his children on two more occasions. When his only daughter left Tyneside in 1968, Ben and his wife moved to a house in Culceth, Cheshire, to live near one of his two sons; and then two years later, at the age of 83, moved to Ulverston, Lancashire to live with his daughter. He died in his daughter's house six years later. Despite having strong roots in West Yorkshire, where he had lived for 76 years (with an interlude in France during the war), in old age the ties of kinship led Ben to move long distances to unfamiliar parts of the country to be near to, and eventually live with, his children. Even in the 1960s, with a well established welfare state, the need to be near close relatives in old age overcame long-standing ties to a specific community.

The final example illustrates the way in which some people, through a combination of circumstances, moved to live with relatives at much earlier stages of the life-course. Polly P. was born in a Cambridgeshire village in 1858. Her father was a farmer, but at the age of just one year and following her mother's death, she was brought up by her maternal grandfather (also a farmer) in Huntingdonshire. Three years later her grandfather died and she went to live with an uncle and aunt in the same village. She remained with them, moving once in Huntingdonshire until, at the age of 24, she married a tenant farmer in the village of Caxton, Cambridgeshire. She remained in the same farmhouse for the next 54 years, bearing nine children until, following her husband's death in 1935, she moved in with an unmarried daughter who was living in the same village. However, five years later, her daughter married and Polly moved out of her daughter's house and went as a paying guest to live with a friend in Huntingdonshire. Three years later, at the age of 86, she moved to the small town of St Neots, Huntingdonshire to take possession of a house left to her by an aunt. Here she lived with another (unmarried) daughter who cared for her until her death in 1955. Thus Polly P. moved to live with relatives on a number of different occasions during her life (though all within the same area), with most moves stimulated by the death of someone on whom she depended. In this, and other examples, movement to live near or with family was often stimulated by a sudden and dramatic change in personal circumstances, especially the death of a parent or spouse. This is a theme returned to in more detail in Chapter 8 where movements stimulated by a range of personal and external crises are examined.

Migration and the life-course

It is well established in contemporary migration studies that the propensity to migrate is closely related to different stages of the life-course. It has been suggested that there have been significant changes in the nature of the life-course, and associated migration characteristics, over time (Warnes, 1992).

205

However, the precise nature of this relationship and the extent of any changes which occurred have not been demonstrated. There are two principal ways of viewing migration decisions within the context of an individual's life-course. First, behavioural studies tend to ascribe migration decisions primarily to short term equilibrium-seeking adjustments reflecting ambitions and responses stimulated by immediate circumstances (Rossi, 1955; Wolpert, 1965; Golledge, 1980; Cadwallader, 1992). Moves at different stages of the life-course thus relate to factors such as pressure from work and related changes in income (promotion, unemployment, retirement), changes in housing space requirements related to family expansion or contraction (childbirth, children leaving home), or changes in caring reponsibilities for kin (moves to be near an old or infirm relative). Many of these reasons for migration are explored in detail elsewhere in this volume: in this section the ways in which they were specifically related to the life-course are explored.

Second, an alternative interpretation suggests that, although they may be difficult to discern and disentangle from other factors, many migration decisions are based upon a long-term migration trajectory (Davies and Pickles, 1991; Davies, 1991; Davies and Flowerdew, 1992). Many individuals and families have a pattern of moves to which they aspire, which may include improvements in housing or moves to a preferred location over a long period of time. Although progress may be erratic, and interrupted by short-term factors, such migrants have a long-term goal to which they aspire over the life-course, and towards which most migratory decisions are directed. Because such long-term goals were rarely articulated they are particularly hard to discern in the past.

These two views of the migration decision-making process are not necessarily contradictory. It seems likely that elements of both processes were present in most migration decisions. In some instances long-term ambitions would have had to be put aside to deal with short-term crises, whilst migratory reponses to short-term needs could eventually have been steered towards a long-term goal. The balance between the two would have varied over time, and between individuals and social groups. It can be suggested that long-term goals would have been most likely to be achieved by those with most control over their own lives, related to factors such as income, marketable skills and power within the household decision-making process. Those with less income or power were more likely to have their migration decisions dominated by external forces. It also seems likely that in the more distant past, when most individuals had few resources to cope with family and employment crises, migration as a short-term equilibrium-seeking adjustment was probably more common than migration to achieve a carefully-planned ambition. Both long- and short-term strategies could also be affected by other factors, especially the cumulative inertia built up by long-term residence in one community. In such cases ties to places and people could prevent short-term migratory adjustments and lead to a reappraisal

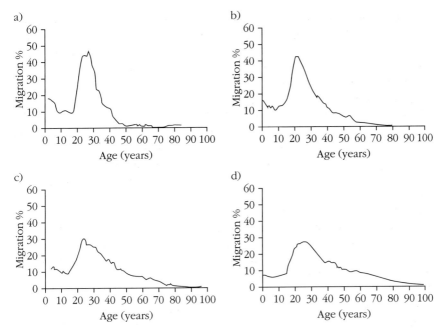

Figure 6.2 Graph of the proportion of moves occurring at different ages, (a) moves before 1800; (b) moves 1800–49; (c) moves 1850–99; (d) moves 1900 and after (moves plotted as five-year running means)

of long-term ambitions. Although difficult to discern in the past, the detailed analysis of migration reasons provided by family historians in relation to stages of the life-course, allows some assessment of these ideas.

Although it is not possible to calculate true migration rates from data provided by family historians, because the sample is drawn from an un-known total population, it is possible to examine the proportion of moves occurring at different ages over the life-course (Figure 6.2). The first point to note from these data is that the graphs are remarkably similar to those produced from plots of migration rates at different ages from late-twentieth century British and American populations (Warnes, 1992). At the aggregate level it would seem that the relationship between migration and the life-course has changed relatively little over the past 200 years. Propensity to migrate was relatively high for young children, who obviously moved with their parents, dropped slightly as children grew older, but increased steeply in the teens and early twenties as children left the parental home. Migration propensity remained high until the early thirties as marriage, new employ-ment prospects and a growing family stimulated moves, but the frequency of movement dropped steadily from the mid-thirties with a slight reduction in the rate of decline around retirement age as people moved to adjustment to a lower income or to be nearer kin in old age.

This basic pattern was quite stable over time and varied little between males and females. The graph for the eighteenth century is more peaked than those for other periods (with most migration being undertaken by those in their late twenties), and the volume of migration later in the life-course (especially after the age of 40) was much less than in later periods. Although this may be due in part to eighteenth-century under-recording of short-distance moves for housing and related reasons, which were more likely to occur later in life, the pattern can also be explained by higher mortality rates and fewer opportunities to move later in life in a more spatially-constrained society. The eighteenth and early-nineteenth century life-course was dominated by the movement of young adults for marriage and work-related reasons, including a small upturn around the age of 12, representing early departure from the parental home for some. By the mid-nineteenth century migration had become more evenly spread over the life-course, with a lower peak for those in their twenties. The peak age for migration also reduced slightly in the early-nineteenth century, rising again in the late-nineteenth and twentieth centuries. In the twentieth century there was also a slight upturn in the volume of migration later in the life-course, reflecting the propensity for retirement migration which has become a feature of the late-twentieth century, with a large population of relatively young and active retired people moving for a combination of family and environmental reasons (Grundy, 1987).

These graphs, in conjunction with the detailed analysis of migration for employment, housing and other reasons examined elsewhere in this volume, suggest that long-term changes in the relationship between migration and the life-course were not only relatively small, but also that they were fundamentally related to major changes in the structure of population, economy and society. It can be suggested that four factors were especially important. First, increased life expectancy which changed the proportion of moves made in old age; second, changes in the age at marriage – related to opportunities for employment – which altered the age at which young adults left the parental home; third, increased housing opportunities in a more highly differentiated twentieth-century housing market; and, fourth, the increased need to move to keep in touch with family members (especially in old age) as kin became more spatially scattered in the twentieth century. The patterns of migration associated with some of these processes, operating at different stages of the life-course, are examined in more detail below. There are certain life-course events which influenced migration, which would have been experienced by most people. Three such events can be identified and examined in more detail: the first move away from a parental home; moves related to marriage and moves associated with retirement. Other, less common events, such as the effects of bereavement or divorce are examined in Chapter 8 which focuses on the relationship between migration and a range of personal crises. The links between migration, housing and the life-course are explored in Chapter 7.

Leaving the parental home was a fairly universal experience with over 90 per cent of men and women in the data set leaving home at some time in their lives. The proportion leaving home increased from the eighteenth to the twentieth century and, especially in the late-eighteenth and early-nineteenth centuries, women were slightly more likely to leave home than men. The mean age at which both men and women left the parental home fell from the eighteenth to the twentieth century, with women leaving home earlier than men for those born 1750–1889, but men leaving home earlier than women for those born 1890–1930 (Tables 6.4a–b). The modal age for leaving home was 20–24 years for all time periods and both genders, but a relatively high proportion of the cohort of men born 1750–1819 remained at home until they were 30 years or more. This suggests that many couples who married in the early-nineteenth century had a period of co-residence with their husband's family prior to establishing a home of their own (see below). In the twentieth century men left home particularly early, primarily due to the impact of two world wars which led to the recruitment of large numbers of young men into the armed forces. One category which seems to be under-represented in the data are moves away from the parental home as a child to take up an apprenticeship. However, it can also be suggested that such moves were, in many cases, temporary with children sent away as apprentices retaining close links to the parental home, returning at weekends, and in some cases moving back to live with parents when an apprenticeship was completed (Pooley and Turnbull, 1997b). Overall, the combination of predominantly short distance mobility, movement as family groups after a period of co-residence with parents, and retention of close links with friends and kin suggest that for most individuals the first move from the parental home was no more disruptive or traumatic than any other form of migration.

The second life-course event which affected a large proportion of the population was marriage. For some this was coincident with leaving the parental home for the first time, but others remained single for varying periods of time. The proportion of the population who never married changed substantially over the period for which data are available. Wrigley and Schofield (1981) estimate that in the late-eighteenth century only seven per cent of the population never married (with a low mean age of marriage for women of 23.5 years), a position which was significantly different from a century earlier when some 15 per cent of the population never married and the mean age of marriage for women was 26.5 years. By 1851, 11.4 per cent of males and 12.4 per cent of females never married, falling to 9.5 and 12.1 respectively in 1881 before rising to 11.9 and 16.0 in 1911. The mean age of marriage for women fell from 25.8 in 1851 to 25.3 in 1881 before

Table 6.4a Characteristics of those undertaking their first move from the parental home by birth cohort (%)

Characteristics of those leaving the parental home	Males				Females			
	1750–1819	1820–49	1850–89	1890–1930	1750–1819	1820–49	1850–89	1890–1930
Per cent of total ever leaving parental home	84.2	93.4	95.9	97.9	90.1	94.5	95.3	95.7
Mean distance moved (km)	37.4	43.0	52.4	66.4	24.8	40.0	45.2	61.9
Age								
% <15	5.8	8.6	8.8	8.4	2.5	7.6	10.5	10.3
% 15–19 years	8.6	13.2	14.4	30.3	13.4	15.0	16.3	17.8
% 20–24 years	30.0	37.1	35.6	33.4	42.9	39.3	35.9	31.6
% 25–29 years	23.9	23.2	25.4	19.3	22.7	22.2	22.8	26.6
% 30–39 years	19.8	13.2	12.4	7.0	12.7	11.4	11.0	9.4
% >39 years	11.9	4.7	3.4	1.6	5.8	4.5	3.5	4.3
Mean age	28.0	24.3	23.9	21.8	25.5	24.1	23.4	23.7
Companions								
alone	28.6	33.2	38.1	46.2	18.2	25.2	30.4	32.6
spouse/partner on marriage*	22.8	24.0	29.2	20.3	48.6	37.3	42.0	37.7
spouse/partner**	14.6	12.1	9.5	4.8	7.8	8.1	6.9	6.6
nuclear family	27.5	19.1	11.5	4.5	19.6	17.6	10.5	9.2
extended family	2.2	4.0	3.6	3.6	3.0	4.6	3.8	5.9
siblings	2.2	3.9	3.9	3.2	1.2	4.3	3.2	4.0
other household members	1.0	1.5	1.0	0.6	1.3	1.6	1.3	0.7
friends/others	1.1	2.2	3.2	16.8	0.3	1.3	1.9	3.3
Reason								
work	38.8	37.0	29.1	15.8	15.8	25.9	24.4	19.1
apprenticeship/training	3.3	3.3	2.8	2.8	0.1	0.1	1.4	2.5
marriage	37.8	38.0	39.2	25.1	74.7	59.6	56.6	44.8
housing	2.8	2.0	3.3	2.8	0.5	1.5	2.6	6.3
family	1.2	1.3	1.5	1.1	1.2	2.5	1.9	2.1
crisis	2.6	3.5	4.2	3.0	2.6	3.2	4.4	6.3
education	1.5	2.5	3.8	5.0	0.5	0.9	1.2	6.2
war service	3.0	3.0	6.7	35.7	0.2	0.7	0.8	4.3
other/combined reasons	9.0	9.4	9.4	8.7	4.4	5.6	6.7	8.4
Total number	3,190	2,516	2,741	1,550	1,041	1,069	1,591	1,162

* Left the parental home to marry and set up home.
** Left the parental home after spending a period of married life co-resident with a parent.
Data source: 16,091 life histories provided by family historians.

Table 6.4b Mean age (years) at first leaving the parental home by migrant characteristics and birth cohort (%)

Migrant characteristics	Males				Females			
	1750–1819	1820–49	1850–89	1890–1930	1750–1819	1820–49	1850–89	1890–1930
Companions								
alone	22.6	20.4	20.5	19.7	23.1	20.6	19.9	20.5
spouse/partner**	26.8	25.0	25.7	26.2	23.5	24.1	24.1	25.3
nuclear family	34.6	31.1	30.7	31.9	31.5	29.5	28.2	29.6
extended family	40.2	26.9	25.3	23.5	34.1	25.1	27.7	22.8
siblings	20.9	17.9	19.3	15.6	32.5	23.2	26.5	19.6
other household members	39.6	27.6	26.9	20.0	33.6	22.2	24.1	18.1
friends/others	20.5	19.3	21.7	19.3	20.0	20.6	21.4	21.1
Reason								
work	27.3	23.5	22.9	22.0	25.3	22.2	20.5	20.4
apprenticeship/training	16.0	16.0	16.1	16.4	*	*	17.9	19.3
marriage	25.6	24.8	25.2	25.9	23.6	23.9	24.2	25.2
housing	34.4	27.0	29.3	28.6	*	29.5	30.7	32.2
family	38.8	28.1	19.9	20.4	42.8	24.3	22.3	21.2
crisis	37.9	21.2	18.9	17.0	40.0	24.3	22.0	22.9
education	15.7	15.8	15.7	16.4	*	*	15.8	16.5
war service	19.4	19.1	23.2	19.7	*	*	*	*
Occupational group of father								
professional	26.2	23.2	24.0	21.6	24.2	23.3	26.3	23.9
landed/farmer	28.5	25.4	25.4	23.9	25.0	23.9	22.8	23.9
intermediate	25.7	24.2	24.1	21.9	26.5	25.4	24.7	24.1
skilled non-manual	26.9	22.2	24.0	21.5	25.0	24.5	24.6	24.7

Table 6.4b (cont'd)

	Males					Females			
Migrant characteristics	1750–1819	1820–49	1850–89	1890–1930		1750–1819	1820–49	1850–89	1890–1930
skilled manual	27.6	24.1	24.3	21.6		25.6	24.3	23.7	24.0
skilled/semi-skilled agricultural	27.7	24.9	22.3	21.1		26.3	24.6	22.3	20.8
semi-skilled industrial	27.6	24.7	23.8	22.3		27.5	22.8	23.3	23.8
unskilled industrial	29.0	23.8	23.2	22.1		23.5	23.9	21.7	21.2
agricultural labourer	28.4	23.8	23.2	22.1		26.1	22.0	20.9	18.0
Occupation group after moving									
professional	27.8	28.5	28.1	25.8	professional	31.1	28.2	24.4	25.5
landed/farmer	30.9	31.5	28.5	32.6	skilled non-manual	*	*	24.7	24.8
intermediate	31.4	27.6	27.7	27.6	skill manual	32.7	23.6	23.2	23.1
skilled non-manual	27.0	24.5	24.5	24.8	semi-skilled manual	*	24.1	22.0	21.9
skilled manual	27.7	24.8	24.9	24.2	domestic service	21.5	18.1	17.5	17.8
skilled/semi-skilled agricultural	27.5	25.0	22.1	21.5	household duties	25.3	25.0	25.0	26.1
semi-skilled industrial	26.7	23.7	24.1	25.8					
unskilled industrial	29.9	25.2	23.1	21.8					
agricultural labourer	27.2	21.9	20.9	19.0					
armed forces	20.7	19.3	23.4	19.8					
apprentice	15.7	15.5	16.3	16.8					

Occupational groups are derived from the OPCS classification of 1951. * Omitted due to very small numbers.

** Includes those leaving home immediately following marriage and those who had a period of co-residence with a parent following marriage.

Data source: 16,091 life histories provided by family historians.

Table 6.5 Mean age of marriage (years) by birth cohort and gender

Gender	1750–1819	1820–49	1850–89	1890–1930
Male	26.0	25.3	26.5	26.9
Female	23.5	24.0	24.6	25.6

Data source: 16,091 life histories provided by family historians.

Table 6.6 Likelihood of selected life-course events producing a move by birth cohort (% based on stated reason for move)

Life-course event	1750–1819	1820–49	1850–89	1890–1930
Marriage	39.5	53.7	66.3	69.1
Retirement	32.7	49.2	53.1	73.6
Death of spouse	7.8	7.8	12.2	10.9
Divorce/separation	14.3	23.5	23.2	33.1

Data source: 16,091 life histories provided by family historians.

rising to 26.3 in 1911. Such variations would have had an influence on the propensity to migrate for marriage, and the age at which young adults left the parental home. However, as shown in Chapter 2, data from family historians were biased towards those who eventually married with only 3.1 per cent never marrying in the whole data set (Table 2.5). This should not affect the reliability of data on age at first marriage in the data provided by family historians and, as can be seen from Table 6.5, the data are broadly consistent with the national trends calculated by Wrigley and Schofield. However, those who never married are under-represented in the data set, and it is possible that their migration patterns were different from those who married (see below).

The broad outlines of the relationship between reasons for migration and migrant characteristics were examined in Chapter 3. The relative import-ance of marriage as a reason for migration decreased over time (due as much to an increase in other reasons for moving as to changes in the proportion marrying), and women were more likely than men to move on marriage, especially prior to 1890. This suggests that where only one part-ner moved, women were most likely to move to their future husband's home, though in the majority of cases both partners moved on marriage. As outlined above, the frequency with which marriage did not lead to a subse-quent residential move changed substantially over time. For the cohort born 1750–1819 only 39.5 per cent of marriages produced a residential move, whereas for those born 1890–1930 a change of home followed marriage in 69.1 per cent of cases (Table 6.6). From these data it can be suggested that in the eighteenth and early-nineteenth century it was relatively common for a woman to move to the home of her husband, but by the late-nineteenth

Figure 6.3 All moves undertaken for marriage, (a) 1750–1879; (b) 1880–1994

and twentieth centuries it was more usual for both partners to move, pos-sibly reflecting the wider range of housing opportunities available and the desire to begin married life in a new home chosen by both partners.

The spatial pattern of movement for marriage is very similar to that for all other moves (Figure 6.3). As shown in Chapter 3, moves for marriage were predominantly over short distances, with a mean distance of 21.9 kms. This was the second shortest mean distance after moves for housing rea-sons and emphasizes the fairly obvious point that most people met and married partners who lived close by (Dennis, 1977b; Dennis and Daniels, 1981). Although mean distances did increase over time in line with the rest of the data set, they remained lower than average. Very short distance moves within settlements are obviously not represented in Figure 6.3, but the pattern of those longer distance moves which did occur for marriage is very like that of all moves, with large towns and especially London dom-inating migration flows. Although there have been changes in the age of marriage and propensity to marry over time, and in the later twentieth century a high incidence of divorce and low rate of marriage has become the norm, the structure and pattern of moves associated with marriage has been quite stable over time. Marriage was a key life-course event which stimulated migration for many. For some it was the only move which was

Table 6.7 Characteristics of all moves by those who never married by birth cohort (%)

Characteristics	1750–1849	1850–1930
Mean number of moves	3.8	4.4
Gender		
female	41.0	67.3
Mean distance (km)	57.8	61.6
Age		
<20	49.6	29.2
20–39	35.2	38.0
40–59	11.6	19.1
60+	3.6	13.7
Companions		
alone	25.9	41.4
couple	2.2	0.8
nuclear family	54.5	34.7
extended family	6.0	5.6
other	11.4	17.5
Position in family		
child	49.5	34.2
male head	28.3	18.5
female head	6.8	26.0
other	15.4	21.3
Reason		
work	51.8	34.9
marriage	3.3	0.8
housing	8.3	18.0
family	5.3	6.2
crisis	10.8	13.0
retirement	1.9	4.6
other	18.6	22.5
Total number	527	1,124

Data source: 16,091 life histories provided by family historians.

made in a life-time. Although numbers are relatively small, it is possible to compare the migration characteristics of those who never married with the majority of the population who did marry at some point in their lives. Those who never married tended to be female, they had a similar level of mobility to all people in the data set but, on average, moved further than most migrants. They were also more likely than other migrants to move as the result of a crisis (either affecting themselves or a relative) and to move alone when young (Table 6.7).

The third life-course event which affected many people was retirement: this section assesses changes in the relationship between retirement and residential migration. The significance of retirement as a life-course event affecting migration was obviously related to longevity and the availability of welfare benefits. In a society where many people died young, and in which it was necessary to remain in the labour market for as long as physically possible in order to avoid destitution, retirement in the late-twentieth century sense of the word was uncommon. However, as people became less competitive in the labour market in old age they may have been forced to change both their employment and home, perhaps moving to somewhere more suited to a reduced income; and before the availability of universal benefits those no longer able to earn a living would have been highly dependent on relatives. A move to be near or live with a relative in old age may have been the only way of avoiding acute poverty.

Crude mortality rates fell dramatically over the 200 years from the mid-eighteenth century from around 29 per thousand in England and Wales in the 1740s to 11 per thousand in the 1950s, and mean life expectancy at birth increased from around 35 years in the mid-eighteenth century to 66 for men and 72 for women born in the early 1950s (Wrigley and Schofield, 1981, pp. 528–9; Tranter, 1996, p. 65). However, most of this change occurred in the twentieth century as a combination of improved nutrition, housing and living standards and the application of new medical knowledge led to a decline in infant as well as adult mortality. In the mid-nineteenth century mean life expectancy at birth was still only around 40 years (much less in large industrial towns), and crude mortality rates in England and Wales remained above 22 per thousand (Wrigley and Schofield, 1981; Woods, 1992). Although many people died young, it should not be assumed that few people lived into old age in the past. High mortality rates and low life expectancy at birth were caused mainly by very high infant mortality rates (in excess of 120 per thousand in the mid-nineteenth century), and many who survived to adulthood lived to a ripe old age. The proportion of the population aged over 60 years actually fell from around 8.5 per cent in the 1750s to around 6.5 per cent in the 1850s, due mainly to changes in fertility (Wrigley and Schofield, 1981). The proportion of elderly in the population remained almost unchanged to the end of the nineteenth century, but with reducing fertility and mortality increased rapidly in the twentieth with 11 per cent of the population aged over 65 in 1951. The extent to which many people did live into old age is amply demonstrated by the data set collected from family historians which, as shown in Chapter 2, is biased towards those who survived into adulthood. The mean age of death amongst the sample population was above 69.2 years in all time periods, and many lived to a stage in the life-course where retirement would have been considered.

The provision of state benefits in old age was also a product of the twentieth century. For most of the period under consideration the elderly who could not support themselves depended on relatives, charity or the Poor Law (which could have led to the workhouse). In 1906 almost six per cent of the population aged over 65 was living in Poor Law institutions (Anderson, 1994). The first old age pensions were provided out of taxation in 1908 in the form of a means-tested non-contributory pension scheme

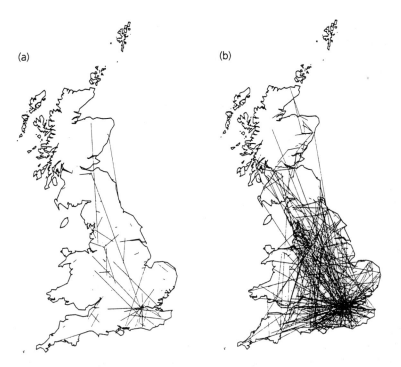

Figure 6.4 All moves undertaken for retirement, (a) 1750–1879; (b) 1880–1994

providing a maximum of 5s per week for those over 70 years. In 1919 pensions were raised and the means test relaxed, and in 1925 a State Pension Insurance Scheme was introduced. However, like other benefits at the time it excluded those who had not paid benefits and it was not until 1946 that a universal pension was established (Fraser, 1984). The ability to retire with some financial security is thus a phenomenon of the second half of the twentieth century, enhanced by the growth of occupational pensions and early retirement schemes for many in non-manual occupations.

These national trends are reflected in data on the relationship between retirement and migration. As shown in Chapter 3 the relative importance of retirement migration increased in the twentieth century (though it still accounted for only 13.7 per cent of all moves), and the likelihood of moving following retirement more than doubled (Table 6.6). Whereas for the cohort born 1750 to 1819 retirement only led to migration in 32.7 per cent of instances, for the cohort born 1890 to 1930 73.6 per cent of those who retired also moved home as a direct consequence. The gender breakdown for retirement migration reflects the composition of the workforce with those moving on retirement being dominated by men (especially for the cohort born 1750 to 1819, and the spatial pattern of retirement moves is similar to that of all migration (Figure 6.4). Whilst most moves were over

217

Table 6.8 Sequential moves by family grouping by birth cohort (%)

Companions	Move 1	Move 2	Move 3	Move 4	Move 5	Move 6	Move 7	Move 8	Move 9	Move 10+
1750–1849										
Alone	21.8	12.7	9.6	8.6	9.4	8.7	10.7	11.8	12.3	13.8
Couple	27.5	24.2	15.2	12.2	12.0	12.9	11.4	12.0	15.6	13.4
Nuclear family	43.3	54.4	65.9	68.6	68.7	67.8	65.7	63.0	59.5	59.6
Extended family	3.0	4.5	5.6	6.7	7.0	8.0	9.3	9.9	7.4	7.1
Other	4.4	4.2	3.7	3.9	2.9	2.6	2.9	3.3	5.2	6.1
Mean age	20.7	30.2	36.9	41.6	45.2	47.7	49.9	51.4	52.5	55.3
Total number	7,619	6,526	5,076	3,681	2,425	1,514	1,014	627	405	674
1850–1930										
Alone	15.7	20.3	20.0	18.1	18.1	17.1	16.4	17.4	18.7	20.8
Couple	12.5	20.5	20.7	20.1	16.8	17.7	19.6	18.9	20.8	23.8
Nuclear family	62.6	48.8	49.2	52.4	55.1	54.2	53.4	52.9	50.3	46.0
Extended family	4.2	4.1	4.4	4.6	5.2	6.8	6.8	6.9	6.0	5.6
Other	5.0	6.3	5.7	4.8	4.8	4.2	3.8	3.9	4.2	3.8
Mean age	12.5	21.3	28.0	33.2	37.8	41.2	44.2	47.0	48.7	51.4
Total number	7,075	6,707	6,132	5,279	4,389	3,449	2,648	1,959	1,426	3,427

The table should be read as follows: for those born 1750–1849, 21.8% of first moves were alone and 27.5% were as a married couple.
Data source: 16,091 life histories provided by family historians.

short distances, the mean distance moved on retirement was 55 kms (less than moves for family reasons but longer than those for marriage). Some people did move much further with London dominating the national migration system. For those who retired movement into London was more common than movement out (see Chapter 8). As with the whole data set distances moved increased somewhat in the twentieth century. Even in the eighteenth century migration following retirement was an important life-course event for some people, and by the twentieth century it had become a majority experience. Some of the case studies cited above give examples of people who moved to live with relatives in old age.

The life-course also influenced the nature and composition of the group of people who migrated together. At different stages of the life-course, people would have been more or less likely to move alone or in a family group, typically moving through a series of stages in which the very young moved as part of a nuclear family grouping; young adults tended to move alone, as married couples or, sometimes, with siblings; more mature adults were most likely to move in a family grouping; but in old age movement alone was again common, especially following the death of a spouse. What these data stress is the fact that the opportunity to move alone was quite limited for most people: for most of their lives they were living as part of a family grouping either as children or parents. Thus it is not surprising that, as shown in Chapter 3, most migration occurred with companions. However, this contradicts the popular conception of migration in the past which, following Ravenstein and others, has stressed the importance of young people moving alone. Although most young people did move alone at some stage of their lives, this was only one of many moves undertaken over the life-course. This is emphasized by Table 6.8 which summarizes migration groupings by successive moves over the life-course. Migration alone was most likely to occur either early or late in a migration life history; movement as a couple was more evenly spread, but also mainly occurred early. Predictably, migration as a nuclear family took place mainly in middle stages of a migration life history (or early as a child), and movement with an extended family mostly occurred towards the end.

In most respects the differences between those moving alone and in family groups were quite small (Table 6.9). All migrants were most likely to move for work irrespective of their companions, with the exception of couples who were most likely to move for marriage (see above). Those moving alone and with siblings were more likely to move in response to crises, especially in old age (see Chapter 8), and nuclear families were more likely to move for housing reasons, in response to changing space demands with a growing family (see Chapter 7). The spatial patterns of moves in different family groupings were similar to each other, and to those in the whole data set, though there was a tendency for nuclear family groupings to move, on average, over shorter distances than those moving on their

Table 6.9 Reasons for migration and distance moved by family grouping and year of migration (%)

1750–1879

Reason	Alone	Couple	Nuclear family	Extended family	Siblings	Other family group	Other
Work	50.4	17.4	62.3	35.3	44.6	43.4	23.7
Marriage	13.5	68.8	4.8	13.2	1.7	6.2	5.0
Housing	0.7	2.1	12.6	13.2	1.0	22.5	1.1
Family	5.0	1.4	1.8	6.0	5.2	2.7	3.8
Crisis	7.6	1.0	6.6	15.1	17.4	7.6	5.4
Emigration	2.0	0.8	2.0	5.4	7.7	0.5	4.3
Retirement	0.8	1.7	1.4	2.5	0.3	1.9	0.3
Education	4.2	0.1	0.2	0.5	8.7	0.3	0.6
War service	4.8	0.5	1.2	0.0	2.2	2.7	51.5
Other	11.0	6.2	7.1	8.8	11.2	12.2	4.3
Total number	3,292	4,577	10,027	838	287	369	317
Distance band (km)							
<5	27.6	53.2	52.9	60.7	28.3	59.9	28.6
5–9.9	11.9	11.0	10.6	8.5	7.3	8.3	7.6
10–19.9	11.6	9.3	9.3	7.6	9.4	6.2	9.8
20–49.9	16.1	11.7	11.6	9.8	13.6	11.2	16.8
50–99.9	13.2	6.3	6.8	6.7	15.3	5.7	8.6
100–199.9	11.4	5.1	5.2	3.6	10.8	4.7	12.4
200+	8.2	3.4	3.6	3.1	15.3	4.0	16.2
Total number	3,350	5,204	16,185	1,225	287	529	185

1880–1994

Reason	Alone	Couple	Nuclear family	Extended family	Siblings	Other family group	Other
Work	37.3	14.8	40.2	20.1	23.8	31.1	20.2
Marriage	4.6	45.9	2.4	8.9	1.5	4.6	0.6
Housing	4.7	12.7	30.3	22.0	10.0	25.3	3.7
Family	8.7	3.9	2.5	7.9	7.6	3.0	0.8
Crisis	13.2	2.9	7.2	17.0	26.8	10.0	3.5
Emigration	2.4	0.8	1.2	1.3	8.8	1.2	1.5
Retirement	1.5	10.1	3.1	5.8	4.7	4.0	0.8
Education	4.5	0.1	0.4	0.5	5.6	1.0	1.8
War service	13.0	1.1	1.8	0.5	1.2	4.2	59.2
Other	10.1	7.7	10.9	16.0	10.0	15.6	7.9
Total number	7,298	7,421	17,462	2,011	340	495	1,077
Distance band (km)							
<5	29.3	51.9	56.3	55.7	42.5	52.0	26.5
5–9.9	6.8	9.2	8.2	8.4	5.1	8.6	7.6
10–19.9	8.3	7.6	7.5	7.2	4.8	7.8	5.7
20–49.9	13.0	8.7	9.3	8.3	17.1	9.7	11.5
50–99.9	13.6	8.2	6.9	6.7	7.0	10.4	14.0
100–199.9	14.5	7.4	6.4	6.5	13.7	4.5	15.5
200+	14.5	7.0	5.4	7.2	9.8	7.0	19.2
Total number	5,730	7,732	21,277	2,279	315	556	407

Data source: 16,091 life histories provided by family historians.

own or with siblings. As with marriage and the first move from the parental home, migration following retirement demonstates very consistent trends over both time and space. It can be suggested that the constraints of the life-course operated in very similar ways for most people and, although changes in longevity and the availability of welfare had an influence, similar sets of responses can be seen over a long period of time.

Individual perspectives on the migration experience

It can be suggested that attitudes to and the impacts of migration varied considerably for different family members depending on their input to the migration process and their stage in the life-course. In most family moves there would have been some who gained more than others. Such issues cannot be explored adequately through aggregate data alone. This section identifies five categories of migrant: dependent children, working-age men, working-age women, elderly men and elderly women and examines their migration experiences using both the aggregate data set and more personal and revealing evidence from two diaries provided by family historians. This section also allows us to return to some of the hypothesized relation-ships between migration and the family outlined in the introduction, and to consider in more detail the interaction between short term expedients and longer term family trajectories in particular migration decisions taken at different stages of the life-course.

The aggregate characteristics of the five different groups are predict-able, and relate closely to factors explored elsewhere. Only key points are outlined here (Table 6.10). The reasons ascribed to moves made by children are obviously mostly the reasons why their parents moved, with work far and away the most important followed by housing and family crises. There were some differences between the reasons why working-age men and women moved, with men more likely to move for work and women for marriage. More marked differences occur between elderly men and women. Women aged over 65 years were much more likely than men to move for family reasons and as the result of a crisis whilst, as shown above, retire-ment motives were mainly ascribed to men. Patterns were stable between different parts of Britain, and there were few differences in either the spatial pattern associated with moves or the distances over which different groups migrated, though the elderly were a little less likely than other groups to move over long distances, especially before 1880.

Exploration of the motives behind migration, and the ways in which migration affected individuals at different stages of the life-course, can be explored through diary and life history evidence. Information from two sources (both referred to elsewhere in the volume) is used: the life history of Benjamin Shaw which covers the moves of the Shaw family in Yorkshire,

Table 6.10 Reasons for migration and distance moved by life-cycle stage and year of migration (%)

	1750–1879					1880–1994				
	Children (0–13 years)	Working age men (14–65 years)	Working-age women (14–65 years)	Elderly men (66+ years)	Elderly women (66+ years)	Children (0–13 years)	Working age men (14–65 years)	Working-age women (14–65 years)	Elderly men (66+ years)	Elderly women (66+ years)
Reason										
Work	59.3	49.8	37.2	18.5	10.1	43.2	36.3	30.7	7.1	4.6
Marriage	3.2	20.7	37.0	2.2	3.6	1.9	13.0	17.2	2.6	1.0
Housing	9.9	7.7	6.8	9.5	5.0	26.0	18.6	20.7	16.3	18.2
Family	3.1	1.4	2.7	18.2	29.5	3.4	1.8	4.1	17.0	22.5
Crisis	10.7	4.2	5.8	20.7	36.0	10.2	4.8	7.7	19.9	31.9
Retirement	0.5	1.0	1.0	23.3	6.5	0.7	3.3	3.6	22.0	9.6
Other	13.3	15.2	9.5	7.6	9.3	14.6	22.2	16.0	15.1	12.2
Total number	2,980	12,007	4,556	368	139	3,656	18,153	11,115	1,753	1,672
Distance band (km)										
<5	48.0	48.8	50.9	64.8	58.8	55.9	50.9	49.6	50.8	46.0
5–9.9	10.9	10.5	11.0	10.6	15.3	7.4	8.3	8.3	8.2	8.5
10–19.9	10.2	9.2	10.0	5.6	8.3	7.9	7.6	7.3	8.7	8.5
20–49.9	13.1	12.4	11.7	8.8	7.4	9.5	9.6	9.9	9.4	10.8
50–99.9	7.9	8.0	6.8	5.3	3.7	7.0	8.2	8.8	9.3	8.6
100–199.9	5.9	6.3	5.7	3.0	4.6	6.8	7.8	8.3	7.4	9.4
200+	4.0	4.8	3.9	1.9	1.9	5.5	7.6	7.8	6.2	8.2
Total number	4,541	16,207	6,108	625	216	4,504	17,992	12,211	2,149	1,877

Data source: 16,091 life histories provided by family historians.

Westmorland and North Lancashire in the late-eighteenth and early-nineteenth centuries, and the life history of Henry Jaques which chronicles the moves of the Jaques family in London during the second half of the nineteenth century. Both manuscripts take the form of life histories written up from personal diaries and, in both cases, rich detail is provided about the relationship of migration to family circumstances and the life-course (Pooley and D'Cruze, 1994; Pooley and Turnbull, 1997a). In this concluding section the migration experience is examined through the eyes of various members of each family, comparing experiences in very different time periods and settings.

Both authors report residential change affecting children whilst living with their parents in the family home, though the impact of parental decisions concerning both family migration and the future careers of children was rather different. The young Joseph Shaw lived with his parents until the age of nine, when he was apprenticed to a weaver in Westmorland and moved away from the family home in 1756. In doing so he was following his father's trade (he had previously worked at home in the domestic woollen industry), but rather than be apprenticed to his father he was sent elsewhere to learn his craft. It seems likely that overcrowding and poverty at home was a factor in sending Joseph away at the age of nine. There is no evidence that the child had any say in the decision to leave home and it is notable that in 1767, when he had finished his apprenticeship, he returned to the family home. He thus retained strong links with his family and returned as soon as he could, working as a weaver in the Shaw household economy until he married in 1771.

The young Henry Jaques, brought up in the Islington district of London, moved frequently with his parents as a child, but always within the same neighbourhood. Ties to both people and places were clearly important for Henry, as he had regular contact with some relatives and had a strong attachment to the school he attended. Though having no say in this family mobility, it clearly did not disturb Henry's life because he remained within the same area of London. At the age of 12 in 1854 Henry reluctantly left school and began work, but in contrast to Joseph Shaw he did not leave home. Again family poverty played a part in the family decision, but operated in a different way. Henry's father intended his son to be apprenticed to his own watchmaking trade, but he could not afford a dependent apprentice (an elder son was already apprenticed elsewhere), and so Henry lived at home but worked as an errand boy for a local clockmaker, clearly a contact of his father. Henry Jaques did not in fact leave home until the age of 16 when he rejected the watchmaking trade and followed in the footsteps of an elder brother by being apprenticed to a draper. Here he lived on the premises but retained close links with his parental home, returning the short distance at weekends. It is clear from the life history that this was a career and residential move in which Henry had some say – his mother was reluctant for him to leave home – and marks the transition from dependence to independence in migration decision making.

It is usually assumed that most married women were dependent on their husbands and that migration decisions were taken primarily by the male household head with relatively little input from other family members. However, evidence from the two life histories suggests that this was not always the case, and that married women could have a considerable say in both the domestic economy and migration decisions (D'Cruze, 1994; Roberts, 1994). In both families women played an essential role in the domestic economy. Isabella Shaw assisted her husband (Joseph) and sons in the domestic weaving that they carried on from their homes in and around Dent (Yorkshire), and Elizabeth Jaques assisted her husband in running the small retail drapers that Henry took when he married in 1864. Subsequently Henry established himself as a home-based shirtmaker, and his wife played a crucial role in running this domestic industry. Both life histories suggest that ties to family and a familiar home area were important for the women. Mothers were reluctant to allow their children to leave home and keen to retain contact with them, and Isabella Shaw had especially close ties to her home settlement of Dent. On one occasion, after the family had moved to Kendal (Westmorland), they returned to Dent specifically because Isabella could not settle in a new community. In this case the migration decision was clearly taken by the mother of the household. In both family histories there were periods when a male household head was incapacitated and their wives played a key role in maintaining income and keeping the household together. Thus Betty Shaw nursed her husband (Ben) through serious ill-health in 1808 and Lizzie Jaques ran the family business on more than one occasion when her husband was ill. Evidence in the life histories suggests that not only were married women an integral part of the household economy, but that they could have major inputs into household decisions about a range of issues, including migration.

In keeping with evidence from the aggregate data, most of the migration decisions undertaken by adult men of working age in the two life histories were related to employment, though they were often interlinked with attempts to improve family circumstances and the residential environment. The importance of the whole family in some migration decisions can be illustrated from both family histories. In 1791 Joseph Shaw took the quite radical decision to leave the familiar area of Dent and move with his wife and seven children to the factory village of Dolphinholme (Lancs) some 50 kms away. The mill owner was recruiting labour in the Dent area and several families moved together to Dolphinholme. Handloom weaving was in recession and there was little work for Joseph and his children, Isabella suffered badly from arthritis, her contribution to the domestic economy was limited, and there was little sign that conditions would improve. The mill at Dolphinholme offered the opportunity of secure employment for Joseph's children, though his employment prospects were probably worsened, and with several members of the family in factory employment the demands on the ailing Isabella would have been lessened. In practice things did not turn out as intended, but the decision to migrate, taken primarily by Joseph Shaw, was clearly designed to benefit all members of the family and was taken in the knowledge that his own employment opportunities were increasingly limited.

The welfare of all family members was also uppermost in the mind of Henry Jaques in 1877 when he moved with his wife, seven children and two elderly parents from the Mile End area of East London to the more suburban location of Forest Gate. The reason given for this move was the health of his family. His wife and children had suffered various bouts of illness and, on the recommendation of his doctor and following the example of other families that he knew, he somewhat cautiously took the step of moving away from the area with which he was familiar to a better environment in a newly expanding suburb. The move also matched his increasing prosperity and status at work, he was supervising the cutting room for a wholesale draper in the City, but he also continued to carry on his domestic shirtmaking from his new home. Although clearly having environmental and status benefits, the move to Forest Gate was not necessarily beneficial to Henry Jaques' employment or welfare. He now had a much longer journey to work, although there was good public transport in the form of a tram, and the combination of long hours in the city and the lengthened journey to work meant that he spent little time at home. Much of the time he did have at home was spent at the cutting board for his domestic shirt making.

Both life histories also provide examples of the impact of ageing on migration decisions as people became increasingly dependent on relatives in old age. Two years after his marriage in 1793 Benjamin Shaw established his home in Preston, Lancs, and this industrial town increasingly became the focus for the Shaw family. One of Ben's sisters followed him to Preston and, although their father Joseph attempted to retain his links with the Dent area for as long as possible, in 1795 he too moved to Preston. Although continuing to live independently following his wife's death in 1798, due to ill-health he moved in with his daughter's family in 1822. This experience was typical of the life pattern of many elderly people, with daughters usually taking the main burden of caring responsibilities within the family, although the life history suggests that Ben was somewhat aggrieved that his father had gone to live with his sister rather than himself. There may also have been a certain element of jealousy as Ben's sister had married well and she was more prosperous than himself. Clearly the Shaw children felt a strong obligation to care for their father in old age, and the moves of Joseph Shaw late in his life were all designed to bring him closer to relatives and to ensure that he could be supported when he could no longer earn his own living.

225

Henry Jaques was also obliged to care for his parents in their old age, even though this clearly placed a considerable strain on the household. In 1876 the Jaques household already consisted of seven children and, despite a reasonable income (over £200 per annum) Henry complained both of poverty and overcrowding at home. However, due to failing health his father could no longer earn a living as a watchmaker, and the elderly couple were living in one room in a state of near destitution. Although there were other siblings, Henry clearly felt a strong obligation towards his parents and despite the lack of space at home took them into his own household. He notes that for several years this imposed a significant extra burden but clearly viewed caring for his parents in old age as a duty that he could not shirk. For the elderly people concerned, movement into their children's home must have caused some concern. Both the Shaws and the Jaques were clearly independent by nature and avoided dependence on their children for as long as possible. The loss of their own home, and the inevitable feeling that they were imposing on their children, must have caused some anguish, though the only alternative was destitution.

Examination of these different perspectives on the migration process emphasizes the ways in which family connections affected a large number of migration decisions, and the extent to which family, employment, housing and other factors were closely inter-connected in the migration process. The case studies also suggest that simple notions of dependence and independence in family migration decisions may be misplaced. In both households the needs of 'dependent' children and wives were clearly given considerable prominence in some migration decisions and the interests of the whole family were frequently placed above those of any one individual. The relative importance of different family members in the migration process also varied at different stages of the life-course, affected by a combination of economic opportunities, housing aspirations and obligations to kin. Evidence from both the life histories and the aggregate data suggests that, although the role of the family in the migration process was affected by structural changes in the nature of society and economy, there was considerable stability over both time and space in the ways in which the family and household operated as a migrating unit.

It is difficult to assess the extent to which the migration histories of the families considered in this section followed a clear life-time trajectory, or were simply a series of pragmatic responses to crises or other life events. Some evidence of both trends can be discerned from the sources, and it can be suggested that for most people migration decisions represented a balance between short-term expedients and longer-term aims. Thus the life of Henry Jaques was affected by a series of crises or other events which precipitated migration as a short-term response: he was dismissed from his apprenticeship on suspicion of stealing and thus had to change both his home and workplace; his small retail business failed due to lack of capital and the family had to find a new home; and he was made redundant on more than one occasion leading to loss of income and, in some cases, a residential move. Yet throughout the life history there is also a clear sense of purpose with many moves designed to provide more space, a better residential environment, and an improved position in the world for both Henry and his family. Henry Jaques was certainly intent on self-improvement through education, religion and employment, he sought and attained respect-ability, provided good educations for his children, and moved to a reasonably desirable suburban location in middle age. As such there were a clear set of aims underlying the employment and residential decisions which Henry Jaques made and, despite some set-backs and moves forced upon him by short-term crises, the life history does give the impression of a life trajectory which was, at least in part, achieved.

CHAPTER 7

Migration and the housing market

Introduction: housing market change in Britain

The 200 years from 1750 brought fundamental changes to the British housing market in terms of tenure, housing quality, architecture, finance and access. This chapter, first, summarizes some of these trends and relates them to the individual migration process, before examining in detail the ways in which residential mobility has been related to housing type and tenure. Housing is a particularly important variable to consider because not only does a residential move, by definition, require a change of house (whereas, for instance, migration does not always lead to job change), but also housing itself can be seen as symbolic of broader social, economic and cultural changes within society. Thus housing choice is in part a statement about an individual family, and their choice of a home, and in part a reflection of wider societal trends for which housing provision provides a good barometer.

In the second half of the eighteenth century most families, and especially those living in towns, rented their accommodation from a private landlord. Only the urban elite owned property and they were not necessarily owner-occupiers. It was, for instance, quite common to own property as an investment and live in a rented house. Renting was also common in the countryside, although in some areas there were substantial numbers of freeholders who owned property. In addition, many people accessed housing by virtue of their employment. This group included domestics, farm servants and apprentices who lived with their employers; most agricultural workers who lived in estate cottages, and some industrial workers especially in mining and textile communities where employers commonly provided accommodation for at least some of the workforce (Chalklin, 1974; Corfield, 1982; Mingay, 1990).

During the nineteenth century the tenure structure of Britain's housing stock changed only slowly, with private renting continuing to dominate in both urban and rural areas, but this stock did become increasingly diversified.

Rapid population growth in rural and urban areas in the first half of the nineteenth century put pressure on the housing stock and, especially in the countryside, this led to massive overcrowding and a decline in housing quality as demand was not met by an increased supply. Indeed, as fewer agricultural workers were employed as farm servants, and the supply of estate cottages was not increased, housing shortage was a factor which pushed some families from the countryside to towns. In urban areas rented accommodation took a variety of forms. Many of the poorest families rented one or two rooms in a multi-occupied house, often centrally located and previously vacated by higher-status occupants. Such accommodation was supplemented by new speculative working-class housing, built at high density close to central urban areas in the first half of the nineteenth century. However, from the 1850s, most new building was increasingly concentrated in the suburbs, and took the form of decent quality by-law terraces. These were rented mainly by skilled workers, the poor continued to live in rented rooms close to the city centre (Gauldie, 1974; Burnett, 1986; Rodger, 1989b).

In both rural and urban areas the proportion of the population that gained housing through their occupation declined during the nineteenth century, although there were some notable exceptions such as in Barrow-in-Furness where the railway, iron and shipbuilding companies provided substantial numbers of houses to attract labour to the town (Marshall, 1958; Roberts, 1978). By the end of the nineteenth century it is estimated that around ten per cent of the population owned the property in which they lived. In addition to the rich who could buy their property outright, an increasing number of skilled workers were moving into owner-occupancy by forming terminating building societies through which they financed and built their own homes. Such activities were encouraged by the Freehold Land Society, with property ownership a means of extending the franchise, but owner-occupancy remained very much a minority experience in the nineteenth century (Burnett, 1986; Daunton, 1987). By 1900 there was also a small amount of municipal housing (located in the largest cities) and some accommodation provided by philanthropic trusts (especially in London), but renting from a private landlord continued to be the most common form of accommodation (Daunton, 1983: Dennis, 1984b; Pooley, 1985).

The structure of housing tenure in Britain changed most rapidly in the twentieth century. This was achieved both by the transfer of rented stock into owner-occupancy (as houses were bought by sitting tenants), and through a massive expansion in the construction of homes that were sold to private owners and municipal housing. The processes involved in this tenure transformation are too complex to discuss in detail. Many had their origins in the nineteenth century, but there were also new factors (including the effects of the First World War) which stimulated change in the twentieth. Owner-occupancy expanded due to both positive and negative factors. Increased affluence, rising aspirations and the activities of permanent building societies who, with a surplus of funds, were for the first time lending to private individuals for owner-occupation (rather than to owners who bought property as an investment) all stimulated the private housing market. But at the same time privately rented housing was becoming increasingly unattractive as an investment for landlords, little new property was constructed to rent, and families perceived renting as an increasingly marginal and inferior form of housing. As the costs of a mortgage and the rent for an equivalent rented house were similar, many chose to move into owner-occupancy, leaving the private-rented sector increasingly as a marginalized tenure for the poor (Burnett, 1986; Pooley, 1992).

From 1919 Local Authorities were required to survey housing need and draw up plans for the provision of council housing, although not all authorities built under the 1919 Housing Act. Although the legislation of 1919 was seen as a temporary measure to meet short-term housing need, and was aimed at the provision of good quality general-needs housing, it set the tone for later developments. Throughout the 1920s council house provision was predominantly suburban and aimed at those on regular incomes who could pay reasonably high rents. The poor were expected to benefit by filtering into vacated inner-urban properties. During the 1930s the subsidy for general-needs housing was removed and municipal housing provision focused on slum clearance and the provision of, usually poorer quality, housing for those dispossessed (Swenarton, 1981; Daunton, 1984; Pooley and Irish, 1984). After the Second World War the same cycle repeated itself, with an initial investment in general-needs housing to meet short-term housing shortage but, by the mid-1950s, the pendulum was increasingly swinging back to slum clearance. During the second half of the twentieth century council housing has been increasingly perceived as a marginal and residual tenure; a trend exacerbated from the 1980s as expenditure has been reduced and stock sold into private ownership. Housing Associations now provide the bulk of new social housing (Merrett, 1979; Forrest and Murie, 1988; Rodger, 1989a).

There are a number of persistent regional variations in the distribution of housing tenures, although there are no national data to give a total picture before the mid-twentieth century. In 1961 there were high proportions of local authority housing in most parts of Scotland, and in most large cities in England and Wales. Small towns and rural areas had higher levels of owner occupancy, but the declining proportion of private renting was concentrated mainly in large urban areas. The 1961 pattern can be related to what is known of the spatial pattern of housing market change in the past. The largest cities, with particularly acute housing crises, were the first to build council housing in the late-nineteenth century (especially London, Glasgow and Liverpool), and acute levels of housing poverty in Scotland led to greater levels of intervention in housing than occurred south of the border (Daunton, 1987; Rodger, 1989a). Although in the nineteenth century the highest levels of owner-occupation were found in industrial communities with a substantial skilled labour force on regular incomes (especially the West Riding, East Lancashire and South Wales), in the twentieth century the expansion of home ownership occurred earliest and most rapidly in Southern and Middle England, in areas least affected by the inter-war depression (Boddy, 1980; Swenarton and Taylor, 1985) Throughout the twentieth century owner-occupation has been associated with affluence and particularly high proportions were found in expanding southern towns and smaller residential communites. By 1971, 50 per cent of households lived in owner-occupied accommodation and this tenure dominates the late-twentieth century housing market.

There have also been significant regional variations in building materials and architectural styles (and associated housing quality), although there has been a trend towards greater uniformity in the twentieth century. For much of the eighteenth and nineteenth centuries local building materials were used for house construction and many parts of the country had distinctive vernacular styles. The availability and nature of local stone, brick clay and wood affected both building materials and styles and led to considerable regional diversity (Tarn, 1971; Brunskill, 1992). However, in the twentieth century a combination of increased regulation of the building industry and the industrialization of housing production has led to a convergence of styles and building materials. This is particularly true of local authority housing where national guidelines on style and space standards have been imposed. Although in England and Wales the traditional form of all private housing, and some council housing, has been low-rise, in Scotland the construction of tenements by both private and public owners has been common, especially in large cities. Whereas in England and Wales living in flats has been associated with poor quality housing, influenced by local authority blocks built in some cities in the 1890s and 1930s, and in most towns from the 1950s, in Scotland (in keeping with much of continental Europe) tenement living has been the norm for most groups of the population (Sutcliffe, 1974; Rodger, 1989a).

Whilst it is true to say that overall housing quality has increased massively since the eighteenth century, differentials between different types of accommodation have probably widened. Moreover, there is not a simple relationship between tenure and quality: each tenure type has had, and continues to contain, both good and poor quality accommodation. For instance, a cross-section of housing quality in a city such as Liverpool in the 1950s would have revealed owner-occupied housing which ranged from new and well built suburban property to older inner-area housing built in the 1890s and bought by sitting tenants in the 1920s. Many elderly people became effectively trapped in decaying owner-occupied stock, located in undesirable areas and which they could not afford to maintain. Likewise, local authority housing consisted of good quality suburban property built at low densities to high design standards in the 1920s and 1940s, but also contained high density tenements constructed cheaply in the 1890s and still lacking modern amenities. Although the privately-rented stock increasingly contained poor quality housing, it was still possible for families to rent reasonable accommodation in some parts of the city (Muchnick, 1970; Pooley and Irish, 1984). Thus although tenure differences provide a neat way of categorizing British housing, they cannot be easily equated with housing quality.

These changes in the British housing market had a large impact on the structure of urban areas, and on the associated quality of life of urban residents. From the mid-nineteenth century most new house construction took place in suburban locations. This contributed to urban sprawl which engulfed villages on the urban fringe and transferred large numbers of people from the city centre to the suburbs either by voluntary out-movement or relocation following slum clearance schemes (Hall, 1973; Cullingworth, 1994). Although both housing and environmental quality was improved in the suburbs, living on new public or private housing estates was not without its problems. Initially, many lacked adequate facilities, people missed the convenience and neighbourliness of their previous communities and, where employment remained centralized, suburban living could mean long and expensive journeys to work. With the possible exception of London, most suburban developments were poorly served by public transport, with tram lines following some time after housing developments. Many poor families moved to corporation estates in the 1930s removed themselves back to privately rented housing closer to workplaces because of the cost and inconvenience of commuting (Barker and Robbins, 1963–74; Jackson, 1973). From the late-nineteenth century very little new private house building has taken place in inner urban areas, and the corporation flats that have been developed (notably in the 1930s and from the 1950s to the 1970s) have been stigmatized by their design and location. A combination of lack of maintenance and location in a decaying urban environment has meant that from the 1970s many have been hard to let and, for tenants, have been equally hard to live in (Ravetz, 1980; Coleman, 1985; Nuttgens, 1991).

As the structure of the housing market has changed, so too have the means by which families gained access to housing. In general, these have become increasingly formalized and subject to regulation in the twentieth century. Throughout the eighteenth and nineteenth centuries most people accessed rented accommodation through relatively informal mechanisms. Although some property was handled by estate agents (who mainly managed rented property), most vacancies were more likely to become known by word of mouth within a community, by an advertisement in the corner shop, or through direct contact with the property owner or his agent who collected the rent. As most people had limited possessions moving between similar rented property was easy, although most landlords would require references and a rent book from a previous landlord as security. The very poor and recent migrants were disadvantaged in the housing market if they could not provide references or evidence of regular income. In the twentieth century, access to housing has been controlled and constrained by an increasing number of agencies. Most people moving into owner-occupied housing have had to gain a mortgage from a building society and meet the society's financial requirements, and those entering council housing were subject to assessment by a local authority housing official. Similarly, Housing Associations have formal mechanisms for the allocation of their property. All systems can discriminate against those perceived as a bad risk or unsuitable tenants (Handy, 1993). Those excluded by more formal mechanisms have had to resort to the declining privately rented sector.

There are many ways in which housing and housing market change can impact upon the process of migration. The desire to improve housing, or to adjust housing provision to changing family circumstances, is obviously an important stimulus to movement. It also clearly interacts with movement for family and employment reasons as life-course adjustments may link family, housing and employment change. However, the operation of the housing market can also act as a constraint on residential movement. Studies from the late-twentieth century show that families in owner occupation have particularly low rates of residential mobility, with the capital investment in buying and the inconvenience of selling, deterring movement. Mobility rates are also low in some council housing, but the privately rented sector is characterized by high levels of mobility (Champion and Fielding, 1992). However, we do not know the extent to which such trends have been consistent over time, and this theme will be explored later in this chapter. Other tenures may also have deterred mobility in the past. It was obviously more difficult to leave accommodation occupied by virtue of employment than it was privately rented housing as space needs changed, and the same may have been true for council housing. The rest of this chapter will explore in detail the links between housing and migration, examining the pattern and process of housing-related moves, individual experiences of gaining access to new housing, and the extent to which levels of mobility were related to housing tenure and housing type in the past.

The structure of migration for housing reasons

The quality of information on movement for housing reasons is likely to be more problematic than much of the data provided by family historians. Because a move to improve housing or to adjust housing to match changed family circumstances did not normally leave any particular record, positive identification of such moves depends heavily on family letters, papers and other documents or, for the more recent past, on oral evidence. It is likely

Table 7.1 Comparison of housing tenure characteristics in sample and census* (%)

Housing tenure	Census*	Sample	Sample size
1890			
Owner-occupied	10.0	13.0	
All other tenures	90.0	87.0	
			7,457
1938–9			
Owner-occupied	25.0	31.0	
Privately-rented	56.0	56.0	
Council housing	10.0	4.0	
Other	9.0	9.0	
			5,400
1961			
Owner-occupied	41.0	35.0	
Privately-rented	27.0	50.0	
Council housing	26.0	6.0	
Other	6.0	9.0	
			3,071

Data source: 16,091 life histories provided by family historians.
* 1961 census data. Earlier data from various sources

that housing related moves are under-represented before the 1850s, though in later periods many respondents did provide convincing evidence (drawn from family documents) of moves to improve housing. For the more distant past it is also often difficult to get firm documentary evidence of property tenure. However, it is likely that home ownership and property linked to employment can be identified and, for the period before about 1900, it is reasonable to assume that almost all property for which a tenure cannot be positively ascribed was rented from a private landlord. This assumption is made in some of the ensuing analysis.

The extent to which information on housing tenure in the sample is reasonably accurate can be assessed by comparing cross-sections of the data with census information on tenure for 1961, and with reliable national estimates for earlier periods (Table 7.1). The split between owner-occupied and privately rented housing seems consistent with national data around 1900, but in the twentieth century there is a clear under-representation of families living in council housing. This should not be due to lack of information as most twentieth century data are drawn from oral evidence. There are, however, two possible explanations which relate to the nature of the sample. First, as indicated in Chapter 2, poorer families are probably under-represented in the data and, in some periods at least, such families would have been more likely to enter council housing. Second, and probably more important, our sample deliberately contains only people born before 1930. Thus a cross-section of our sample in 1930 or 1960 will necessarily

235

Table 7.2 Housing-related moves by distance, migrant characteristics and year of migration (%)

Migrant characteristics	1750–1839	1840–79	1880–1919	1920–94
Mean distance moved (km)	12.6	4.2	3.7	8.1
Distance band (km)				
<5	79.6	87.5	88.8	76.8
5–9.9	7.2	5.9	5.0	9.2
10–19.9	4.2	1.4	2.8	6.0
20–49.9	4.9	2.8	1.9	4.4
50–99.9	1.5	1.7	0.8	1.9
100–199.9	0.8	0.5	0.6	1.0
200+	1.8	0.2	0.1	0.7
Gender				
female	18.1	27.3	39.0	43.9
Age				
<20	23.8	25.0	28.8	10.4
20–39	51.3	44.9	40.5	41.6
40–59	21.2	23.8	25.1	30.1
60+	3.7	6.3	5.6	17.9
Marital status				
single	28.8	30.1	34.7	16.6
married	67.2	67.2	62.9	75.1
other	4.0	2.7	2.4	8.3
Companions				
alone	3.0	1.1	1.7	6.8
couple	7.9	5.5	5.5	18.0
nuclear family	83.8	79.0	83.8	66.3
extended family	3.4	7.7	5.9	6.3
other	1.9	6.7	3.1	2.6
Position in family				
child	25.2	26.5	32.2	12.9
male head	60.2	53.0	42.0	48.3
wife	13.4	16.9	21.6	28.4
female head	0.4	1.0	2.1	7.6
other	0.8	2.6	2.1	2.8
Total number	270	1,322	2,875	4,358

Data source: 16,091 life histories provided by family historians.

contain a high proportion of elderly people. Given the fact that large scale council house development did not take place until the 1920s, and that families moving into council housing were predominately young, it is perhaps not surprising that we have picked up relatively few people moving into the council sector. This bias must be borne in mind when interpreting data for different tenure groups.

Moves undertaken primarily for housing reasons were overwhelmingly short distance (over three quarters were less than 5 kms in each time period), and many were contained within one urban area (Table 7.2;

(a)   (b)

Figure 7.1   All moves for housing reasons, (a) 1750–1879; (b) 1880–1994

Figure 7.1). Very short moves cannot, of course, be shown in Figure 7.1. The longest mean distances moved for housing reasons were in the period 1750–1839, but this is probably due to an under-representation of very short distance moves in the more distant past. Mean distances are stable until 1919, but double in the twentieth century reflecting both the larger size of urban areas and the willingness and ability of people to commute over longer distances. There are some long distance moves which were ascribed to housing reasons, though it is likely that in many such cases migration was undertaken for a variety of reasons of which housing change appeared to be the most significant when the data were recorded. Apart from the slight increase in mean distances moved in the twentieth century, the pattern is very stable over time, although the proportion of all moves due to housing reasons does increase dramatically in the twentieth century.

The characteristics of those moving for housing reasons also vary little from the eighteenth to the twentieth centuries, and for the most part reflect the characteristics of the whole data set (Table 7.2). Movement for housing reasons was most likely to be undertaken by a nuclear family grouping in the young adult age range (20–39 years), presumably reflecting housing adjustment during the family-building stage of the life-cycle. In the twentieth century an increasing proportion of moves for housing reasons were

Table 7.3 Housing-related moves by region, distance, migrant characteristics and year of migration (%)

Region	Average distance (km)	Gender female	Age <20	Age 20–39	Age 40–59	Age 60+	Marital status single	Marital status married	Marital status other	Total number
1750–1879										
London	7.0	21.6	22.1	50.8	21.3	5.8	26.4	67.8	5.8	245
South East	8.3	22.0	19.5	47.6	24.4	8.5	22.0	71.9	6.1	242
South Midland	10.4	26.7	25.7	48.6	17.6	8.1	29.7	66.2	4.1	75
Eastern	14.6	19.5	23.7	39.5	23.7	13.1	28.9	71.1	0.0	77
South West	15.2	17.1	28.9	39.5	28.9	2.7	33.3	65.3	1.4	76
West Midland	4.0	34.0	19.0	41.0	33.0	7.0	24.0	74.0	2.0	100
North Midland	2.3	20.0	21.3	49.3	18.7	10.7	25.7	70.3	4.0	75
North West	2.6	28.6	23.5	46.3	25.3	4.9	30.7	67.7	1.6	294
Yorkshire	9.8	21.7	31.6	44.5	21.3	2.6	38.2	58.8	3.0	157
Northern	8.8	27.3	26.6	42.2	26.0	5.2	34.1	62.6	3.3	154
North Wales	7.0	21.7	13.0	60.9	21.7	4.4	14.3	85.7	0.0	23
*South Wales	2.1	27.1	27.1	58.3	10.4	4.2	30.6	66.7	2.7	48
**Northern Scotland	16.8	38.0	17.6	40.7	31.5	10.2	25.0	71.3	3.7	108
***Southern Scotland	6.8	25.6	31.6	50.4	12.8	5.2	37.6	60.4	2.0	117
1880–1994										
London	10.1	39.0	21.1	48.4	23.2	7.3	29.9	67.4	2.7	783
South East	13.2	43.2	13.4	39.3	30.2	17.1	19.1	72.7	8.2	827
South Midland	11.5	43.2	17.8	41.7	27.0	13.5	24.8	68.8	6.4	824
Eastern	12.2	39.7	19.9	40.8	25.5	13.8	26.2	69.0	4.8	436
South West	19.2	42.8	16.3	33.9	29.7	20.1	17.1	74.7	8.2	313
West Midland	12.9	44.1	14.8	39.0	26.4	19.8	19.7	73.4	6.9	569
North Midland	7.9	41.7	16.6	37.7	28.9	16.8	23.1	69.3	7.6	494
North West	5.6	42.6	17.7	43.4	28.2	10.7	25.7	69.1	5.2	1193
Yorkshire	6.7	44.5	17.5	38.7	29.7	14.1	22.5	69.8	7.7	723
Northern	8.7	40.7	19.0	40.6	29.6	10.8	23.6	71.3	5.1	567
North Wales	22.8	45.1	17.6	45.1	23.5	13.8	29.4	70.6	0.0	51
*South Wales	3.6	42.4	20.6	38.5	34.6	6.3	23.9	70.4	5.7	257
**Northern Scotland	10.0	27.8	12.9	44.9	27.9	14.3	19.1	76.1	4.8	275
***Southern Scotland	5.0	43.6	20.8	44.1	25.6	9.5	25.0	69.4	5.6	390

* Includes Monmouth and Glamorgan.

** Includes Highland, North-East Scotland and Scottish Midland census regions.

*** Includes South-West Scotland, South-East Scotland and Southern Scotland census regions.

Data relate to all moves with an origin or destination in a given region.

Data source: 16,091 life histories provided by family historians.

undertaken by older people (mostly as married couples), reflecting the fact that not only did more people live longer but also that many couples in late-middle age experienced increasing affluence and were able to improve their housing situation and/or move on retirement. With these exceptions there is very little change over time.

As with other aspects of the data set, variations between different parts of Britain were also very small (Table 7.3). The longest mean distances moved for housing reasons were recorded in Northern Scotland, Eastern England and South-West England in the period 1750 to 1879, and in South-West England and North Wales after 1880. These were all predominantly rural regions where substantial settlements were widely spaced, and this may account for the slightly longer distance over which housing-related moves were undertaken. In all other respects the characteristics of migrants were very similar between British regions and reflect the patterns already described for the full data set (Chapter 3). This stability emphasizes the way in which moves for housing reasons formed part of a continuing life-cycle of mobility and often interacted with mobility for other reasons.

More insight into the pattern of housing-related moves, the ways in which they interacted with moves for other reasons, and the strategies adopted for gaining access to housing can be illustrated through a series of case studies spanning the eighteenth to the twentieth centuries and drawn from a series of diaries and life histories. The examples are concerned entirely with movement between privately rented property – the most common experience for most people in the past. Collectively the case studies demonstrate an essential similarity of experience, albeit in rather different circumstances, stretching over three centuries. They also emphasize the importance of local networks and personal contacts in gaining access to a new home. More detailed examination of variations between tenure groups is presented later in the chapter.

The Shaw family (mentioned in previous chapters) moved frequently in Yorkshire, Westmorland and North Lancashire from the 1740s to the 1820s, and their life history was recorded by Benjamin Shaw in 1826. Their moves were stimulated by a combination of housing, employment and family reasons, with the three often interacting to produce a particular outcome. Thus, on more than one occasion the move was stimulated by a desire for more space for both a growing family and the workshop and tools that Joseph Shaw had accumulated as a self-employed weaver and clock repairer. Whilst the house was chosen to match the space demands of work and family, and on at least one occasion to provide a larger garden so that vegetables could be grown, for much of their life the general area in which they lived was largely determined by the family attachments of Isobella Shaw to the area around Dent. The search process for a new home was clearly a complex interaction of a number of factors and, although on some occasions it may be possible to say that housing improvement was the main reason for moving, it was rarely the only reason. As far as can be determined from the life history, all houses were gained without difficulty and knowledge about the availability of rented accommodation would be easy to come by in a mainly rural area where the family was well known.

Many of the same themes recur in the life history of Henry Jaques (also mentioned previously), but set in the very different context of London in the second half of the nineteenth century. The Jaques family also moved frequently, mostly in East London, after Henry's marriage in 1864. Initially the Jaques' lived and worked in the same place, running a small retail clothiers in the Mile End Road and working from home as shirt makers. However, they were forced to leave the business due to bankrupcy and rented first rooms and then various houses in the same area of London, before moving further east to the suburb of Forest Gate. Each of the moves was an attempt to improve the housing space available at the lowest possible cost. Space was at a particular premium as there were a total of 14 children born to Lizzie Jaques, Henry's parents lived with the family for part of the period, and he required room for a workshop and machine room for his home-based shirt making. Throughout the period he combined (in varying proportions) working from home with employment in various city clothiers. Thus although improving housing space was a major consideration, employment and family considerations were also important. All homes, including the initial shop, were rented following word-of-mouth enquiries and scouring the area for suitable vacant premises. The Jaques' lived in the same area for most of their lives, and had a good knowledge of the local housing market and a network of friends, relatives and business contacts from whom they could get information.

The trade-off that was frequently made between the desire to improve housing and constraints of income and convenience to work is illustrated from the diary of David Brindley, covering his life in Liverpool in the 1880s (Lawton and Pooley, 1975). Brindley was a porter at Canada Dock Goods Station, and needed to be within walking distance of his work. Although not poor, his income was limited (24s per week in 1883), and for most of his life he lodged or rented a terrace house in the same area of north Liverpool. Although starting married life in lodgings in 1887, very soon the Brindley's rented a property at 5s 6d per week and they moved three times to very similar property over the next three years. On one occasion they did consider moving to a larger property with outbuildings, a greenhouse and chicken pens. This was clearly a very attractive property, which would have enabled Brindley to produce some of his own food, but the increased rental of the property coupled with the fact that it was further from his place of work meant that in the end they declined the opportunity. As with the other examples, all the Brindley's homes were, as far as can be discerned, obtained through local contacts, in one instance moving to another property owned by his current landlord.

The final example relates to the movement of Rhona Little who migrated alone from Londonderry (Northern Ireland) to London at the age of 18 to take employment as a secretary in the Civil Service, and subsequently moved lodgings on two occasions between 1938 and 1942. In each case her move was stimulated by factors to some extent beyond her control, and housing improvement was certainly not the most significant factor in any move. But the example does demonstrate the way in which access to rented property continued to be determined by local knowledge and contacts in the mid-twentieth century. When she initially arrived in London Rhona Little lived in a YWCA hostel for girls in Earls Court. This accommodation had been arranged for her by her parents using contacts that they already had in London. After less than a year Rhona was joined in London by her sister (L.) and she used this as an opportunity to move to a less restricted environment. The two girls rented a room in a family house in North London: the accommodation had been advertised on a noticeboard in L.'s workplace and the owner of the house also worked for the Civil Service. Following the outbreak of war, the girls were forced to move because their landlady evacuated herself to Wales, but they (and a friend) easily found a room in another house just around the corner. During the war there was much vacant property in London and finding a room to rent was very easy.

There is thus a degree of continuity in each of these case histories. The desire to improve housing, or to make it fit more closely to current space requirements, interacted with employment and family factors. This was especially important where at least some work was carried out at home. In every case rented property – be it a room or a house – was gained using local networks of information, and there is no evidence of use of official agencies. In this sense many people determined their own housing strategies and the outcomes of these individual decisions were remarkably similar over a long period of time and between different parts of the country.

The influence of housing tenure on residential mobility

There are some reasonably well established links between housing tenure and residential mobility in late-twentieth century Britain. In general, it has been found that owner-occupiers tend to stay in the same house for longer than those who rent from a private landlord. On average council tenants have lower rates of mobility than private tenants, but the distribution is often bi-modal with some council tenants moving frequently and others staying in the same house for very long periods. Recent studies of council tenants who have bought their property suggest quite low rates of mobility amongst this group, and Housing Association tenants also stay in the same accommodation for longer than private sector tenants (Hughes and McCormick, 1987; Champion and Fielding, 1992; Boyle, 1993, 1995).

Table 7.4 Persistence at the same address (number of years) by housing tenure and year of migration

Housing tenure	1750–1839	1840–79	1880–1919	1920–94	All
Privately-rented	15.3	10.0	9.9	7.1	9.6
Owner-occupied	23.1	17.5	16.5	12.7	14.7
Tied house	15.5	10.7	9.4	7.8	10.2
Council house	–	–	14.9	11.5	11.7
Lodger	9.3	7.0	6.2	3.9	5.4
Board and lodging provided with employment	7.9	7.9	5.3	3.8	5.5
Tenant Farmer	21.8	15.9	14.9	12.8	17.3
Other	7.2	7.5	5.4	3.9	4.9
All	16.1	10.8	10.2	8.8	10.3

Data source: 16,091 life histories provided by family historians.

These patterns are usually explained in relation to the characteristics of both housing tenure and occupants. It is suggested that the cost and inconvenience of moving in the owner-occupied market acts as a deterrent to mobility, despite the fact that owners may have higher than average incomes and thus greater opportunities for mobility. Home owners are also often able to adjust their housing to changing family circumstances by extending or altering their home. For many this may be cheaper and more convenient than moving. The high rates of mobility in the privately rented sector relate to the inflexibility of this tenure – it is not usually possible to alter it to suit housing needs – the relative simplicity of moving from one tenancy to another, and the fact that private renters are most likely to be young and therefore more mobile than mature families that have moved into owner-occupation. Low rates of mobility in council property relate to the relative lack of good quality property available and the relatively bureaucratic procedures involved in transferring from one council property to another. However, there have been no studies to date which examine the extent to which such trends have been stable over time. The nature and significance of different housing tenures has changed dramatically in the past 200 years, and it might be expected that the relationship between housing tenure and mobility has also altered.

As explained in Chapter 3, overall persistence rates were notably higher before 1840 than they were in later time periods. This could reflect greater opportunities for mobility from the late-nineteenth century, but it could also be an artefact of the data with the likelihood that some short distance mobility has not been recorded for the eighteenth century (see Chapter 2). Whatever the reason, it should not affect relative differences between the various tenure categories: there is approximately a threefold difference in persistence rates between housing tenures with a very stable pattern across each time period (Table 7.4). Owner-occupiers and tenant farmers were most likely to remain in the same house for long periods of time (in excess of 14 years for the whole data set); those in privately rented accommodation, housing tied to an occupation and council housing were close to the average (around ten years) although council renters tended to stay in the same house rather longer than those in the private sector; whilst boarders and lodgers had the shortest average tenancies (five years). These figures thus closely reflect trends found in late-twentieth century studies, and suggest that the factors producing particular mobility rates in different housing tenures have been very stable over a long period of time. Possible reasons for this stability are explored in detail below, but they must be related to the characteristics of particular households and their propensities to migrate, as well as to the attributes of particular tenures.

There are some significant variations in persistence rates between different tenures, the distance moved, and the characteristics of migrants. As these relationships were also very consistent over time (and numbers are quite small in some sub-categories), they are presented only for the whole data set (Table 7.5). Analysis of the distance moved to housing in different tenures is instructive. In all tenures there is a clear tendency for short distance moves to be followed by a relatively long period of residency, and for people who moved over longer distances to move home again relatively quickly. Thus amongst owner-occupiers, whereas the average persistence for the whole data set was 14.7 years, those that moved less than 5 kms to their owner-occupied house remained at the same address on average for 16.2 years, and those moving over 100 kms only stayed in the same place for 11.3 years. It can be suggested that these patterns reflected varying degrees of knowledge of the housing market. Those moving short distances were presumably moving within an area which they knew well and they could choose their housing carefully with reasonably full knowledge of alternatives. Such moves could be seen as a considered adjustment based on good information. In contrast, those moving over longer distances would be likely to have had much less knowledge of the area they were moving to and, particularly in the more distant past when communications were more difficult, would have had limited opportunities to look around and appraise housing alternatives. It is thus not surprising that, having moved to a new area, such households looked around and fairly quickly moved to more suitable accommodation based on a fuller knowledge of local housing markets.

Variations in persistence by family characteristics were relatively small and were consistent between tenure groups. Married couples and nuclear families tended to remain longer in the house to which they moved than single people who moved alone. Thus amongst owner occupiers whereas single migrants remained in the same house for 10.1 years, married couples and nuclear families persisted for over 15 years. Following from this there was also a tendency for younger migrants to stay in the same house for a shorter period than those in more mature stages of the life-cycle. Those

243

Table 7.5 Persistence (number of years) by tenure, migrant characteristics, reason, distance and region

Migrant characteristics	Privately-rented	Owner-occupied	Tied house	Council house	Tenant farmer	All other tenures
Gender						
male	9.6	14.7	10.6	11.5	17.3	5.4
female	9.6	14.6	9.4	12.1	17.3	5.4
Age						
<20	9.3	12.5	9.1	8.1	14.3	6.5
20–39	9.5	17.1	10.8	12.8	19.3	4.6
40–59	11.2	15.7	10.3	14.0	16.4	6.9
60+	6.8	9.4	7.5	8.2	10.4	4.7
Marital status						
single	8.4	12.0	8.6	8.6	15.7	5.5
married	10.2	15.9	10.8	12.5	18.1	4.9
other	8.3	9.9	9.1	9.9	13.2	5.7
Companions						
alone	7.4	10.1	8.9	9.5	20.0	5.4
couple	10.0	15.1	11.9	9.5	22.5	5.1
nuclear family	9.9	15.5	9.9	13.1	15.6	6.2
extended family	9.1	12.3	9.2	9.4	13.5	6.1
other	8.6	13.6	11.2	6.6	16.3	4.7
Position in family						
child	8.7	12.0	8.4	8.5	14.4	7.2
male head	9.8	15.3	10.9	12.0	18.0	5.5
wife	10.6	16.5	10.2	13.4	19.1	5.2
female head	8.1	10.5	6.8	10.1	14.7	5.3
other	7.8	10.4	9.1	7.4	12.2	4.9
Reason						
work	9.2	14.6	9.4	9.0	16.4	5.9
marriage	10.8	18.7	15.1	10.5	21.8	6.3
housing	10.4	15.8	14.6	13.0	24.2	5.4
family	8.9	8.6	6.3	8.1	16.8	6.3
crisis	8.5	11.8	9.0	8.3	12.4	5.1
Distance band (km)						
<5	10.1	16.2	12.4	13.1	17.8	6.5
5–9.9	10.2	14.6	11.0	9.2	18.0	6.4
10–19.9	9.8	13.9	9.2	7.5	16.9	5.6
20–49.9	9.6	13.8	9.6	8.4	16.4	5.9
50–99.9	8.5	12.0	8.1	9.0	15.3	4.8
100–199.9	7.4	11.3	8.7	7.1	13.3	4.1
200+	7.4	11.3	6.9	7.5	18.8	4.2
Census region						
London	8.3	14.5	8.5	13.6	9.9	5.2
South East	8.8	13.1	10.0	12.8	13.3	4.6
South Midland	9.6	14.3	9.6	11.8	14.9	4.6
Eastern	10.8	15.0	11.5	9.6	14.9	5.2
South West	10.2	14.6	10.0	8.6	19.0	5.4
West Midland	8.7	12.3	9.8	9.6	16.4	5.0
North Midland	10.8	15.7	10.8	11.8	15.9	5.5

Table 7.5 (cont'd)

Migrant characteristics	Privately- rented	Owner- occupied	Tied house	Council house	Tenant farmer	All other tenures
North West	9.6	14.2	8.5	12.3	15.8	5.4
Yorkshire	9.9	15.0	11.6	10.6	18.0	5.0
Northern	9.7	14.2	9.5	9.9	18.2	5.1
North Wales	9.7	15.4	11.8	9.3	10.4	6.7
*South Wales	9.5	15.6	10.0	13.0	22.0	4.8
**Northern Scotland	10.3	16.2	10.4	11.6	20.9	5.9
***Southern Scotland	8.4	13.9	9.3	14.9	16.5	5.8
Total number	20,642	9,776	3,952	973	1,015	7,596

* Includes Monmouth and Glamorgan.
** Includes Highland, North-East Scotland and Scottish Midland census regions.
*** Includes South-West Scotland, South-East Scotland and Southern Scotland census regions.
Data source: 16,091 life histories provided by family historians.

over 60 obviously had relatively short persistence rates because of higher mortality. However, tenant farmers and owner occupiers recorded particularly high levels of persistence amongst the over 60s, which may relate to greater longevity amongst this group. Such patterns can obviously be explained by the life-course patterns of migrants moving through the housing system. In all periods, young single migrants were most likely to move frequently as they adjusted their housing to suit changing employment and personal circumstances. Such people were also likely to have fewer ties and possessions and thus mobility was relatively easy. More mature families were not only constrained by the inertia created by children and increased material possessions, but also were more likely to have had the opportunity to move to a home which suited their longer-term needs.

The length of time a family stayed in the same house may also have been related to the reason why they moved to that house in the first place. Once again, variations in persistence by reasons for migration are very stable across different tenure groups with tenure differentials reflecting the overall pattern In almost all tenure categories families who moved to a house for marriage or housing reasons subsequently stayed longer in that property than if they had moved for employment, family or other reasons. The only exception was amongst lodgers and boarders where persistence rates were uniformly low and variations with different reasons were small. This was probably because boarding and lodging was clearly related to particular stages in the life-course and persistence was much more strongly related to life-course factors than to the reasons why movement occurred. Explanation of the other patterns is relatively straightforward. Moves for housing reasons and on marriage were most likely to be undertaken after careful consideration and with time to seek out alternative accommodation. Such moves were also most likely to be over short distances and thus there should have been a reasonable knowledge of alternatives. It is thus not surprising that such moves were followed by a relatively long period of residency in the same house (though those moving to council housing on marriage had a much shorter residency than those making the same move for housing reasons). In contrast, at least some moves for employment and family reasons were undertaken at relatively short notice – possibly because of unemployment or the death of a relative – and many entailed moving over long distances to a relatively unknown area. Such migrants may thus have needed to readjust their housing situation relatively quickly after a move.

Variations in residential persistence between regions of Britain were relatively small and, for the most part, reflect patterns in the whole data set. The detailed data are not presented here. Highest levels of persistence in most tenure groups were recorded in Eastern England, the North Midlands, North Wales and Northern Scotland. Although quite a varied group these regions do all exhibit quite high degrees of rurality and/or marginality and this may have contributed to above average residence periods in most tenures. Mean length of residence in council housing was above average in London and South-East England, possibly reflecting a region with high housing demand and thus a lack of alternative accommodation, and tenant farmers had higher than average persistence in South-West England, Northern England and South Wales. Again these include more rural and/or peripheral regions, although the pattern may also reflect differences in farming patterns and tenancy agreements in different areas. It may be that there was greater stability in pastoral farming areas with well established family farms (Armstrong, 1988; Beckett, 1990; Overton, 1996). However, as with other aspects of the data set, the most striking feature is the lack of substantial regional variations. The characteristics of migration and subsequent persistence at an address were very similar in all parts of Britain.

In conclusion to this section, it is necessary to assess the extent to which the variations in persistence recorded were reflections of the characteristics of particular housing tenures, of the families who moved, or of the interaction between these and possibly other factors. Tenure characteristics were clearly of some significance. It can be argued that the expense, time and commitment required for a skilled working man to gain access to owner-occupancy through a terminating building society in the nineteenth century meant that home ownership may have been an even greater brake to mobility than it was in the twentieth century. By the 1920s mortgages were relatively easily available and affordable by many skilled working-class families, and there was a more differentiated private housing market in which it was possible to quickly and easily find a home to suit particular needs. By contrast, in the nineteenth century becoming a home owner could be a much more protracted business as freehold properties were developed slowly by a group of individuals in a terminating building society, constrained both by resources and (where a substantial amount of building was done by the eventual occupier) by time (Price, 1958; Boddy, 1980). Tenant farmers would likewise invest a large amount of time and capital into their enterprise, and it is thus not surprising that rates of persistence were consistently high for both groups.

The privately rented housing market was both larger and more differentiated in the nineteenth century than in the twentieth, but a majority of the population continued to live in this tenure until after the Second World War. It was thus a tenure in which there was a relatively wide range of choices and little need for long-term commitment. Indeed, many landlords preferred the flexibility of relatively short lets, and thus families in the privately-rented sector could move easily to suit their particular needs. In contrast municipal landlords were more likely to encourage stability, especially amongst the relatively skilled and affluent tenants that moved into suburban housing in the 1920s (Daunton, 1987; Kemp, 1988).

Tenure characteristics can account for some of the variations, but this is only part of the story. We should beware of giving too much autonomy to housing as such because, although particular tenure characteristics may have encouraged stability or mobility, of greater importance were the requirements of families in different stages of the life-course. Thus the principal reason why persistence rates were lower in rented than in owned property was because most families were only able to aspire to home ownership in a relatively mature stage of the life-course, and at a time when they would have been likely to stay in the same house for a long period whatever tenure they had occupied. Thus life-course and tenure characteristics interacted to produce particular patterns. What has been clearly demonstrated in this analysis is that they interacted in similar ways, and produced convergent outcomes, in all regions of Britain and at all periods in the past 200 years.

The experience of moving home in different housing tenures

This section focuses on the extent to which the characteristics and experiences of migration varied as people moved through different housing tenures. Living in privately rented housing was clearly the most common experience overall, but the relative importance of owner occupation and, to a lesser extent council housing, did increase dramatically in the twentieth century (Table 7.6a). Tenant farmers and those living in tied housing declined over time, but the proportion of lodgers and boarders remained

Table 7.6a Housing tenure after migration by year of migration (%) (Includes only those moves where the tenure was clearly specified)

Housing tenure	1750–1839	1840–79	1880–1919	1920–94	All moves
Privately-rented	44.9	57.2	56.0	33.2	46.9
Owner-occupied	15.4	11.4	14.9	36.5	22.2
Tied house	14.3	13.5	9.2	5.2	9.0
Council housing	–	–	0.4	5.6	2.2
Tenant farmer	9.9	3.9	1.7	0.5	2.3
All other tenures	15.5	14.0	17.8	19.0	17.4
Total number	3,257	9,402	14,935	16,360	43,954

Table 7.6b Housing tenure after migration by year of migration (%). (Includes all moves and assumes that all those not otherwise identified are privately-rented)

Housing tenure	1750–1839	1840–79	1880–1919	1920–94	All moves
Privately-rented	79.8	81.0	72.4	45.4	68.4
Owner-occupied	5.7	5.1	9.3	29.9	13.2
Tied house	5.2	6.0	5.7	4.2	5.4
Council housing	–	–	0.2	4.6	1.3
Tenant farmer	3.6	1.8	1.0	0.4	1.4
All other tenures	5.7	6.1	11.4	15.5	10.3
Total number	8,895	21,149	23,810	20,010	73,864

Data source: 16,091 life histories provided by family historians.

more stable. Analysis and explanation now focuses, first, on the aggregate characteristics of movement into different housing tenures and, second, on the individual experiences of moving between different housing tenures. The cumulative effect of these data is to reaffirm the essential similarity of experiences between places and across time periods.

For the eighteenth and nineteenth centuries there is relatively little direct archival information on housing tenure, and thus quite a large number of family historians did not complete this section of the form. As most people rented from a private landlord before the twentieth century, this probably means that private renting is relatively under-represented and other tenures (which in any case are more likely to have left documentary evidence) over-represented. This is borne out by the relatively high proportion of moves into owner occupation in the eighteenth century. However, if we assume that all missing data is in the privately rented sector, a much more realistic pattern emerges (Table 7.6b). Results from subsequent analyses are essentially similar whichever definition of private renting is used, though the inclusion of missing data as being in the private sector makes the characteristics of this tenure more like those of lodgers and boarders. It seems

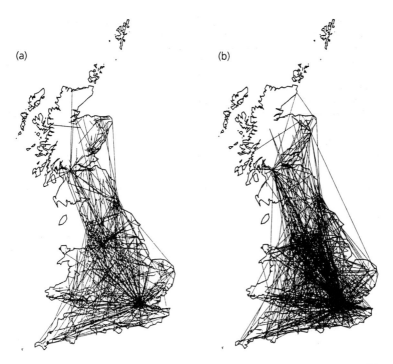

(a)   (b)

Figure 7.2 All moves between privately rented housing, (a) 1750–1879; (b) 1880–1994

likely that some of the missing data relates to those who lodged as well as those who rented a house or room. In practice a clear distinction between the two may be hard to draw.

Mobility within the privately rented sector was obviously the most common experience and, not surprisingly, the spatial pattern of such moves reflects that of the whole data set (Figure 7.2). In both time periods the mean distances moved and the characteristics of migrants were quite similar to those for all migrants, particularly if unidentified tenures are assumed to be privately rented. Although moves for employment reasons were the largest single category for all tenures, there was a relative over-representation of moves for marriage and housing reasons amongst those moving into privately rented accommodation. Variations between regions were small, but moves to privately rented housing in South-West England were noticeably longer than average (Table 7.7). For people born before 1930 the experience of moving within the privately rented sector was both common and similar in all places and time periods studied. Evidence cited above from four diaries illustrates some of these common elements in more detail.

As the second largest tenure group, owner-occupiers were also similar to the whole data set but, especially after 1880, accommodated rather fewer

Table 7.7 Housing tenure after migration by migrant characteristics, reasons, distance and region by year of migration (%). (Includes only those moves where the tenure was clearly specified)

Migrant characteristics	1750–1879						1880–1994					
	Privately-rented	Owner-occupied	Tied house	Council house	Tenant farmer	All other tenures	Privately-rented	Owner-occupied	Tied house	Council house	Tenant farmer	All other tenures
% by column												
Gender												
female	26.1	25.3	23.8	–	24.6	22.2	38.9	41.5	36.9	43.6	38.2	35.3
Age												
<20	23.2	17.9	21.3	–	18.4	41.7	18.8	9.8	18.9	9.6	21.7	25.9
20–39	56.5	50.3	60.2	–	56.4	46.1	52.9	36.8	48.1	45.7	43.9	49.0
40–59	15.9	23.1	15.8	–	20.7	5.4	19.9	30.4	27.7	22.8	26.0	8.2
60+	4.4	8.1	2.7	–	4.5	6.8	8.4	23.0	5.3	21.9	8.4	16.9
Marital status												
single	29.4	28.0	27.8	–	26.4	68.7	26.9	17.1	26.5	13.7	31.6	57.8
married	67.2	67.2	70.2	–	70.2	23.9	67.7	74.8	70.2	75.4	65.3	27.6
Companions												
alone	7.1	6.9	8.9	–	7.3	53.8	9.2	8.5	9.8	11.0	7.0	58.2
couple	21.0	20.4	19.3	–	22.8	6.8	21.4	27.1	17.7	19.5	15.0	6.4
nuclear family	64.3	57.9	66.9	–	60.3	16.8	61.3	54.1	66.5	62.5	67.3	13.9
extended family	4.7	5.3	2.8	–	6.3	5.4	5.5	6.6	3.8	5.5	7.2	5.7
other	2.9	9.5	2.1	–	3.3	17.2	2.5	3.7	2.2	1.5	3.5	15.8
Position in family												
child	21.1	19.1	17.9	–	16.7	6.5	19.6	12.0	17.2	10.7	23.9	4.9
male head	57.4	60.3	60.9	–	60.2	56.5	48.1	50.8	52.1	48.4	46.3	47.4
wife	16.5	15.1	18.2	–	17.0	5.1	23.4	26.8	26.1	29.0	22.4	5.8
female head	2.5	2.7	1.1	–	2.6	11.9	5.5	6.6	2.9	8.9	2.8	22.0
other	2.5	2.8	1.9	–	3.5	20.0	3.4	3.8	1.7	3.0	4.6	19.9

	Average distance (km)											
	27.8	23.6	41.6	—	21.4	61.5	35.0	45.2	56.8	15.5	27.7	91.2
Distance band (km)												
<5	59.2	60.4	38.4	—	43.3	30.9	60.1	49.6	31.2	74.6	39.4	26.9
5–9.9	8.5	11.1	12.4	—	16.0	10.0	8.1	9.0	10.8	6.6	12.0	6.6
10–19.9	7.4	7.3	12.4	—	14.3	10.8	6.2	8.7	11.3	6.8	15.4	9.2
20–49.9	10.2	9.4	15.2	—	15.5	16.3	7.9	9.4	16.3	4.0	18.7	12.8
50–99.9	6.2	4.7	8.6	—	6.3	11.6	6.3	8.2	12.1	2.6	8.4	13.4
100–199.9	5.0	4.2	7.7	—	3.3	12.2	5.7	7.9	10.8	3.2	4.8	14.6
200+	3.5	2.9	5.3	—	1.3	8.2	5.7	7.2	7.5	2.2	1.3	16.5
% by row												
Region												
London	70.2	10.9	4.6	—	0.3	14.0	56.8	21.2	2.8	1.4	0.1	17.7
South East	45.7	13.3	15.5	—	5.6	19.9	36.2	33.6	7.8	2.6	0.8	19.0
South Midland	50.9	14.6	11.6	—	5.7	17.2	39.6	32.0	5.6	2.8	1.7	18.3
Eastern	45.6	13.5	20.6	—	6.6	13.7	40.0	32.3	8.0	1.6	0.7	17.4
South West	47.2	9.3	22.3	—	5.0	16.2	36.4	28.5	10.8	3.2	1.3	19.8
West Midland	55.5	13.9	11.9	—	4.5	14.2	41.4	27.9	8.2	3.6	1.8	17.1
North Midland	47.4	14.4	13.6	—	6.4	18.2	43.6	28.2	6.5	2.9	1.5	17.3
North West	70.0	10.7	5.8	—	4.3	9.2	51.0	26.8	4.4	3.2	0.3	14.3
Yorkshire	62.9	11.6	8.6	—	3.6	13.3	49.2	25.5	6.2	3.7	0.3	15.1
Northern	53.6	11.8	19.8	—	4.4	10.4	47.2	21.0	12.8	4.4	0.6	14.0
North Wales	41.0	19.0	14.0	—	10.6	15.4	38.5	31.7	11.0	2.2	1.4	15.2
*South Wales	54.8	16.0	9.8	—	8.2	11.2	44.1	28.0	8.8	2.9	0.6	15.6
*Northern Scotland	42.3	9.0	26.7	—	10.4	11.6	37.6	20.7	15.1	3.5	3.5	19.6
*Southern Scotland	49.8	8.5	21.9	—	7.0	12.8	50.0	21.4	9.1	3.5	0.4	15.6
Reason												
work	41.4	10.4	24.9	—	7.9	15.4	40.3	20.2	16.9	1.3	2.2	19.1
marriage	62.7	12.8	11.7	—	6.7	6.1	63.5	23.4	5.2	1.4	0.9	5.6
housing	64.4	26.7	5.6	—	0.2	3.1	44.2	39.9	2.5	9.1	0.1	4.2
family	52.4	12.4	6.2	—	2.2	26.8	33.0	30.2	1.6	2.0	1.0	32.2
crisis	51.5	12.9	3.9	—	2.3	29.3	40.6	23.0	2.6	3.3	1.1	29.4

* Regions as Table 7.3.
Data source: 16,091 life histories provided by family historians.

single migrants than the private sector. Financial constraints of entry to owner-occupation meant that it was usually achieved later in life. Migration for housing reasons was over-represented in movement to owner-occupied property, emphasizing the extent to which home ownership was viewed as a means of housing improvement and upward mobility, and the longest migration distances again occurred in South-West England together with Northern Scotland. Tied housing provided with employment was also mainly occupied by nuclear families, but such families had travelled rather further than the average to take up the accommodation. Thus in the period 1750–1879 the average distance moved into owner-occupied housing was 23.6 kms, but moves into tied houses were on average 41.6 kms. It can be suggested that the offer of work and accommodation attracted migrants over longer distances than would normally be the case, with the longest moves occurring in London, South-West England and the West Midlands. Not surprisingly, migration to tied housing was almost totally dominated by movement for employment reasons (Table 7.7).

In contrast to accommodation provided with employment, movement into council housing was over a much shorter distance than that for any other tenure, with movement dominated by families moving for housing reasons. This obviously reflects the allocation policies of particular housing authorities, with most housing going to local families who could demonstrate housing need. The longest distance moves into council housing were in North Wales, presumably reflecting housing problems in a dispersed rural area. Those tenants who lodged or boarded in someone else's household (either privately or by virtue of their employment) were the most distinctive group. In both time periods they tended to travel much further before entering their accommodation, and were overwhelmingly young and single. Although migration for work was the most common reason for moving, such migrants were also over-represented amongst those who moved for family reasons and as the result of crises. This implies that such migrants consisted both of young single people leaving home and travelling some distance away for work, and migrants who were forced to move due either to family circumstances or some other crisis, and for whom lodgings were a temporary expedient.

The way in which movement between these tenures was linked together is best illustrated through a series of case studies. The first two examples illustrate the extent to which similar constraints operated in both the eighteenth and twentieth centuries. John H. was born in Markeaton, Derbyshire, in 1773, the son of a shoemaker. He moved once with his parents before joining the army in 1795 at the age of 22. However, five years later he bought himself out of the army with money provided by an uncle in order to get married. John and his wife then lived on his uncle's rented farm and John worked as an agricultural labourer. In 1808 the lease expired on his uncle's farm and an extended family of nine moved from

Belper to nearby Holbrook (Derbyshire) to live with John's parents on the farm which they now owned. He worked here as an agricultural labourer until in 1814 he inherited two houses in Derby and moved there with his wife and four children to set up business as an urban cowkeeper. Though obviously coming from a relatively affluent farming family, the uncertainty of being a tenant farmer could disrupt life patterns. However, his inheritance allowed him to move into owner occupation in the early-nineteenth century, and he effectively transferred a rural occupation to an urban setting (Grundy, 1982).

The importance of inheritance in allowing movement from the privately rented sector to owner occupancy is a recurring theme, and one which has become increasingly important in the twentieth century (Hamnett, Harmer and Williams, 1991). This is illustrated by the example of Henrietta M., born in 1897 in Tottenham, Middlesex. Her father was a tinsmith, and she moved three times with her family between rented property in the same neighbourhood. In 1918 she married and moved with her husband to a rented property also in Tottenham. She was at this time working as a telephonist for a furniture company in Tottenham, and thus her work and family were all concentrated in the same neighbourhood of north London. She stayed in the same house for 17 years, but in 1935 after inheriting some money from her father, she moved with her husband and only child to a more modern owner-occupied house elsewhere in Tottenham. Many people were moving into owner-occupation at this time, but the relatively small sum that she inherited (£50), was enough to provide a deposit for a house and to move out of the privately rented sector. She moved twice more between owner-occupied property in 1963 and 1985, on both occasions to be nearer to her son.

Many people received some assistance with the move from renting to owning, as illustrated by the example of Harold G. who was born in Beighton, Derbyshire in 1895. Harold's father was a coal miner, but in 1900 the family moved some 10 kms to the city of Sheffield where his father took work as a wagon agent. The main motive for the move was to provide a better education for their son in Sheffield. Harold remained at home until 1914 when, on the outbreak of war, he joined the army and spent most of the next four years in France. On his discharge in 1919 he lived at home in the same rented house that his parents had moved to in 1900, and trained as a teacher at Sheffield University, beginning his teaching career in Sheffield in 1921. Following his marriage in 1924 he left home, but only moved a short distance within Sheffield to another rented house. Harold moved twice more with his wife and four children in 1931 and 1935, on both occasions to get a larger house or garden. The first move was within Sheffield and the second to nearby Dronfield, and both properties were privately rented. However, in 1938 the family (now with five children) moved again to a larger house in Sheffield and this time they became owner-occupiers. The immediate reason why they were able to buy was the offer of financial assistance from Harold's mother-in-law, but the move also coincided with him becoming headmaster of a Sheffield secondary school and, presumably, the increase in salary also encouraged the commitment to owner occupation. Harold moved once more to a smaller house in Sheffield in 1960, and lived there until his death in 1983. This example illustrates that even families with relatively secure (though not particularly well paid) occupations such as school teaching often remained in rented property during the first half of the twentieth century, and that financial assistance from a relative was often the stimulus which led to a move into owner-occupation. All Harold's moves were very short distance and most were taken to improve his housing situation.

The other common tenure change in the twentieth century was movement into council housing. Two examples demonstrate how this tenure change fitted into life-time mobility patterns, and also suggest that for some council housing was a relatively temporary period in individual housing careers. Emily C. was born in Birmingham in 1890, the daughter of a railway porter. She left home at the age of 14 to work as an unpaid domestic servant with her aunt and uncle in Birmingham, and a year later moved to another family in the same city as a trainee nanny. At the age of 20 (in 1910) she moved families again and took a new position as a nanny looking after one child, also in Birmingham. She remained with this household until 1920 when, at the age of 30, she returned to her parents who were now living in rented property in Wilmcote, Warwickshire. The main reason for this move was her pregnancy. Emily married in 1921 and (with her husband and baby son) lodged with relatives of her husband also in Wilmcote, whilst awaiting a council house. Later the same year Emily and her family moved into a new council house in the nearby village of Binton, Warwickshire, where she remained for 16 years until in 1937 she moved with her husband and three children into privately rented property in the same village. In fact the house they moved into had previously been occupied by her father-in-law, and her husband was offered the tenancy after his father had died. She lived in the same house until her death in 1970. Thus Emily C.'s housing career included living with various families in connection with her work, lodging with relatives whilst awaiting council housing (which must have led to considerable inconvenience), and moving back into the privately rented sector when a house became available in the village in which they lived. For this family, at least, the privately rented sector had not totally lost its attractions in the 1930s.

The housing and migration career of William S., born in Bothwell, Lanarkshire, in 1891 shows some similarities with the previous family, but with an eventual move into owner-occupancy. William's father was a bunkersman on a mineral railway, and the family was lodging with relatives. However, in 1903 they moved to their own rented accommodation in the nearby town of Motherwell, and in 1905 moved again into Glasgow where William's father took a new job as a dairyman and William himself began work as a milkboy. However, in 1907 the family moved south to Crewe, leaving William (now an apprentice warehouseman for a drapery company) lodging with relatives in Glasgow. William continued as a lodger until 1914 when he was called up and served in the army until his discharge in 1919 when he resumed his work as a warehouseman. However, he was now again living with his parents and three siblings who had returned to Glasgow. In 1920 William married and rented a house in Glasgow until some five years later he took a new post with the drapery firm he worked for and moved to Leeds as a traveller in the drapery trade. The family lodged in Leeds, but when he was moved by his employers to Liverpool in the late-nineteen twenties they initially rented a house in Waterloo (north Liverpool), before gaining access to a council house in Eastham on the Wirral. However, William did not stay long on Merseyside because in 1931 he negotiated a move back to Glasgow (still working for the same Glasgow based drapers), because his wife was homesick. They initially lodged with in-laws in Glasgow, but within a year moved into owner occupation and bought a house in Paisley, where he died in 1959. Thus William lived variously in lodgings, privately rented housing, and (briefly) council housing before moving into owner-occupancy. The case study shows the ease with which families moved in and out of different tenures, and the role of kin in providing accommodation at crucial times. Only the last move into owner-occupation was stimulated mainly by a desire to improve housing.

Conclusion

This chapter has examined the links between migration and housing from three perspectives, focusing on the characteristics of all moves made mainly for housing reasons, on varying levels of persistence in different housing tenures, and on the characteristics and experiences of families moving into different housing tenures. The individual case studies vividly illustrate the way in which families could occupy a variety of housing tenures over their life-course, often choosing tenures to suit their employment and life-cycle circumstances, but with migration specifically to improve housing most likely to occur later in the life-course. In the twentieth century this frequently culminated in a move into owner-occupancy.

Research on migration in the late-twentieth century has explored the concept of housing and migration careers, suggesting that although many moves are necessarily determined by employment and family constraints, most households have a housing trajectory which includes a reasonably clear view of the type of housing towards which they are aiming. Not all moves fulfil this aim, but when the opportunity arises such families will move to improve their housing situation in a way which fits in with their housing career aspirations (Davies, 1991; Davies and Pickles, 1991). There is some evidence that this may also have been the case in the past. In the nineteenth century there was only limited scope for movement between housing tenures. Most people lived in privately rented housing, or in tied property occupied by virtue of their employment. For such families housing improvement was achieved by moving within the highly-differentiated privately rented sector, although some households did aspire to home ownership.

In the twentieth century there was much more tenure choice and the relationship between tenure change and housing careers is clearer. For most households migration in the early stages of the life-course was mainly stimulated by employment and family factors, with housing choices quite severely constrained by finance. Many people spent a period of their lives lodging with relatives. However, later in the life-course moves to improve housing became more common, and it is at this stage that it can be suggested that households were moving in line with longer term housing career aspirations. Of course, it is impossible to tell for how long such aspirations had been held, but in the twentieth century it is clear that for most people the desire to move into owner-occupation became a well defined aspiration during the inter-war period. Although some remained in privately rented or council rented housing, for others these were interim tenures on the way to becoming home owners.

As with other aspects of migration analyzed in this book, the relationship between housing and migration has proven to be extremely stable both over time and between regions. The same sorts of trends were occurring in all parts of the country, and the persistence characteristics associated with particular tenures in late-twentieth century studies have been shown to be applicable to earlier time periods. However, it is important not to overemphasize the significance of housing in structuring migration. In many ways decisions to move to different types of housing simply reflect opportunities and constraints provided by employment and life-course factors. Housing rarely acted as an independent variable, but interacted with and reflected other migrant's characteristics. Movement solely to improve housing occurred relatively rarely and for many reflected a comfortable stage of the life-course when other constraints had been minimized.

CHAPTER 8

Migration as a response to crisis and disruption

The role of crises in migration decisions

Some of the most visible migration flows in history have occurred as a response to crisis situations. Thus the removal of crofters from the Scottish Highlands following the clearances (Richards, 1982–5; Craig, 1990), Irish famine migration in the mid-nineteenth century (O'Grada, 1988; O'Sullivan, 1997) the mass movement of Jews from Central and Eastern Europe following persecution (Dotterer, Dash-Morre and Cohen, 1991) and, more recently, the exodus of refugees from the former Yugoslavia (Glenny, 1993; Black and Robinson, 1993) have all produced highly emotive pictures of migration as a traumatic and disruptive event stimulated by crisis conditions beyond the control of individual migrants. Fortunately, in the context of all migratory movement, such events are rare, but their impact on both individuals involved and on society in general can be out of all proportion to the numbers of migrants. It can also be suggested that many families and individuals experience their own personal crises due to events such as illness, the death of a partner, divorce or bankruptcy. Though much less public, these events too can stimulate migration and produce residential change which was much more traumatic for the individuals involved than movement for most other reasons. This chapter focuses on both personal crises and the much rarer externally generated events, and examines the extent to which they influenced migration patterns.

It must be stressed that the definition of a crisis is not straightforward. At a personal level different people may respond to particular situations in highly varied ways: an event which is a crisis for one person may have a much more minor effect on the life of another individual. The link between migration and personal crises is thus partly linked to individual personalities, a theme which is returned to in the conclusion to this chapter. Although major events such as war are likely to be widely viewed as producing crisis conditions, their very universality may make the impact of the event less severe. The fact that most people were in the same situation was likely to affect reactions to the crisis and, as demonstrated below, it was possible for many people to continue their lives much as normal during wartime and even find some enjoyment in the situation. Personal and family crises were also likely to have been viewed in rather different lights by various family members. Thus a planned move by a family to improve the career prospects of the main wage earner may be viewed as a major personal crisis for a teenage son or daughter who is loath to change schools and leave behind friends. The definition of a crisis is thus a personal and subjective process which varied between individuals and may have changed over time.

There are a number of factors that may influence the extent to which one or more events may become a crisis which stimulates migration for individuals and families. In some cases it may not be any one individual event which is crucial, but rather a combination of events which may cumulatively create a situation which is viewed as a crisis and which may lead to migration. Alternatively an unexpected crisis may act as a tipping point which stimulates a move which had been previously considered but not acted upon. It can also be suggested that reactions to particular situations may have changed over time. In the late-twentieth century we live in a highly regulated society which, despite recent changes in welfare provision, provides a wide range of support mechanisms. There are many different organizations designed to provide support for all manner of personal crises. In this context it can be suggested that many potential crises can be avoided and that help in dealing with misfortune is always available. In contrast, two centuries ago life was much more uncertain. Most families had few personal resources, they depended on kin and neighbours for support in times of trouble, and strategies for dealing with crises were more limited.

Exactly how these changes have affected the relationship between crises and migration is uncertain. It may be that in the past personal and other crises were a more normal and expected part of life and thus people adapted to them more easily than in the late-twentieth century when there are much stronger expectations for life to go according to a preconceived plan. The remainder of this chapter explores some of these issues by, first, focusing on personal crises that affected families and stimulated migration; and, second, through an analysis of the impact of two world wars on population migration. In each case analysis of aggregate patterns is followed by exemplification through a series of case studies. Other externally generated crises, including the effects of Highland clearances, Irish famine migration, and persecution of Jews in Central Europe are dealt with more briefly.

Migration due to personal crises

Not only is the definition of personal crises subjective, but the origins and impacts of crises overlap and interact with other causes of migration considered in earlier chapters. Thus many events which may have been viewed as crises could relate to family, employment or housing circumstances explored in Chapters 5, 6 and 7. In this section we focus on seven specific sets of events which occurred fairly commonly and which, under at least some circumstances, could be viewed as crises that may have stimulated migration. These events are severe personal illness, redundancy or bankruptcy, the death of a partner, divorce or separation, the death of one or more parents, the need for long-term care in old age, and severe family disagreements. The basic characteristics of residential moves associated with these causes are summarized in Table 8.1.

It is notable that in relation to the characteristics of the data set as a whole, migration as a result of personal crises was heavily dominated by women, particularly in the period after 1880. The total volume of movement due to crises also increased markedly after 1880 and the most important factors also changed. Before 1880 the crises most likely to stimulate migration were the death of a parent or the death of a partner; after 1880 personal illness was the most significant factor followed by the death of a partner. These trends obviously reflect changing levels of mortality from the eighteenth century, with fewer parents of young children dying in the twentieth century but more chronic illness as a higher proportion of the population lived into old age. But they may also be affected by the nature of the historical evidence. For the more distant past it is certainly easier to gain information on deaths than on illness, and thus the relative impact of mortality may be exaggerated. It is important to stress that particular personal crises did not always stimulate migration. Death of a spouse led directly to a move in about one quarter of cases where this information was recorded, death of a parent produced a move in about one fifth of cases, but divorce/separation generated migration in two thirds of instances recorded (Table 8.2).

Focusing on those cases where a personal crisis did stimulate migration, there are some interesting variations in the characteristics of people affected (Table 8.1) Women were more likely than men to migrate following the death of a partner, especially in the period before 1880, but they were less likely to move due to illness or redundancy/bankruptcy. These trends clearly reflect the fact that, for much of the period under study, most women were dependent on men (Dyhouse, 1989; Rendall, 1990). Thus in some cases the death of their husband could mean giving up the family home if it was linked to their partner's employment, and in many instances was likely to lead to a severe drop in income which could necessitate moving. As men were mostly the main wage earners their migration was more likely to be stimulated by long-term illness or unemployment. After 1880, women were more likely than men to move due to the need for long-term care in old age. This obviously reflects the differential effects of increased longevity, with women living, on average, longer than men.

259

Table 8.1 Migration for selected personal crises by migrant characteristics, distance and year of migration (%)

Migrant characteristics	Illness	Redundancy/ bankruptcy	Death of partner	Divorce/ separation	Death of parent	Care in old age	Left home after family dispute
1750–1879							
Gender							
female	21.7	19.5	59.1	44.8	34.0	30.4	34.6
Marital status							
single	27.1	36.4	0.4	35.7	90.4	8.7	75.0
married	58.8	57.4	8.0	28.6	9.0	34.8	20.8
other	14.1	6.2	91.6	35.7	0.6	56.5	4.2
Age							
<20	18.3	27.2	7.9	28.6	74.7	0.0	50.0
20–39	35.5	37.3	25.3	50.0	21.2	8.7	38.5
40–59	25.6	29.6	30.4	21.4	3.9	4.3	7.7
60+	20.6	5.9	36.4	0.0	0.2	87.0	3.8
Companions							
alone	27.8	6.8	28.5	35.8	15.7	22.7	42.3
couple	9.1	4.9	0.8	0.0	1.4	9.1	11.5
nuclear family	53.4	77.1	53.3	46.4	55.9	31.8	30.8
extended family	6.3	6.8	13.8	7.1	12.2	31.8	0.0
other	3.4	4.4	3.6	10.7	14.8	4.6	15.4
Distance band (km)							
<5	33.3	41.1	61.4	40.0	50.3	73.9	30.8
5–9.9	15.1	6.9	10.2	12.0	10.7	17.4	3.8
10–19.9	10.8	10.8	5.3	8.0	9.8	0.0	11.5
20–49.9	13.8	19.0	11.4	16.0	9.5	0.0	0.0
50–99.9	11.9	6.4	6.1	0.0	7.6	8.7	11.5
100–199.9	9.4	8.2	4.0	0.0	6.8	0.0	19.2
200+	5.7	7.6	1.6	24.0	5.3	0.0	23.2
Total number	184	169	254	29	365	23	26

1880–1994						
Gender						
female	49.4	41.6	62.2	51.0	49.2	67.6	54.3
Marital status						
single	22.6	27.2	0.4	13.7	77.6	6.0	56.9
married	48.7	69.5	7.1	26.1	20.4	20.3	26.4
other	28.7	3.3	92.5	60.2	2.0	73.7	16.7
Age						
<20	10.6	20.4	0.7	12.3	53.3	0.5	35.8
20–39	24.2	35.3	15.3	25.8	31.0	0.9	32.1
40–59	17.9	33.7	31.4	48.4	11.4	1.8	11.1
60+	47.3	10.6	52.6	13.5	4.3	96.8	21.0
Companions						
alone	39.9	7.5	43.0	41.4	17.5	65.8	46.3
couple	12.4	8.5	0.8	1.3	3.1	14.0	7.5
nuclear family	30.6	74.1	38.8	42.1	52.4	4.0	27.5
extended family	13.1	6.9	14.4	9.9	11.6	11.7	7.5
other	4.0	3.0	3.0	5.3	15.4	4.5	11.2
Distance band (km)						
<5	38.9	44.4	56.0	46.2	59.1	49.3	52.1
5–9.9	10.6	8.3	8.1	12.2	8.2	13.2	5.5
10–19.9	10.1	9.7	7.3	9.5	5.6	14.2	8.2
20–49.9	11.0	12.8	9.4	8.8	9.1	7.8	15.1
50–99.9	11.5	9.2	6.7	12.3	7.5	7.8	9.6
100–199.9	10.1	8.1	5.4	5.5	5.6	4.5	4.1
200+	7.8	7.5	7.1	5.5	4.9	3.2	5.4
Total number	860	377	539	155	465	222	81

Data source: 16,091 life histories provided by family historians.

Table 8.2 The frequency with which selected personal crises led to migration by birth cohort (% based on all recorded crises)

Personal crisis	1750–1849	1850–1930
Death of spouse	21.2	26.2
Death of parent	14.0	23.9
Divorce/separation	69.1	63.6

Data source: 16,091 life histories provided by family historians.

Variations by age and marital status are mostly obvious. The young and dependent were clearly most likely to move following the death of a parent, though the age range affected by such crises increased after 1880 reflecting the fact that many children remained dependent on their parents for longer in the twentieth century due to the demands of education. Migration following the death of a partner mostly affected older people, especially after 1880, reflecting greater longevity in the twentieth century, but migration following illness mainly affected young adults before 1880 but those over 60 after 1880. This is a graphic illustration of the changing impact of mortality as infectious diseases, especially tuberculosis (which mainly affected the young adult population in the nineteenth century) declined markedly from the 1880s, to be replaced by chronic illnesses which mainly afflicts the over-60s in the twentieth century (McKeown, 1976; Woods and Woodward, 1984). Whereas before 1880 only 20 per cent of those who moved due to illness were over 60, after 1880 over 47 per cent were 60 years old or over.

As with migration for other reasons there were very few variations between different parts of Britain, with each county having very similar trends and no obvious or consistent pattern to the variations that did occur. It is likely that any differences are due to chance factors reflecting the relatively small sample sizes in some categories when broken down by counties and specific crises. The distances over which people moved following a crisis were however distinctive, with all crisis moves taking place over rather longer distances than the average for the whole data set. There was also relatively little change over time (although crisis migration formed a particularly high proportion of long distance moves before 1880), and it can be suggested that in both time periods migration due to crises was a major factor stimulating longer distance moves. Whilst illness consistently generated the lowest proportions of short distance moves the same trends were present for all types of crises. It would thus seem that most crises that stimulated migration necessitated removal to a new community. These and other trends can be explored through a series of individual case studies of migrants affected by various personal crises.

John G. was born in Kirkcudbright (Scotland) in 1893 and lived with his parents in a tied cottage on the estate where his father was employed as a hedger. From the age of 18 he was employed as a solicitor's clerk, but suffered sporadically from ill-health which required a kidney operation in 1906. In 1913 at the age of 20 John left home and took a new post (as a clerk) in the small town of Sanquhar (Dumfries) some 80 kms from his parental home. However, after two years his illness forced him to return to his parents – he was unable to cope on his own – where he was unemployed for almost two years before again taking up clerical work in Kirkcudbright. Eventually in 1926 he married and established his own home in Kirkcudbright. This case demonstrates the importance of family in times of ill-health, and the way in which illness could stimulate return migration due to the inability to cope alone. More frequently, illness promoted migration later in life. For Agnes N. her declining years were affected by illness and the death of her partner, both of which influenced her migration patterns. Born in Aberdeen in 1870, Agnes left home and married at the age of 24. She had trained as a teacher, and initially taught in Aberdeen, but gave up her work on marriage. During her married life she lived in Barrow-in-Furness, Glasgow and Hull until in 1958, at the age of 88, she moved with her husband to Bristol. The reason for the move was to be near her daughter due to her husband's ill-health. In 1960 Agnes was widowed, and due to this and her own ill-health she moved into her daughter's home, dying there the following year. This example represents the increasingly common twentieth-century pattern of the 'crisis' of old age and ill-health which leads to, often long distance, migration to be near relatives who can provide appropriate care.

Men as well as women were affected by the death of a partner as illustrated by the case of Donald M. Born in the village of Rafford (Elgin) in 1874, Donald first assisted his father (who was a shepherd) before leaving home in 1900 to work for a grocer in Edinburgh. Seven years later he married and moved twice in Edinburgh to better housing before being called up in 1914. Donald served in France with the Royal Scots Regiment, but before his return in 1918 his wife died and he was left with three young children. He returned to work for the same Edinburgh grocer, but moved with his children to live with his mother-in-law in Edinburgh. The early death of a spouse created a crisis that could only be resolved by moving in with relatives who could care for Donald's three children.

In some instances the death of a parent could lead to small children leaving their other natural parent and going to live, for a time at least, with a more distant relative. Edith S. was born in Otley, Yorks, in 1888 the daughter of a railway clerk. By 1895 the family was living in Derby, but at the age of nine her mother died and she moved with her siblings to live with relatives in a Derbyshire village. Her father did not move with her and in 1904 he too died leaving her an orphan at the age of 16. Edith married at the age of 19 and lived in various houses in Rochdale and Congleton before her death in 1941. Elizabeth B. was affected by the death of a parent at an even earlier age. Born in 1898 in Binfield, Berkshire, Elizabeth's mother died two weeks after her birth when she moved with her father (a brick-maker) to live with her aunt also in Binfield. Elizabeth remained with her aunt until her father remarried and she moved to Windsor with her father and stepmother in 1906. Kezia W. was born in a Huntingdonshire village in 1757, the daughter of a butcher. However, when she was only seven years old both parents died and she moved with four siblings to live with a guardian in a nearby village. Here she remained until her marriage. In the absence of close relatives a guardian or family friend provided a secure home for the orphaned children. The pattern of personal crises created by the death of a parent or spouse leading to migration to live with a relative's family is repeated frequently in the life histories provided by family historians, and further emphasizes the importance of kinship links in migration (see Chapter 6).

Business failures were another frequent cause of migration, as illustrated by the life histories of Amos Kniveton and Henry Jaques related elsewhere in this volume (Chapter 5). One further example is given by Charles B. whose movement following business failure again illustrates the importance of kinship ties in helping people in times of crisis. Charles was born in Theale, Berkshire, the son of a master baker in 1864. He followed in his father's footsteps and in 1890 married and took over his father's bakery business near Theale. However, after ten years the business failed, and he was forced to move with his wife and seven children to the nearby town of Reading. Here they rented a house, but he was given employment in a bakery run by his father and brother. He continued to work at the same bakery until his retirement, living in various houses in Reading. Although Charles did not live with relatives, his choice of destination when he had to give up his home at the bakery was clearly affected by the offer of employment within the family.

As noted above, although personal crises often stimulated migration – much of which was assisted by other family members – crises could occur without migration taking place. For instance, migration immediately following the death of a spouse occurred in only a minority of instances. Whether or not migration occurred depended very often on the stage of the life-course in which bereavement took place, the number of dependent children to be considered, the health of the remaining spouse and, for women in particular, the impact of widowhood on income. By and large, men were much less likely to move following the death of a partner than women, as illustrated by the case of Isaac C. Born in a small village in Kent in 1804, Isaac married at the age of 23 and lived in a tied cottage working on an estate in Kent. He had been married for 19 years when his wife died in childbirth, leaving him with several small children. However, he did not leave his secure job and home on the estate and remarried the following year. He remained in the same house until the age of 62 when, in 1866, he moved with his second wife to an owner-occupied cottage in the nearby village of Worth. The combination of secure work and the opportunity to remarry and provide his children with a new mother meant that, for Isaac C., the death of a spouse caused less disruption to his life than it did for many others.

Migration due to external disruption

The two world wars in the twentieth century had a massive demographic and social impact on the population of Britain. Their effect on mortality can be seen in aggregate statistics, with crude mortality rates rising between 1914 and 1918, and the disruption to employment, housing and family structures caused major changes to British society (Winter, 1986; Wall and Winter, 1988; Pope, 1991). Although some changes, especially those during the First World War, were temporary – for instance, many women who entered the labour force between 1914 and 1918 gave up their jobs when men returned from the army (Braybon and Summerfield, 1987) – other changes may have had more permanent effects. It can be suggested that wartime conditions both stimulated migration at the time and contributed to longer-term change as people's horizons and attitudes towards travel were broadened. The extent to which wartime moves caused permanent or temporary change is assessed below. The migration patterns of many groups were affected by wartime conditions. For those in the forces, especially in the First World War, enlistment to fight in France was often the first time that they had travelled a substantial distance from home. Members of the armed forces moved around Britain and Europe to an extent which was previously unimagined. However, the two world wars also affected those left at home. The effects of bomb damage, especially during the Second World War, caused many to move, whilst others were evacuated from urban areas threatened with air raids. Many women migrated to take work in munitions factories or to work on the land, replacing male labour which had been called up, whilst in some families kin moved to share housing, thus easing the burden of missing male relatives (Summerfield, 1989). The extent and impact of such factors is examined through the aggregate statistics provided by family historians, and through a series of individual case studies.

The characteristics of those who moved due to wartime conditions are quite predictable (Table 8.3). Whilst members of the armed forces were mainly male, those forced to move due to bombing or evacuation were predominantly female. Whilst those evacuated or who moved due to bomb damage usually migrated in a family grouping, and members of the armed forces either moved alone or with others who were joining up, both men and women who moved within Britain for reasons connected with the war (but not in the forces) often migrated alone. This category included those

Table 8.3 Migration stimulated by World War One and World War Two by migrant characteristics and distance (%)

Migrant characteristics	Soldiers/armed forces personnel	For war reasons but no connection with armed forces	Bombed out/ evacuated
World War One			
Gender			
female	2.2	34.8	58.5
Marital status			
single	71.0	53.5	8.5
married	28.0	46.5	39.6
other	1.0	0.0	1.9
Age			
<20	30.7	15.2	35.8
20–39	64.2	71.7	47.2
40–59	5.0	13.1	15.1
60+	0.1	0.0	1.9
Companions			
alone	55.4	52.3	19.3
couple	0.6	6.8	1.9
nuclear family	3.0	29.5	59.6
extended family	0.2	0.0	11.5
other	40.8	11.4	7.7
Distance band (km)			
<5	7.3	13.8	20.8
5–9.9	0.9	3.4	4.2
10–19.9	4.5	6.9	8.3
20–49.9	18.2	13.8	14.6
50–99.9	19.1	10.3	27.1
100–199.9	19.1	27.6	10.4
200+	30.9	24.2	14.6
Total number	801	46	53
World War Two			
Gender			
female	10.6	32.7	52.1
Marital status			
single	58.2	42.6	23.6
married	41.1	48.9	66.2
other	0.7	8.5	10.2
Age			
<20	19.6	19.2	18.6
20–39	72.5	55.8	27.1
40–59	7.9	21.2	33.5
60+	0.0	3.8	20.8
Companions			
alone	54.6	52.9	12.5
couple	1.0	3.9	10.9
nuclear family	4.8	23.5	52.9
extended family	0.0	3.9	16.6
other	39.6	15.8	7.1

Table 8.3 (cont'd)

Migrant characteristics	Soldiers/armed forces personnel	For war reasons but no connection with armed forces	Bombed out/ evacuated
Distance band (km)			
<5	4.9	5.3	15.6
5–9.9	1.6	2.6	4.6
10–19.9	8.1	5.3	10.7
20–49.9	8.1	15.8	17.0
50–99.9	18.7	31.6	16.9
100–199.9	17.9	18.4	15.3
200+	40.7	21.0	19.9
Total number	509	52	582

Data source: 16,091 life histories provided by family historians.

moving to take new war-related employment or as a result of major disruption to household structures. All those moving as a result of the effects of war moved much further than most migrants, and it can be suggested that wartime conditions were a major cause of long distance migration in a period when the most common experience for all people remained short distance mobility. Differences between regions were relatively small, but the disproportionate effect of air-raid damage in London and South-East England is clearly reflected in the high rates of migration due to evacuation and bomb damage in these regions. These themes are illustrated through a series of individual examples.

Henry H., born in 1896 to a farm labourer living in Wargrave, Berks, was probably typical of many men called up during the First World War. He had moved twice, very short distances, with his parents and was employed as a men's hairdresser in Henley when he was called up in 1915 at the age of 19. He left home for the first time and served with his regiment (7th Royal Warwickshire) in France and Belgium before being wounded and hospitalized to Newcastle in 1918. Here he met his future wife, whom he married in 1919; he remained in Newcastle until 1921. However, he could not get employment in the North East and, in 1921, moved with his wife and daughter back to the same house and work in Henley that he had left before the war. Although moving three more times he remained in the same part of the country for the rest of his life. For Henry H. the effect of the First World War was to both remove him from home for the first time and to take him to the other end of the country as an invalid. However, despite his marriage in the North East, he returned to his home area due both to employment opportunities and family ties. Thus, in this case, the impact of the First World War was quite temporary as by 1921 Henry had returned to a pattern of living that he might well have followed had the war never occurred.

Although few women joined the forces during the First World War, many did move around the country to take work vacated by men. Kate H. was typical of such women. Born in 1897 in Southampton, the daughter of a diver, she moved four times with her parents (due to her father's work) to Kent, Gloucestershire and Monmouthshire (where she was employed as a domestic servant). In 1914, at the age of 17, she left home with her sister to move to Birmingham to work in a munitions factory during the First World War. In 1919, she returned to her parental home (now Newport), but married soon after and lived in various houses in South Wales for the rest of her life. She did not take any further paid employment after her marriage. Kate H.'s move to Birmingham was clearly stimulated by the recruitment of women to work in munitions factories during wartime, but the effect of this was tempered by the fact that she moved and lived with her sister, and at the end of the war she not only returned to her home area but left the labour force. Wartime conditions had quite a temporary impact on this individual's life patterns.

The residential moves of Percy W. also suggest that for many the effects of the First World War were a temporary interruption to more usual migration patterns, although this example was very much a minority experience. Percy W. was born in Battersea, London, in 1893, the son of a bricklayer's labourer. He moved very frequently during his life, mostly in and around London, South Wales and the Welsh Borders (his parents moved to Herefordshire in 1907). However, two of the longest distance moves of his life were made during wartime. At the outbreak of war in 1914 Percy W. was living at home in Glamorgan, working for a grocer. However, as a conscientious objector he refused to be called up and in 1917 was imprisoned for six months in Wormwood Scrubs. Following his release he was rearrested in 1918 and sentenced to two years in Walton Gaol, Liverpool. However, he was released in 1919 and returned to live with his family in Monmouthshire where he gained work as a ledger clerk. Although Percy W. moved frequently during the rest of his life, including moves to Birmingham and London, most of his migration was relatively short distance. His wartime imprisonment took him away from his family and home area, but on his release he returned immediately to the area and people that he knew.

Although some families experienced bomb damage during the First World War, this was much more common during the conflict of 1939–45, particularly in London. Doreen G., born in Poplar, London, in 1930 moved twice as a child due to the effects of air raids. In 1940 her parents' rented house in Poplar was destroyed by a bomb, and the family moved slightly further out of London to another rented property in Ilford, Essex. However, the following year that house was also damaged by bombs, and they moved again to a house in a nearby street. Although this was obviously temporary accommodation gained at short notice, the family remained in this rented house until 1947 when it was required by the owners. She then moved with her parents into an owner-occupied house on the outskirts of London. For this family although bomb damage forced movement, they remained in much the same area and it did not significantly disrupt patterns of living and employment. Many adults and children left London during the Second World War, either moving to live with relatives or, and especially in the case of children, to be billeted with families in the countryside. Patricia H. had this experience between the ages of 11 and 15. Patricia's father had died when she was six, prompting a move to a smaller flat in the Highbury district of London. However, on the outbreak of war in 1939 the 11-year-old girl was evacuated with other pupils at her school and lodged with a family in Ashstead, Surrey. The following year she moved to new lodgings in Ashstead but, soon after, all pupils from the school were moved to what was considered to be a safer location in Keighley, Yorkshire. She remained in Keighley until 1943 when she returned to her mother in Highbury, London, where she remained until leaving home for higher education at the age of 18. For this teenage girl the war had a major effect on her life, removing her from her family and home area, but she too returned to pre-war patterns of life as soon as it was deemed safe to do so.

The impact of the Second World War on a London resident, and particularly the extent to which normal patterns of behaviour were retained despite the occurrence of air raids, is illustrated by the detailed diary entries of Rhona Little (mentioned in Chapter 4). On the outbreak of war Rhona was living with her sister in lodgings in the Canonbury district of London, having migrated to London from Londonderry at the age of 18 in 1938. The war did have some immediate effects on her life and residential location as her landlady evacuated herself to Wales, leaving the house empty, and Rhona, her sister and a friend were forced to find new lodgings. In fact this was very easy as there were many empty properties and they rented a room in an adjacent street from a landlady who had previously taken only male lodgers but realized that these would be in short supply during the war. Soon after this move Rhona's sister was evacuated with her office to Morecambe (Lancashire), but Rhona herself was adamant that she would stay in London. This was a consistent view which she first recorded in her diary in September 1938:

> Nearly all the Irish and Scotch girls say that if there is a war they are going home. Did you ever hear of such a cowardly lot! There are, however, a few people like myself who would stay here.

Apart from the disruption caused to transport in the city due to bomb damage, Rhona Little did not seem to find the effects of air raids too disruptive or disturbing. She describes with 'great excitement' the night a bomb fell close to her house, and she seems to enjoy the sense of community engendered through her voluntary work as an air raid shelter warden. She also seemed to enjoy fire fighting following air raids recording in March 1942:

> Lovely day of putting out fires, crawling about and holding branches from the hydrants and motor pumps. Beautiful time.

Although recording some sense of fear and weariness as the war progressed, it is clear from the diary that for Rhona and many others in London the impact of air raids was minimized by a determination to ensure that most aspects of life went on as normal. Although forced to move lodgings, Rhona remained in the same part of London and showed no interest in returning to Londonderry, even when the opportunity was presented to her.

In all of these individual examples the theme which comes through most strongly is that, although both the First and Second World Wars fundamentally changed people's lives and frequently stimulated unusually long distance migration which almost certainly would not have taken place in other circumstances, for most people this pattern of disruption was temporary. Despite experiencing new places, and in theory having much broader horizons, many people returned to their home areas following wartime migrations; many women who had worked during the First World War left the labour market when the war ended; and very often movement during wartime was the only long distance migration undertaken during a lifetime. Although external crises such as war clearly were disruptive and stimulated unusual migration patterns, their long-term impact should not be over-emphasized.

Whereas the impacts of the two world wars in the twentieth century were, to some extent, felt by all people in all parts of Britain, other external crises had more local significance. Due to the smaller numbers affected, and the fact that (as explained in Chapter 2) migrants from Ireland to Britain were under-represented in the sample, numbers are too small to present meaningful aggregate statistics relating to events such as the Highland clearances or Irish famine migration. However, four selected case histories illustrate some individual impacts of such moves. Angus M. was born on Barra, in the Outer Hebrides in 1819 but in 1830 the family were forced to move when the Eoligarry peninsula was cleared for a sheep farm. Initially they migrated only a short distance and his father took a croft elsewhere on Barra, but in the 1840s Angus left home and gained work as a farm servant in Perthshire, living with his employer. Alexander M., born in Clyne, Sutherland in 1807, was also affected by clearances in 1825 when his family was forced to give up their farming tenancy. However they moved only about 20 kms down the coast to Dornoch where he eventually became a tenant farmer. Many families affected by the Highland clearances did move over long distances to more prosperous farming districts such as Perthshire or to urban areas such as Glasgow, and many others emigrated (Richards,

1982–5; Craig, 1990; see also Chapter 9). But as these examples show, for at least some families, the response to adversity was to move, at least initially, only a relatively short distance. Ties of family and locality were clearly strong and such commonplace short distance migration can easily get lost amongst more dramatic accounts of long distance movement following events such as the Highland clearances.

The impacts of Irish famine migration have been well-researched, even if the precise causes and extent of the famine are still debated (O'Sullivan, 1997). The population of Ireland fell by 20 per cent 1841–51 and some 1.5 million people emigrated between 1845 and 1852, mostly bound for Britain or North America (Davis, 1997). Two case studies illustrate the way in which this localized but large scale crisis emigration interacted with life-time migration patterns in Britain. Jeremiah N. was born in Newcastle West, Limerick, in 1817, the son of a smallholder. He continued to live in the same place, employed as a farm worker, until in 1844 he was pushed by the first year of potato crop failures to emigrate alone to England. He probably sailed from Limerick to Bristol and gained work as a navvy constructing the Bristol to Taunton railway through Somerset. However, after a year he married and moved to London where he joined the Metropolitan police. He remained as a police constable for the rest of his life, living in nine different houses in and around the Poplar district of London until his death in 1885. Myles D. was born in Blackwater, County Wexford, in 1830, the son of an agricultural labourer. In 1847 he sailed with his family to Liverpool to escape the effects of the potato famine and, like many Irish emigrants, gained work on the Liverpool docks. Myles D. remained in Liverpool all his life, continuing to work as a dock labourer and living in at least four different houses (probably more) in central Liverpool until his death in 1912. What these two examples clearly demonstrate is that although the move from Ireland to England as the result of successive failure of the potato harvest was a disruptive long-distance move stimulated by a crisis event, thereafter the residential life history of such migrants was very like that of all other migrants. Once settled in England they moved over short distances within a single community and, as such, crisis migration again had only a temporary effect on the long-term residential history of individuals.

Conclusion

Whilst major crises such as wars or famines, fortunately, occurred infrequently, most families have been afflicted by personal crises of some sort during their life-course. Although the perception of what constituted a crisis, and hence the reaction to that event, varied from individual to individual, crises often interacted with other factors to stimulate migration in the past. For many people such moves were also over much longer distances than the residential changes which more frequently occurred. However, what emerges most clearly from an analysis of the ways in which crisis migration fits into life-time residential histories, is the temporary effects of such events. Even the disruptions caused by war, often leading to long distance migration to places which would not otherwise have been visited, did not seem to have any long-term effects on subsequent migration patterns. Once the crisis was over, the residential history of those affected was much the same as that of all other families: dominated by short distance moves within particular neighbourhoods.

It seems likely that individual personality was one of the most important factors in explaining reactions to particular crisis situations. Whereas some people would have had a natural inclination to be mobile, and thus reacted to adversity by seeking to improve their circumstances elsewhere; others would have had a preference for stability and were thus more likely to react to a crisis by remaining in the place with which they were familiar in the hope that conditions would improve. Such personality differences may go a long way towards explaining why various families faced with similar circumstances reacted in rather different ways. Unfortunately such behavioural factors are extremely difficult to discern in the past, and explanation based on such factors is often little more than speculation. However, it is possible to gain fuller insights into the migration decision from some diary evidence.

As has been demonstrated already (Chapter 5), the life history of Benjamin Shaw provides detailed information about mobility of members of the Shaw family. One of the themes which emerges from a detailed analysis of the life history is the extent to which various members of the Shaw family reacted to events in different ways. Joseph Shaw, father of the author of the life history, was clearly a person who reacted to adversity by moving elsewhere. He moved house a total of ten times during his life and, whenever trade was bad or work was not available he was happy to move elsewhere (albeit within a fairly circumscribed area of Westmorland, Yorkshire and North Lancashire). However, Joseph's wife (Isabella) did not share this willingness to move, and even a relatively short distance move from Dent to Kendal (25 kms) led to her feeling homesick and unsettled. After a short time the family returned to her home settlement of Dent at her request, despite the fact that employment prospects were probably better in the small town of Kendal. In many respects Benjamin Shaw took after his mother. After he left home he only moved when he absolutely had to and, even when made redundant from a rural mill in Dolphinholme and with no real prospect of alternative work in the neighbourhood, his initial reaction was to stay in the village in the hope that work would become available. In the event he gained work in the Lancashire town of Preston (where his wife had worked for a time), and apart from a short period when they were removed by the Overseers of the Poor, Benjamin Shaw and his wife remained in the same house in Preston for the rest of their lives. This stability occurred despite considerable personal difficulties including unemployment and serious illness.

The implication to be taken from the life history is that Benjamin Shaw and his mother both shared a preference for stability, and thus their reaction to most crises was to sit tight and hope that things would improve. In contrast, Joseph Shaw had a preference for mobility. At various periods of his life he worked as a weaver and travelling watch and clock repairer, which required him to move over considerable distances, and his reaction to most problems seemed to be to get up and move and try something different. Clearly it would be over-simplistic to either read too much into one life history, or to extrapolate to other families from such limited evidence. However, it does seem likely that some people were more prone to mobility than others, and that such personality characteristics had a bearing on how they responded to particular crisis situations.

CHAPTER 9

Overseas migration, emigration and return migration

Introduction to the history of emigration

Migration from Britain to another country could occur for many different reasons: employment prospects, family ties, marriage and the impact of specific crises could all stimulate such moves. The factors considered in Chapters 5 to 8 are thus all relevant to overseas migration which, in effect, was no more than an extension of long distance internal migration. However, the nature, characteristics and impacts of overseas migration are sufficiently distinctive to merit separate consideration. The term emigration is not easy to define precisely. It is usually taken to imply a move to another country in which there was an intent to settle permanently. However, the definition of permanent settlement is always nebulous in migration studies and intentions are very hard to discover for past events. Furthermore, the reasons given by family historians for movement overseas – which were gleaned from a range of public and personal documentary records – are also confusing. In some cases 'emigration' was stated as a reason for moving, in other cases more specific reasons such as work or to join other family members were stated as the reasons for migrating overseas. Whilst some such moves were permanent, and in any definition would be classed as emigration, in other cases they included temporary (though sometimes prolonged) periods of overseas employment during which there was always an intention to return to Britain. Moreover, some 'emigrants' also returned to Britain and others who initially moved temporarily eventually settled overseas. To avoid making arbitrary distinctions, in this chapter all migration from Britain to other countries is included, with the exception of moves undertaken when employed in the armed services. Clearly movement from Britain to France with a regiment during the First World War was not emigration! Return migrants are all those (excepting people moving with the armed services) who moved from Britain overseas and, at some time, returned to live in Britain for a period of time. The life-time migration histories collected by family historians allow overseas migration to be placed within a broader context, relating it to both the pattern of internal migration which occurred before an overseas move and to the incidence of return migration.

In many respects the history of emigration from Britain (and from the rest of Europe) is rather better researched than the history of internal migration. One reason for this is the survival of a much more complete range of sources for the study of emigration. The combination of ships' passenger lists, immigration records, the archives of emigration companies, and personal letters and accounts from emigrants, together with more conventional census and related sources, has provided rich raw material for a range of detailed studies which include complex econometric analyses and richly-detailed individual accounts of the emigration experience (Baines, 1985, 1994; Erickson, 1972, 1994). Because migration overseas was both a major event for the individual emigrant, and had implications for the governments of countries sending and receiving such migrants, movement overseas was recorded much more fully in official documents and personal papers than the more commonplace and everyday events of internal migration. It is not the purpose of this chapter to duplicate this research but, rather, to use the material provided by family historians to examine some aspects of overseas migration and emigration from new perspectives. Following a brief review of current research on emigration, and the formulation of hypotheses to be tested in this chapter, we focus particularly on the characteristics of overseas migrants and return migrants, the relationship of emigration to the internal migration process, and the nature of the emigration decision.

There has been a long history of overseas migration from Britain and the rest of Europe, linked to exploration, settlement, colonization and exploitation of different parts of the world (Lucassen, 1987; Baines, 1991; Moch, 1992; Hoerder and Moch, 1996). However, the volume of overseas migration and the range of destinations increased dramatically during the nineteenth century, with migration overseas a major feature of Victorian society. From the 1840s to the 1930s Britain regularly lost more people through migration overseas than it gained through immigration, with the peak decades for out-migration occurring towards the end of the nineteenth century. North America attracted by far the largest volume of emigrants due both to its relative accessibility and the long history of British and Irish settlement in the USA and Canada. However, from the 1840s movement to Australia and New Zealand was increasingly important, with South Africa also attracting substantial numbers of migrants by the late-nineteenth century. In the twentieth century immigration controls have restricted entry to many traditional destinations, but overseas migrants from Britain have gone to wider range of destinations including South America, Asia and, increasingly, continental Europe (Carrier and Jeffery, 1953; Baines, 1985, 1994; Erickson, 1989, 1990; Constantine, 1990; Devine, 1992; Johnston, 1993; Haines, 1994).

276

Although emigrants came from all parts of Britain, some counties lost far more people than others. Remote rural districts which were experiencing economic decline (especially the Highlands of Scotland and South-West England) lost a large proportion of their populations, but the greatest numbers of emigrants came from large urban areas with surplus labour (Baines, 1985). Most emigrants responded to a combination of economic and social push factors and the allure of imagined opportunities elsewhere, but emigration was also encouraged by government with migration overseas seen both as a means of establishing control in the colonies and of relieving economic and social pressures at home (Constantine, 1991; Erickson, 1994; Baines, 1994). Whilst most movement to North America was undertaken by individual families making their own arrangements for transport and eventual settlement, the greater distance and lesser knowledge of Australia and New Zealand meant that much emigration to these countries was organized through emigration companies who could provide assisted passages and organize settlement in the Antipodes. These companies could select emigrants and regulate the characteristics of those who moved (Richards, 1993; Haines, 1994; Hudson, 1996).

Given the large numbers of emigrants that left Britain during the nineteenth century, it is not surprising that they included people of many ages and from all walks of life. However, there was a tendency for emigrants to be relatively young adults with emigration companies seeking to avoid transporting large families. Young couples with skills relevant to the society to which they were moving were the most cost-effective emigrants from the point of view of emigration companies moving people to Australia and New Zealand. Migration to North America included a wider cross-section of society, with a large volume of relatively poor and unskilled migrants from Ireland during the famine years, but even in these circumstances it can be suggested that those Irish who made the crossing were rather better off than the Irish who moved to English or Scottish cities. The migration of family groups was important for all destinations, and there is also evidence of chain migration with many people who apparently moved alone actually migrating to join relatives (Carrier and Jeffrey, 1953; Erickson, 1990; Haines, 1994). Whilst many families who emigrated settled quickly in a new society, others were disappointed, and there is considerable evidence that a proportion of emigrants returned to Britain after relatively short periods. For those who moved to New Zealand or Australia, the cost of returning and the thought of a second uncomfortable sea journey of some six months must have been daunting. However, the nature and extent of return migration in the past is a relatively under-researched topic (Shumsky, 1995).

Whilst the broad dimensions of emigration from Britain in the nineteenth century are now quite well known, with good data on the volume, direction, characteristics and experiences of emigrants, there are a number of themes on which the data provided by family historians can provide fresh evidence. Although Baines (1985) has used published census data to try to link internal migration with emigration, most studies have examined emigration in isolation without linking it to the life-time migration experience of emigrating families. This chapter therefore addresses a series of specific questions about the nature of overseas migration and return migration, and their links to the life-time migration experiences of individual migrants. First, how did the decision to emigrate relate to the previous migration history of individuals? Were those who moved overseas more or less likely to have been highly mobile within Britain prior to their emigration? Second, did those who moved overseas have different characteristics from internal migrants? It can be suggested that emigration was simply an extension of long distance internal migration and thus the characteristics of emigrants and long distance internal migrants should have been similar. Third, was the decision to emigrate seen as a more significant decision, and taken in a different way, from decisions about internal migration, or were emigration decisions taken for much the same reasons as long distance internal movement. Fourth, did those emigrants who returned have any special characteristics, or were they a cross-section of all overseas migrants? Finally, what were the factors that contributed to the decision to return to Britain after emigration? It is sometimes suggested that returned emigrants had somehow 'failed' in their emigration venture, but it is rarely implied that a migrant who returned to a previous home after an internal move had failed. It seems likely that the decision to return was as likely to be taken for positive as negative reasons, and that rather than viewing return emigrants as failures, they were simply part of a global migration system in which some people circulated around the world as easily as others moved frequently within a circumscribed local area.

There is no a priori reason why internal migration, overseas migration, and returned emigrants should be treated as separate and distinct. Migration theory concerning the decision to migrate and the way it is related to distance, opportunities and family life-cycle stage is as relevant to overseas migration as it is to short distance internal movement. Analysis in this chapter is designed to stress the links between different types of migration, and place the emigration decision within the context of a life-time of residential migration.

The characteristics of emigrants and return migrants

Analysis presented in this section serves two principal purposes. First, the characteristics of overseas migrants in the sample collected from family historians can be compared with the evidence (outlined above) from other aggregate studies to assess any biases in the data. Second, and most import-

Table 9.1 Emigration and overseas migration by destination and year of migration (%)

Destination	1750–1839	1840–79	1880–1919	1920–94	All moves
Ireland	10.3	5.7	5.8	4.7	5.8
USA & Canada	39.8	34.6	44.7	19.5	34.5
South/Central America	6.8	1.7	4.2	6.9	4.5
Africa	4.1	3.5	8.3	17.3	9.2
Australia/New Zealand	16.4	35.9	16.5	18.8	22.6
Asia/Middle East/India/ Russia	5.5	6.5	8.0	13.3	8.8
Eastern Europe	0.7	2.2	0.8	1.3	1.3
Western Europe	12.3	6.8	9.8	15.1	10.6
Other	4.1	3.1	1.9	3.1	2.7
Total number	146	459	589	451	1,645

Data source: 16,091 life histories provided by family historians.
All moves made by armed service personnel are excluded.

ant, the characteristics of emigrants and return migrants are compared with each other, and with the characteristics of all internal migrants to examine the extent to which emigrants, return migrants and internal migrants were similar to one another. It should be stressed again that the data relate to all overseas moves (excepting those made with the armed forces) and, at this stage, no assumptions are made about intentions for permanent or temporary settlement. It should also be emphasized that the data set included information from family historians currently living overseas, and thus the sample includes the descendants of those who settled and remained overseas as well as those who returned.

The distribution of overseas migration by time period and destination (Table 9.1) broadly reflects the pattern demonstrated in other studies of emigration from Britain. Overall, North America was the most important destination (34.5 per cent of all overseas moves), followed by Australia and New Zealand (22.6 per cent). However, migration to the Antipodes peaked in the period 1840 to 1879, but movement to North America in the period 1880 to 1919. In the twentieth century overseas migration destinations were much more widely dispersed with movement to the former colonies greatly reduced but with increased migration to Africa, Asia and Western Europe.

The characteristics of overseas migrants varied a little between different destinations, and differed in some respects from the characteristics of all migrants (Table 9.2). Overseas migrants to most destinations were more likely than all migrants to be male, with the exception of those moving to Asia and Western Europe. Movement to these areas increased significantly in the twentieth century suggesting that whereas nineteenth-century colonial emigration was male dominated, twentieth century movement to other

Table 9.2 Characteristics of emigrants and overseas migrants to different destinations (%)

Migrant characteristics	Ireland	USA/Canada	Central/South America	Africa	Australia/New Zealand	Asia/Middle East/India/Russia	Eastern Europe	Western Europe
Gender								
female	25.0	30.0	26.5	29.5	27.5	39.3	27.3	38.3
Age								
<20	27.7	18.1	6.8	9.4	19.7	13.1	13.6	23.4
20–39	50.0	56.3	67.6	65.8	55.1	65.5	81.8	60.0
40–59	19.1	20.7	18.8	21.5	20.6	18.6	4.6	11.4
60+	3.2	4.9	6.8	3.3	4.6	2.8	0.0	5.2
Marital status								
single	49.4	38.6	39.3	41.0	41.2	43.4	55.0	58.9
married	48.2	57.1	57.4	54.5	55.9	53.7	45.0	38.7
other	2.4	4.3	3.3	4.5	2.9	2.9	0.0	2.4
Companions								
alone	28.7	27.5	38.9	34.9	26.7	43.8	25.0	49.1
couple	6.4	9.9	19.4	26.0	10.4	17.4	20.0	14.0
nuclear family	48.9	43.3	29.2	29.5	48.9	26.4	30.0	23.4
extended family	4.3	9.2	4.2	0.0	4.8	0.7	5.0	1.2
other	11.7	10.1	8.3	9.6	9.2	11.7	20.0	12.3
Position in family								
child	16.8	12.8	8.3	6.8	12.9	6.9	9.5	8.2
male head	57.9	55.0	59.7	58.9	56.0	45.8	57.1	49.1
wife	8.4	14.6	16.8	16.4	13.4	18.8	14.3	11.1
female head	5.3	6.2	6.9	10.3	6.7	16.7	0.0	20.5
other	11.6	11.4	8.3	7.6	11.0	11.8	19.1	11.1
Total number	96	568	74	149	371	145	22	175

Data source 16,091 life histories provided by family historians.

parts of the world was not. Migration to all overseas destinations was more likely to consist of young adults (age 20–39) than all migration and was much less likely to include those aged over 60 years. In most cases children under 20 years were under-represented, the only exceptions being movement to Ireland and the rest of Western Europe. This again suggests that migration to European countries was more similar to internal migration than emigration to more distant locations. However, all overseas migrants were more likely to be single and to have moved alone than internal migrants, with single migrants most dominant (and thus different from internal migrants) amongst movement to Western European destinations. Lone migrants were also most common to Western Europe (excluding Ireland), Asia and South America, with nuclear family groups most commonly moving to Ireland, the Antipodes and North America. Thus the large scale emigrations of the nineteenth century (to the USA, Canada, Australia and New Zealand) were most likely to consist of family groups, though slightly less so than all internal migrants. Their characteristics were, however, quite similar to those of longer distance internal migrants (Chapter 4).

Although in some cases sample sizes become rather small when emigrants are broken down by their regional origins, none the less some interesting patterns emerge (Table 9.3). There were marked regional differences in the relative importance of emigration and internal migration. Emigration was most significant amongst movement from the Scottish Highlands, but the ratio of emigration to internal migration from a region was also high for North-East Scotland, South-East Scotland and North-West England. These figures reflect both levels of economic prosperity and the availibility of sailings from west coast ports. However, it should not be assumed that all those who emigrated from the North West had lived there previously for long periods as some people may have moved to Liverpool and lived there for a short period before embarking for North America. There were also some regional differentials in the destinations to which people sailed. North America was over-represented amongst emigrants from North-West England, Yorkshire, South Wales and all of Scotland, whereas the Antipodes were over-represented among emigrants from South-West England, Highland and South-West Scotland, and Northern England. Mainly twentieth century destinations such as Asia and Western Europe were particularly common amongst migrants from the whole of South-East England.

Examination of the stated reasons for overseas migration provide some additional explanation of these trends (Table 9.4). The main distinction is between those destinations where the main reason was simply 'emigration', and those where movement was primarily for employment reasons. As discussed above, although much emigration was undertaken to improve employment prospects, the decision to move permanently to North America, Australia or New Zealand was likely to have been related to more than a simple job opportunity, possibly reflecting broader dissatisfaction with life

Table 9.3 Regional variations in emigration and overseas migration (%)

Region of origin	Ireland	USA/ Canada	Central/ South America	Africa	Australia/ New Zealand	Asia/Middle East/India/ Russia	Eastern Europe	Western Europe	Other	Total number	Ratio emigration to internal out-migration
London	12.0	19.1	6.0	8.7	20.8	7.7	3.3	18.0	4.4	183	8.0
South East	7.4	15.7	8.4	7.9	21.3	20.2	0.0	15.7	3.4	178	9.4
South Midland	2.6	24.8	6.8	11.1	25.6	12.0	0.9	13.7	2.5	117	7.1
Eastern	4.7	17.2	1.5	9.4	29.7	10.9	3.1	18.8	4.7	64	6.1
South West	3.1	34.6	2.3	11.5	37.7	3.8	1.6	3.8	1.6	130	12.2
West Midland	10.9	18.5	2.5	15.1	31.1	12.6	0.0	9.3	0.0	119	9.7
North Midland	0.0	35.1	0.0	12.2	19.3	5.3	3.5	22.8	1.8	57	6.3
North West	4.1	48.6	4.9	6.8	20.2	4.1	2.3	6.3	2.7	222	17.8
Yorkshire	2.0	58.8	1.0	6.9	12.7	11.8	2.0	3.9	0.9	102	8.9
Northern	4.2	34.7	2.1	9.5	28.4	7.4	0.0	12.6	1.1	95	12.6
Wales*	4.2	42.3	11.3	4.2	18.3	5.6	0.0	5.6	8.5	71	12.1
Highland	0.0	50.0	4.2	4.2	33.3	0.0	0.0	8.3	0.0	24	41.7
NE Scotland	1.8	63.6	5.5	9.2	14.5	3.6	0.0	1.8	0.0	55	22.6
Scottish Midland	2.4	63.4	2.4	0.0	7.3	12.3	0.0	7.3	4.9	41	6.1
SW Scotland	7.5	47.5	2.4	7.5	32.5	1.3	0.0	0.0	1.3	80	13.4
SE Scotland	11.9	52.9	0.0	2.9	8.9	2.9	0.0	17.6	2.9	34	19.6
Southern Scotland	25.0	41.7	0.0	8.3	8.3	0.0	0.0	16.7	0.0	12	5.0

* All census regions in Wales are combined as the numbers involved are very small.
Data source 16,091 life histories provided by family historians.

Table 9.4 Reasons for overseas migration and emigration by destination and year of move (%)

Reason	Ireland	USA/Canada	Central/South America	Africa	Australia/New Zealand	Asia/Middle East/India/Russia	Eastern Europe	Western Europe
1750–1839								
Work	73.3	5.2	20.0	16.7	4.2	87.5	100.0	50.0
Marriage	0.0	0.0	0.0	0.0	0.0	0.0	0.0	0.0
Family	0.0	0.0	0.0	0.0	0.0	0.0	0.0	0.0
Crisis	0.0	3.4	10.0	0.0	0.0	0.0	0.0	0.0
Retirement	6.7	0.0	0.0	0.0	0.0	0.0	0.0	0.0
Education	0.0	0.0	0.0	0.0	0.0	0.0	0.0	11.1
Emigration/better prospects	20.0	91.4	70.0	83.3	95.8	12.5	0.0	38.9
Total number	15	58	10	6	24	8	1	18
1840–79								
Work	53.8	10.1	75.0	31.3	4.2	63.3	60.0	71.0
Marriage	7.7	0.6	0.0	0.0	0.6	0.0	10.0	0.0
Family	7.7	0.6	0.0	6.3	0.0	6.7	10.0	0.0
Crisis	15.4	1.3	0.0	6.3	3.6	0.0	0.0	0.0
Retirement	0.0	0.0	0.0	0.0	0.0	0.0	0.0	0.0
Education	3.8	0.0	0.0	0.0	0.0	0.0	0.0	9.7
Emigration/better prospects	11.6	87.4	25.0	56.1	91.6	30.0	20.0	19.3
Total number	26	159	8	16	165	30	10	31

Table 9.4 (cont'd)

Reason	Ireland	USA/Canada	Central/South America	Africa	Australia/New Zealand	Asia/Middle East/India/Russia	Eastern Europe	Western Europe
1880–1919								
Work	61.8	9.5	68.0	24.5	11.3	57.4	60.0	44.8
Marriage	0.0	1.9	4.0	4.1	2.1	12.8	0.0	3.4
Family	5.9	1.9	4.0	6.2	4.1	4.3	0.0	1.7
Crisis	5.9	2.7	16.0	2.0	5.2	4.3	0.0	5.3
Retirement	8.8	0.4	0.0	0.0	0.0	2.1	0.0	3.4
Education	5.8	0.8	0.0	2.0	0.0	0.0	0.0	20.7
Emigration/better prospects	11.8	82.8	8.0	61.2	77.3	19.1	40.0	20.7
Total number	34	263	25	49	97	47	5	58
1920–94								
Work	71.3	13.6	73.6	73.2	18.7	73.3	83.3	42.7
Marriage	0.0	2.3	5.9	3.8	0.0	6.7	0.0	1.5
Family	9.5	4.5	2.9	3.8	1.2	3.3	0.0	4.4
Crisis	4.8	2.3	0.0	0.0	2.4	1.7	0.0	4.4
Retirement	4.8	0.0	0.0	0.0	1.2	1.7	0.0	8.8
Education	4.8	0.0	0.0	0.0	0.0	0.0	0.0	13.2
Emigration/better prospects	4.8	77.3	17.6	19.2	76.5	13.3	16.7	25.0
Total number	21	88	34	78	85	60	6	68

Data source 16,091 life histories provided by family historians.

in Britain. Thus it can be suggested that whereas most moves to Ireland, South America, Africa, Asia and Europe were clearly work related and had varying degrees of permanence (especially in the twentieth century), moves to North America and the Antipodes were undertaken for a more complex range of factors, and there was at least the intention of permanent settlement. It is also notable that, in relation to internal migration, movement overseas was much more likely to be stimulated by family related factors, especially in the case of movement to Africa, Australia, New Zealand and North America.

The amount of return migration also relates to the extent to which movement was stimulated by job opportunities or was part of a broader process of emigration and settlement (Table 9.5). Thus those who migrated to Ireland, South America, Africa, Asia and the rest of Europe had a very high rate of return (in excess of 63 per cent of overseas migrants returned to Britain), whereas return migration from North America and the Antipodes was much lower (20–25 per cent). Although distance and communications were clearly factors which affected the rate of return, the reasons why people moved overseas were also important. Thus the mainly work related moves all over the globe in the twentieth century had high return rates both because there was little intention of permanent settlement, and because twentieth century communications made return that much easier. Perhaps most striking is the fact that up to a quarter of those who emigrated long distances to North America and the Antipodes in the nineteenth century also returned, with the decades of most frequent return migration coinciding with the periods of greatest outflow. This suggests that some intending emigrants stayed overseas for relatively short periods. Although emigrants to New Zealand/Australia and Africa had the longest mean lengths of residence prior to return, for most destinations (including North America and the Antipodes) the modal length of residence prior to return was only two to five years, although in the case of Australia/New Zealand over one fifth also remained in the country for over 20 years before they returned (Table 9.6). Clearly distance and accessibility were important factors, but a substantial minority of long distance emigrants failed to settle in their new home and returned to Britain. Whilst it may be suggested that a British based study is likely to produce more information on those who returned than on those who remained overseas, as outlined above some information was received from family historians overseas, and figures for return migration are comparable with estimates produced by Baines (1985) who suggests that 20–25 per cent of transatlantic migrants returned.

When the intervening time period is taken into account, the characteristics of return migrants were broadly similar to those of emigrants to different regions (Table 9.7). Thus return migrants were, on average, older than emigrants; they were more likely to be married, and were less likely to move alone. They were thus more like internal migrants than the original

Table 9.5 Return migration by country and year of move (%)

Area of origin	1750–1839	1840–79	1880–1919	1920–94	% of total	Return migrants as % of all emigrants
Ireland	18.8	11.7	12.5	5.3	9.2	69.8
USA & Canada	9.4	26.5	28.3	13.4	20.1	25.7
South/Central America	6.3	5.1	4.3	8.4	6.5	63.5
Africa	6.3	3.7	11.6	20.3	14.2	69.0
Australia/New Zealand	3.1	11.1	8.5	12.2	10.5	20.5
Asia/Middle East/India/Russia	9.4	9.4	11.2	16.6	13.4	66.9
Eastern Europe	0.0	7.7	1.7	1.7	2.6	86.4
Western Europe	34.2	19.7	17.2	18.0	18.7	77.7
Other	12.5	5.1	4.7	4.1	4.8	77.8
% of all emigrants	21.9	25.5	39.6	76.3	44.1	
Total number	32	117	233	344	726	

Data source 16,091 life histories provided by family historians.

Table 9.6 Length of residence overseas prior to return by area of residence (%)

Length of residence	Ireland	USA/ Canada	Central/ South America	Africa	Australia/ New Zealand	Asia/Middle East/India/Russia	Eastern Europe	Western Europe
Mean length of residence (number of years)	6.6	8.5	8.6	10.5	12.4	9.0	6.8	5.3
< 1 year	5.0	2.5	0.0	1.9	0.0	3.2	0.0	4.8
Between 1 & 2 years	8.3	13.0	8.7	7.7	5.6	10.4	14.3	23.2
Between 2 & 5 years	40.0	31.7	43.5	21.2	29.2	32.3	21.4	36.0
Between 5 & 10 years	23.3	28.0	15.2	32.7	24.7	22.9	35.7	18.4
Between 10 & 20 years	20.0	16.1	17.4	22.1	18.0	17.7	28.6	12.8
20 years+	3.4	8.7	15.2	14.4	22.5	13.5	0.0	4.8
Total number	67	146	47	103	76	97	19	136

Data source as 16,091 life histories provided by family historians.

Table 9.7 Characteristics of migrants and reasons for return migration by area of residence (%)

Migrant characteristics	Ireland	USA/Canada	Central/South America	Africa	Australia/New Zealand	Asia/Middle East/India/Russia	Eastern Europe	Western Europe
Gender								
female	16.4	26.0	36.2	22.3	21.1	44.3	26.3	35.3
Age								
<20	20.9	6.2	6.4	4.9	6.6	3.1	0.0	10.3
20–39	50.7	50.3	46.8	41.7	48.7	46.4	68.4	66.2
40–59	23.9	33.7	29.8	38.8	34.2	41.2	31.6	17.6
60+	3.5	9.8	17.0	14.6	10.5	9.3	0.0	5.9
Marital status								
single	29.3	26.4	32.5	22.0	39.4	20.2	10.5	45.4
married	67.2	63.6	67.5	74.7	56.1	74.5	89.5	53.1
other	4.5	10.0	0.0	3.3	4.5	5.3	0.0	1.5
Companions								
alone	17.9	29.4	36.2	23.8	31.6	24.0	11.0	37.6
couple	19.4	4.9	17.0	21.8	11.8	13.5	22.2	14.3
nuclear family	53.7	55.2	46.8	49.5	52.6	57.3	55.6	41.4
extended family	4.5	7.7	0.0	3.0	1.3	1.0	5.6	0.8
other	4.5	2.8	0.0	1.9	2.7	4.2	5.6	5.9
Position in family								
child	18.5	9.8	4.3	5.0	10.5	3.1	0.0	9.0
male head	64.6	64.3	70.2	72.3	68.4	51.0	72.4	54.1
wife	12.3	14.0	14.8	15.8	10.5	31.3	11.0	13.5
female head	1.5	6.3	6.4	5.9	6.6	12.5	11.0	17.5
other	3.1	5.6	4.3	1.0	4.0	2.1	5.6	5.9
Reason								
work	67.2	24.6	70.2	54.4	27.6	64.9	57.9	48.5
marriage	1.5	2.8	2.1	2.9	2.6	8.2	0.0	1.5
family	4.6	0.7	2.1	3.9	3.9	3.2	0.0	1.5
crisis	5.9	2.1	6.4	1.0	5.3	2.1	0.0	2.2
retirement	3.0	0.0	0.0	0.0	0.0	0.0	0.0	2.9
education	5.9	1.4	0.0	0.0	0.0	0.0	0.0	17.6
emigration/better prospects	11.9	68.4	19.2	37.8	60.6	21.6	42.1	25.8
Total number	67	146	47	103	76	97	19	136

Data source 16,091 life histories provided by family historians.

emigrants, and were simply a sub-set of emigrants at a later stage of the life-cycle. However, there was a tendency for return flows to be more male dominated than emigration from Britain, suggesting that males who moved alone (possibly for work reasons and leaving their family at home) were, not surprisingly, also more likely to return. Variations in the rate of return by origin region within Britain mostly reflect the characteristics of the original flows. Thus those regions which had a high rate of emigration to North America and the Antipodes had lower rates of return than regions such as South-East England from which more workers migrated to Europe and Asia in the twentieth century. These factors are also reflected in the reasons why people returned. Destinations such as Ireland, South America, and Asia which attracted large numbers of labour migrants generated similar work related return moves. The most notable feature of reasons for return migration is the low incidence of return for family reasons. Whereas many people emigrated to North America and the Antipodes for family reasons, very few cited family ties as an important reason for a return move.

Overall, the patterns and characteristics of overseas migrants in the sample were similar to those found in other studies of emigration. More interestingly, the characteristics of overseas migrants and return migrants were very similar to those of internal migrants and, especially, long distance internal migrants. This evidence suggests that the process of emigration and overseas migration should not be viewed as a separate and special event, but rather as an extension of the internal migration process, with the same sorts of factors operating to structure migration decisions. The relative significance of return migration, even from distant destinations where emigration and permanent settlement was common, further emphasizes the extent to which overseas moves were simply part of a continuum of life-time migration experiences. The ways in which such moves fitted into life-time migration patterns will be examined below, using evidence from a series of individual case studies of emigration and return migration.

The links between emigration, return migration and internal migration

In this section the life histories of eight individuals are examined, four of whom emigrated to North America and four to Australia. The individuals were born between 1809 and 1898 with emigration taking place at various times between 1846 and 1925. In each case, as far as can be ascertained, the intention was to settle permanently in a new country though, in the event, six returned to Britain and one emigrant undertook the return journey from Britain to North America twice. The reasons for emigration were quite varied, though crises of one sort or another played a significant role in both emigration and return migration. In this sense the case studies link closely to the material examined in Chapter 8. The case studies are typical of many others, and illustrate the ways in which emigration and return migration fit into a life-time of migration experiences.

Mary B. was born in Runcorn, Lancashire in 1809, the daughter of a shoemaker. She lived at home with her family until the age of 16 when she married and moved to a new home, also in Runcorn. After 21 years in Runcorn the family of husband, wife and four children emigrated to Wisconsin, USA to start a new life as farmers. However, they were clearly unused to such work and after less than a year they all returned to Runcorn. Four years later, in 1851, they attempted to emigrate again, settling on a farm in Ohio. However, once again they found the work unfamiliar and too hard and within a few months returned to Runcorn. Mary and her husband moved at least once more in Runcorn before her death at the age of 58. Although the double emigration is unusual, in other ways this is a typical pattern of family migration from a small Lancashire town, seeking economic improvement and a new life in the USA. However, the failure to settle illustrates that farming in Wisconsin and Ohio presented considerable problems to a family unused to such work and conditions, and their home town of Runcorn clearly exerted a considerable pull in times of difficulty. With the exception of moves to the USA all migration was short distance within the same settlement.

Henry A. had a very different life history. He came from a relatively wealthy and upwardly mobile family, moved very frequently, but still failed to settle after emigration. Born in Blackburn, Lancashire, in 1836 Henry A. was educated at a private school in the town and, at the age of 16, was employed in a clerical capacity in the cotton mill of which his father was part-owner. He moved with his parents in Blackburn to a larger house before marrying in 1859. At the age of 23 he was running the cotton mill with his father and subsequently moved twice with his wife and family in and around Blackburn. All moves were stimulated by housing and environmental reasons, including a move to the nearby village of Chatburn to provide a better environment for his children to grow up in. However, by 1867 the mill was failing – precise reasons are not given but presumably the cotton famine had an impact – and he was forced to move to cheaper housing in Blackburn. The family was effectively bankrupt and they were forced to give up the mill, but Henry was given work by his father-in-law who ran a tea and coffee merchants in the town. He continued to manage this business for some 11 years, moving four times in Blackburn mainly for housing reasons. During this period his wife died, leaving him with six children to care for, but he remarried in 1875 and continued to work for the father of his first wife. However, in 1880 his father-in-law died and the shop was sold, leaving Henry without work again. This time his reaction was to seek his fortune elsewhere and, after a period without proper employment, at the age of 52 he left his second wife and family in Blackburn and emigrated to Manitoba, Canada, to take up farming. The intention was presumably to establish himself before his family followed, but after two years he gave up because farming was too hard at his age and he returned to his wife in Blackburn. Here he moved twice more prior to his death in 1919. Although Henry A. had a privileged life in comparison with many, he also encountered some hardship and crises, and he too was unable to settle in North America. This is perhaps not surprising as managing a cotton mill and tea dealers in Blackburn was a far cry from farming in Manitoba. He would have had few relevant skills and little knowledge of what to expect. Apart from his two years in Canada his life-time migration pattern was very typical of many others and consisted entirely of frequent short distance moves in and around the same community.

James J. was born in Cramond, Midlothian, in 1847 where he worked as a machinist and lived with his wife and children until, at the age of 39, he emigrated alone to Fall River, Massachusetts. No reason is given beyond the desire to improve his employment prospects. He gained work as a blacksmith, thus using some of his skills, and the following year he was joined by his wife and six children. The family now seemed settled in Fall River, they moved house three times and James improved his occupational status eventually teaching engineering. However in 1915, following the death of his wife, James moved into his son's house in Fall River and the following year, at the age of 69, he returned to Cramond in Scotland, the town he had left some 30 years previously. Here he remarried and lived in Cramond until his death in 1925. Once again, apart from emigration, all other moves were within local communities (Fall River and Cramond) and, despite apparently settling in the USA and prospering economically, James still chose to return to his home town of Cramond after the death of his wife. Such examples illustrate clearly the difficulty of assuming permanence in any migration, and further emphasize the extent to which people continued to feel links to their origin communities over long distances and time periods.

The final example of migration to North America is that of John K. who was born in Ormsgill, Barrow-in-Furness, Lancashire in 1898. His father was an iron miner and John lived at home and worked as a riveter in the Barrow shipyard until at the age of 26 he emigrated with a friend's family to Hamilton, Ontario, Canada. Barrow was hard hit by depression in the 1920s and the move was primarily to improve employment opportunities. John gained work in a fruit canning factory in Hamilton, but after only a year he returned to Barrow. This move was for a combination of employment and family reasons. Work in the canning factory was seasonal and he had difficulty gaining employment during the winter, and his father (a widower) was ill and required assistance. He lived with his father in Barrow until 1930 when he married and rented a house on Walney Island, Barrow, moving once more within the same community in 1937. On his return from Canada he was employed first as a fruit salesman and, from 1937, as a meter fixer for the electricity company. He remained on Walney Island until his death in 1973. Once again, all moves apart from the journey across the Atlantic were short distance and within the same community, and ties to home overcame the attractions and opportunities of life in Canada.

Although a considerable journey, movement across the Atlantic was nothing compared to emigration to Australia. For instance, in the 1860s surface mail from London to Sydney took 70 days compared with ten days to New York (Allen and Hamnett, 1995). Migration to Australia or New Zealand meant that communication with home and return migration was especially difficult. However, as shown above, around one fifth of emigrants to the Antipodes did make a return journey. John F., born in Radcliffe-on-Trent, Nottinghamshire, in 1827 was one emigrant who did not return. His father was a carpenter and John too was apprenticed as a carpenter (though not to his father) and by the age of 21 he was married and working as a carpenter in Great Grimsby, Lincolnshire. This was over 100 kms from his parental home in Nottingham. The following year he moved some 75 kms down the Lincolnshire coast to the village of Burgh-le-Marsh, near Skegness, before in 1851 returning to live and work as a carpenter in Nottingham. John and his family thus had a history of migration for employment within Eastern England, and when faced with adversity in Nottingham he decided to emigrate. In 1851 there was a strike and lock out in the building trade in Nottingham and this, combined with stories about finding gold in Victoria, led John, his wife and two children to emigrate to Geelong, Victoria, Australia in 1851. Here he continued to work as a carpenter, but died in Geelong at the early age of 30 years. Although the circumstances of his death are not known, it would seem that emigration did not bring the long-term advantages that had been hoped for.

Many people who migrated to Australia and New Zealand went on assisted passages, often recruited because of their skills to fill particular economic niches in the developing economies. For those employed in declining or uncertain economic sectors such opportunities were attractive. Thomas J. was born in Merthyr Tydfil, Glamorgan in 1828, the son of a coal miner. Thomas worked for the Tredegar Iron Company as an iron miner, living in Rhymney, Monmouthshire, only some 5 kms from Merthyr. However, in 1853 a local agent was recruiting labour to work in the coal mines in New South Wales and, at the age of 25, he travelled with a party of miners from the Tredegar district to work as a coal miner for the Australian Agricultural Company. Thomas settled in Newcastle, NSW, and lived there for the rest of his life, eventually becoming an engine driver when he was too old for mining. Whilst there is a tendency to think of those who stayed (like Thomas J.) as successful emigrants, and those who returned as unsuccessful, there were some who returned directly because of their success overseas. Elizabeth J. was born in 1857, daughter of a tenant farmer near Bridgend in Glamorgan. In 1859 her father decided to emigrate to Australia to look for gold and the whole family travelled to Ballarat. Within five years her father had found gold and had sufficient money to return to Glamorgan to buy a farm in the same village that the family had previously lived in. Success in Australia led directly to return migration and contributed to later financial and residential stability.

Martha Y. also returned from Australia within five years of emigrating, but her story was less successful. Born in Salford, in 1865, daughter of a waterman, Martha moved three times in Salford with her family before her marriage in 1883. Whilst single she worked as a cotton spinner, but gave up work on marriage, though continuing to live in Salford. The family moved twice more within Salford, mainly connected with her husband's work (they lived in tied housing for part of the time), but in 1921, at the age of 56, Martha Y. emigrated with her husband and youngest of 12 children, to Brisbane, Queensland, to join a married daughter who had previously emigrated. Martha and her husband worked as market gardeners in Brisbane, but after five years they returned to Withington, Manchester, because they did not like the Queensland climate. The tropical climate of Brisbane is certainly very different to that of North-West England and it suggests that, despite her daughter's presence in Queensland, Martha was not well prepared for conditions there.

All of these life histories have many features in common. For most people emigration across the Atlantic or to Australia was the only long distance move that they made in their lives. Migration experiences either side of emigration consisted of frequent short distance moves within a community, the type of movement which dominates the whole data set collected from family historians. There is thus no reason to suggest that those who emigrated were any different from other migrants, except that they responded to a particular crisis or opportunity by emigrating. As seen in Chapter 8, many others responded to similar crises by undertaking long distance moves within Britain. It is also clear that many of those who did emigrate were poorly prepared for the experience and had little idea of what to expect. This was especially true of the less controlled migration streams to North America: it is not surprising that urban Lancastrians found farming in central Canada or the USA difficult. More migrants who undertook the longer journey to Australia went on assisted passages and were to some extent screened by emigration companies and recruited because of their skills. Despite a willingness to emigrate, what is also clear from these life histories is the extent to which at least some emigrants retained links with their home communities. When faced with adversity they frequently returned home and remained in the same community for the rest of their lives.

Conclusion: explaining the emigration decision

Evidence provided by family historians allows emigration and migration overseas to be placed within a longitudinal perspective; however, most such life histories do not provide detailed evidence about emigration decisions. Only rarely were sufficiently detailed sources available to probe the reasons why people decided to move overseas. In this concluding section, evidence from three diaries is used to shed some light on the decision-making process. Three distinctively different overseas migration experiences are examined: an emigrant who intended to settle but in fact returned; a man who migrated overseas for work; and a woman whose overseas migration was controlled by others.

George Osborne was born in 1820 in Northampton and, at the age of 16, was apprenticed to an agricultural implement manufacturer in Essex. However, in 1841 (at the age of 21) George decided he did not like the iron trade and determined to emigrate to Australia. He writes of his decision to emigrate in the introduction to his journal kept during the voyage:

> The writer of this served as an apprentice to the ironmongery business at Chelmsford in SX England but not liking the mode of transacting retail business there resolved to emigrate to Sydney New S Wales having heard of a situation there in his own trade which promised well . . . having purchased a turning lathe and a few tools of different sorts being of a rather mechanical turn he sailed from Portsmouth in the brig Sarah Bell . . . for Sydney Cove on 13th of May 1841 ready for any kind of employment for a living and having a desire to see and experience more of the ways and customs of the world than he could do behind a counter at home.

The journey from his home in Northampton to Portsmouth is described in detail. George and two brothers travelled by bus from Northampton to the railway station, then by train to London, stopping overnight in London before travelling on to Portsmouth by coach. George said goodbye to his brothers in Portsmouth full of anticipation for the journey, recording that he was 'All alone . . . full of spirits as a gin bottle with a trifling exception of parting'.

On arriving in Sydney, George started work for a Mr Gordon but was immediately disappointed by the prospects:

> Went to work this morning at 1/2 past 6 had an hour from 8 to 9 for breakfast and an hour to dinner. left off at 6 at night business very dull. Nothing to do. Gordon is about to give up business.

294

Three days later he 'determined to give up the place and go to Port Philip and take Pot Luck.' He then spent several weeks in and around Sydney working on ships and wherever employment was available, but was increasingly unhappy and unwell. By mid-January 1842 George had had enough of Sydney writing in his journal:

> I thought of getting work in an engine factory but found my health so indifferent that my mind was soon made up to leave this horrid hole at once and once more see the only bit of earth in my estimation worth living for and I sincerely hoped there to spend the rest of my days for I felt it was better to live upon bread & water in the midst of one's Friends than to live independent in a country where I had not yet seen a soul that I knew at home.

George could not afford the fare home, so he signed on as a seaman on a cargo boat bound for England and he finally left on 17th February 1842. His journal entry sums up his experience of some four months in Australia: 'This day I took a last look at detestable Sydney'. On his return to England he went first to his home town of Northampton and then, in 1844, to Kinver (Staffordshire) where he married and remained for some 30 years, working as a clerk in an iron works.

By any criteria George Osborne was an unsuccessful emigrant. None of his expectations were fulfilled and, apart from realizing his ambition to see more of the world, he gained few benefits from the experience. From the entries in his journal it would appear that both the decision to emigrate and the decision to return to Britain were taken relatively quickly. Although he had disliked his situation in Essex, he found Sydney even more unpleasant and made little attempt to wait until things improved. Ties of family and friends were also clearly important and one gets the impression that he could have withstood some of the hardships of life in Sydney had he had more support from kin or friends. George clearly had little idea of what to expect in Sydney and was prepared to work his passage home in order to return to a community with which he was familiar.

The experiences of John James, born in Sithney, Cornwall, in 1822, were very different. During his life he worked for long periods in Tennessee, Newfoundland and Ireland, (and elsewhere for shorter periods) but always retained contact with Cornwall which he viewed as his 'home' despite the fact that his children were also scattered around the globe. 'Captain' John James was a skilled Cornish tin miner and mine manager, and all his periods overseas were taken to seek work when the Cornish tin mining industry was in recession. However, despite living for almost ten years in Ireland, he clearly viewed all his overseas moves as temporary migration associated with his work rather than as emigration. John James's diary provides quite detailed evidence about his decision to move to New-foundland in 1869:

> After having been home from Norway near two months with but little prospect of a situation. Trying to help bring out New Rosewarne Mine but without success, everthing in the mining world being dull, I accepted the offer of a situation in this place – Burtins Pond – Notre Dame Bay, Newfoundland. This is through my friend Mr. Pike. Went to Liverpool, engaged with C J Browning Esq. the head of the firm. My salary £250 per annum, taking with me my son & seven others as miners at £8 per month, also a smith at £10 per month. Also took with me wife & three daughters, leaving home three others. This course would never have been adopted if I could have seen any other source. I have left friends whom I hope to meet again. It seems I must follow what I hope are providential openings. We left on 4th August 1869 via Bristol and Liverpool.

Despite moving to Newfoundland with most of his family and with other miners from Cornwall, John James was clearly apprehensive about the ex-perience and would have preferred to remain in Cornwall if there had been work in the tin mines. He stayed in Newfoundland for almost two years but, with others, seized the opportunity to return to Cornwall when the mine was failing:

> The Tilt Bore Mine is failing & the Cornishmen, the Agent included, leaving. We are about to leave all our men except one. We are glad of it.

However, he did not find it easy to gain work in Cornwall and was torn between his desire to remain in Cornwall and his need to find work:

> We have been home about three months. I have had much anxiety. I have had no situation, have travelled many miles & made many inquir-ies after mine situations etc, but have not succeeded to anything as yet. There are some friends who would help me when opportunity offers & I hope to succeed soon. I have no desire to leave home & home com-forts again.

Within a month he did find work as a mine manager in Cornwall, though it was too far from home for him to travel daily and he only returned home at weekends, and the following year he took a better job managing three mines. However, although tin prices were high the mining industry was in turmoil with labour strikes and much financial speculation. John resigned from his post as manager of three mines due to the worry and uncertainty and was again forced to travel overseas to find work, accepting the post of manager of the Ballycastle Coal and Iron Mine in Ireland in 1873. Initially he moved alone, but his wife and some children joined him the following year. The move to Ireland was made with reluctance and he writes in 1873 '. . . was very sorry to have again to leave home, but it seemed I had no alternative', and again in 1874 when he gaves up his home in Cornwall and his wife joined him in Ireland 'It seemed as if I could not stay at home . . . It is a new line of life to me, this management of coal and Iron mines . . . I shall soon get used to it.' He remained in Ireland – though visited his children in England from time to time – until in 1883 when, at the age of 61, he at last returned to Cornwall. He writes in his diary:

> I think it is time for me to leave this strong Northern atmosphere. If my life be spared until the end of March next I hope to go to Cornwall and try the effect of a warmer climate and my native air.

John James died in Cornwall some two years later.

Although John James was highly mobile and lived overseas for substantial periods, all his moves were stimulated by the need to find work and were generated by the precarious employment position of a Cornish tin miner and mine manager. He could earn good money when in employment, but he also had to endure long periods without work. All his moves overseas were undertaken reluctantly, and he always seemed to be working towards the goal of returning to Cornwall. Although some moves were undertaken with his family, he was often away from home for long periods. He was attracted to Cornwall both by his identity with the place and by his attachment to relatives and friends who lived there. Consideration of the diary evidence, with details of why he moved and how he felt about this migration, places the bare bones of his residential history in a different light, and emphasizes the extent to which movement was forced upon him by economic circumstances.

The final example of the way in which decisions to move overseas were taken is very different again. Evidence comes from a family memoir compiled in 1976 by Mollie Potts (born 1902). The family were relatively affluent, living in Cheshire where the author's father worked as a cotton merchant in Manchester. In the 1890s Mollie's Aunt Emily, then aged 18 and living at home, formed a romantic attachment of which the family disapproved – mainly because they considered the couple too young – and to separate the young lovers Emily was sent to a family friend in Dresden, Germany, for a few months to learn German and finish her education. However, this did not have the desired effect and so the 18-year-old girl was sent with a sister and brother on a trip to Australia. The specific reason for the journey was her sister's health – the doctor had suggested a long sea voyage – but they also had a brother living in Australia. Initially Emily's sister returned without her, but Emily booked her own fare and sailed back to London in 1896 where she married the man from whom her family had tried to separate her. This unusual case history demonstrates that some people, and especially young women, were often not in control of their own lives (though Emily did eventually get her own way), and also emphasizes the relative ease with which the rich could move over long distances at whim. Although Emily's parents would have wished her to stay in Australia for a substantial period of time, she rebelled against this forced migration and quickly made arrangements for her return.

Collectively, these three diaries emphasize the diversity of reasons for which people migrated overseas. Only one could be classed as a true emigrant who went with the expectation of settling, but he returned within a few months, whereas the man who reluctantly migrated for work remained abroad for much longer periods. For most people both employment opportunities and family ties strongly affected the decision to move overseas, but most migrants also retained strong links to their home areas. They kept in touch with kin and returned home when the opportunity arose. There is little evidence from either the aggregate data or the individual testimonies that the decision to emigrate was fundamentally any different from the decision to move within Britain. Many of the same sorts of factors were involved and, although the impact on the individual was often considerable, when viewed in the context of the whole life-course an emigration decision was often set within a life-time of migration very like that of many other people.

CHAPTER 10

The role of migration in social, economic and cultural change

Introduction

Analysis presented in this volume has examined the migration process from a variety of different perspectives. Following assessment of the main characteristics of the data set, the ways in which these varied over time and space, and the role of towns in the national migration system, attention has focused on migration undertaken for a variety of different reasons: employment, family motives, housing, as a response to crisis and for emigration. It has been stressed that these were rarely discrete categories – many moves were undertaken for a variety of reasons and employment, family and housing considerations often interacted to produce a migration decision – and that the precise motives for migration were at all times difficult to discern precisely. This chapter uses the evidence so far presented as a starting point to focus more specifically on the wider ramifications of the migration process, examining the relationship of different migration experiences to the processes of social, economic and cultural change which occurred in Britain during the two centuries after 1750.

The central question to be addressed in this chapter can be put very simply: did migration matter? This question can be explored by focusing on the impact on individual people and their families, on the places that migrants left and went to, and on change within the wider structures of economy and society. The main difficulty in examining links between migration and social and economic change at either the individual or societal level relates to the issue of causation. The fact that migration was occurring at a time of economic change in a particular region, or of turmoil in the lives of individual migrants, does not mean that the events were necessarily related. Even if there was a relationship the nature of causation is not always clear. For instance, was migration a response to forces of change operating within society as a whole, or was the process of migration itself an agent of change? Did an individual move because of major difficulties or changes in their circumstances, or did the act of migration itself create new problems and fundamentally change people's lives?

There are, of course, no simple answers to such questions, and in many instances a whole host of causes and effects will have been closely inter-related. Disentangling these for the more distant past may be an impossible task. The aim of this chapter is to use both the longitudinal migration data and evidence from diaries and life histories provided by family historians to begin to explore some of the impacts of migration on individuals, places and wider society. This is achieved by, first, examining the impacts of short distance and long distance mobility, suggesting that the impacts of these events would have been rather different. Second, the impact of migration on places is assessed by reviewing evidence from the aggregate data set for links between migration and processes of local and regional change. Third, the extent to which the meaning of migration, and its links to other processes of change, varied over time is explored by focusing on four cohorts of migrants with similar characteristics to examine the extent to which parallel processes were operating at different periods of time. It is obvious that historical events assume much greater significance if they had an impact on and meaning for the people involved, and if they were related to the broader processes of change that were taking place within society. Because of the volume of migration in the past, and its potentially disruptive effects in moving people from one part of the country to another, it is usually assumed that migration had such significance. This chapter explores this proposition in detail drawing on evidence provided by family historians.

The impact of short distance migration

It has been demonstrated that short distance moves within the same com-munity, or to adjacent settlements which would have been within the every-day action space of most migrants, were by far the most common migration experience at all times in the past. As it is likely that unrecorded moves in the more distant past were overwhelmingly short distance, the volume of local mobility has probably been understated rather than exaggerated. Although the distances over which people moved did increase in the twen-tieth century, short distance mobility of under 10 kms still accounted for well over half of all residential moves. Although residential mobility was a relatively common experience, with individuals moving on average between four and five times during their life-time in the nineteenth century, and over seven times in the twentieth, such moves only rarely removed people from their local communities. Whilst aggregate analyses of migration flows using census and vital registration data emphasize the net effects of migration, and the inter-regional redistribution of population which occurred, the ex-amination of life-time migration histories gives a different perspective. These data do not replace or challenge the picture of net movement between regions given by other studies, but rather emphasize that such moves were a very small part of the total migration experience. By focusing on gross migration patterns over complete life-times, the relative rarity of long distance migration is emphasized.

It can be suggested that most short distance residential mobility would have had only limited impacts on the people or places involved. Such migration could be seen as an individual or local adjustment mechanism, with few implications for wider society or for the lives of people who moved. Migrants travelled through familiar territory, they continued to experience similar social, economic and cultural circumstances, and they could retain contact with friends and kin. In this respect the relative neglect by historians of short distance moves in favour of the long distance inter-regional transfer of population can, perhaps, be justified. However, there is a danger of making over-simplistic assumptions about the relative impact of different types of mobility in the past. As shown in Chapter 8, the majority of moves that occurred as the result of major personal crises (such as bereavement or redundancy) were also over short distances, and there is a danger of transferring late-twentieth century values about distances to the past. It should not be assumed that a move of, for instance, 20 kms, which seems short by present-day standards, was necessarily perceived as a local move in the late-eighteenth century.

Perceptions of relative distance, and of the disruptive effects of mobility, are almost impossible to determine precisely. They varied according to factors such as the nature, cost and availability of transport in particular areas; constraints of local topography; local variations in labour markets; cultural and dialect differences over short distances; and individual life-cycle characteristics. Thus for a small child, with limited mobility, movement of only a few kilometres away from familiar places and people may have been disruptive, whereas an adult who may have undertaken a long daily journey to work may have construed the same move as retaining links with the local community. Although large cities, especially London, had clearly defined local neighbourhood communities, by the late-nineteenth century most urban areas were developing effective intra-urban transport systems and, for those who could afford the fares, movement from one part of the city to another was relatively easy. In contrast, in upland areas, including the Scottish Highlands and the English Pennines, topographical barriers could limit contact between adjacent settlements, and in areas like the Yorkshire Dales individual valleys and villages retained distinctive characteristics with which locals identified well into the twentieth century (Aldcroft and Freeman, 1983; Hallas, 1991).

The residential histories and diaries provided by family historians provide only glimpses of such issues. Most historical sources do not record the feelings of people when they moved and the few life histories that deliberately

301

dwell on the effects of migration may well have been atypical. However, selected evidence taken from diaries and life histories already cited can be used to illustrate particular points, and demonstrate the diversity of responses to short distance migration. It is clear that many people had very strong attachments to particular settlements and neighbourhoods within towns. This is demonstrated by the frequency with which migrants returned to an area in which they had previously lived – including many emigrants as shown in Chapter 9 – and the reluctance shown by families such as the Shaws when they left Dent and moved to Dolphinholme (Chapter 6). The significance of locally differentiated labour markets is demonstrated by the example of Ben M. (Chapter 6) who moved for employment (in wool textile factories) between the adjacent West Riding towns of Holmfirth, Halifax and Huddersfield, but was clearly uncomfortable moving from a worsted to a woollen area and deliberately moved back to work in a worsted mill when the opportunity arose.

The frequent moves of Henry Jaques around London in the second half of the nineteenth century also demonstrate equivocal reactions to short distance mobility (Chapter 4). As a child he was clearly upset when his aunt moved away from the neighbourhood and he could no longer visit her alone, and when he first left home at the age of 16 to take an apprenticeship his mother felt a great sense of loss even though he was only moving some 6 kms across the river. During his life Henry Jaques travelled over most parts of London (and elsewhere) in connection with his work, and when he was seeking employment he was clearly prepared to consider almost any location in the metropolis. Yet underlying these searches for work and housing was a strong preference to be in a part of the city with which he was familiar, where he could retain contact with kin, continue friendships and business associations, and maintain an established pattern to his limited leisure activities based, especially, around particular churches. Even after his move from the Mile End Road to Forest Gate he returned once a month to the church he had previously attended.

In the eighteenth and nineteenth centuries, when most ordinary working people had relatively few possessions and most homes were rented, movement within a town or to an adjacent village would have been relatively easy. Transport would have been on foot and by horse and cart, and the costs in time, trouble and finance would have been limited. Even in the late-twentieth century when most households have many material possessions and buying and selling a home can take many months, a local move is much less disruptive than migration over a long distance. Such moves also reinforced local and regional social, economic and cultural structures as people with particular employment or cultural characteristics were retained within local economies. However, at the individual level, even short distance moves could induce some trepidation, affect employment prospects and alter established patterns of social interaction. Although such impacts may have been relatively minor and short-lived, the limited evidence available suggests that whilst short distance movement was easy and frequent, it was not without meaning for people's lives. Any residential move is followed by a period of readjustment to changes circumstances. In those cases where a local move was the result of personal disaster, such as the time when the Jaques family were forced to quit their home and business due to bankruptcy and move to two rooms in the same locality, coming to terms with decreased circumstances whilst living in the same community must have been especially difficult.

The impact of long distance migration

It is difficult to determine a meaningful definition of long distance migration. Not only would perceptions of distance have changed over time in line with alterations in transport technology (Chapter 3), but also as suggested above the friction of distance would be felt differently according to age, local topography, labour market characteristics and many other factors. As shown in Chapter 3, the proportion of moves over 100 kms was relatively stable (between ten and 12 per cent of all moves) during the eighteenth and nineteenth centuries, but increased rapidly in the twentieth century (18.7 per cent of all moves after 1920). However, if long distance moves are defined as being over 50 kms for the period before 1920 these also account for just over 18 per cent of all moves recorded. After 1920 such moves form 27.7 per cent of the total. In the period before motor vehicles, 50 kms was clearly a substantial distance representing over a day's walk and several hours travel by horse and cart. Although from the 1840s the railways provided relatively rapid transport over long distances, many people could not afford railway fares until at least the late-nineteenth century (Dyos, 1982). In the mid-twentieth century a journey of 100 kms could be accomplished in a few hours by automobile or train and such transport has become increasingly accessible to most people. It seems reasonable to suggest that at all times from the mid-eighteenth century something under one fifth of all moves within Britain have been over what might have been perceived as long distances.

Although existing migration studies which focus on the net effects of inter-regional migration flows emphasize such long distance mobility, they still only give a partial picture of the totality of long distance migration. As demonstrated in Chapters 3 and 4, the volume of movement in and out of regions within Britain, and the amount of movement up and down the urban hierarchy, was quite evenly matched at all times in the past. There were small differences at particular times and between some regions, and these were sufficient to produce the well known net migration flows shown by census and vital registration evidence (for instance net movement to larger towns in the eighteenth and nineteenth centuries, away from towns in the twentieth century, towards the industrial northern counties of Britain in the nineteenth century, but away from such areas and towards midland and southern Britain in the twentieth). The data collected from family historians does not challenge such evidence of net migration – and as a relatively small sample of the total population the data are not designed to address such questions which are much better researched from census and registration evidence – but they do provide some qualifications and fresh perspectives on long distance mobility. In particular, they reinforce the observation made by Ravenstein (1885, 1889) that every flow produced a counterflow, and from the perspective of investigating the impact and meaning of migration it is the gross number of flows in all directions which is of most importance rather than the net differences between inter-regional moves.

It is often assumed that long distance migration was disruptive to the individual, removing them from familiar people and places, and that such moves often occurred as a response to major external stimuli. The effects of long distance migration could also disrupt local economies, for instance by removing skilled labour, and altering the age and social structure of particular communities. However, there is increasing evidence from previous studies that this was not necessarily the case. Much long distance migration took place to take up particular employment opportunities, people often moved in family or larger groups, and networks of friends or kin could provide a structure to assist the process of long distance migration (Darroch, 1981; Schurer, 1991). Evidence from family historians (presented in previous chapters) reinforces this picture and suggests that movement over long distances was not necessarily a great deal more disruptive than short distance moves. Whilst moving over long distances clearly required more effort and organization than a move across the road, and such a move also necessitated the reorganization of many aspects of an individual's life, there were also many mediating factors which meant that migrating individuals retained some familiar elements within their lives. The individual who moved alone and speculatively over a long distance to an unfamiliar place in which they had no friends or relatives, and no previously arranged employment or accommodation, was relatively rare.

As demonstrated in Chapters 4 to 8, long distance moves within Britain were undertaken for a wide variety of reasons, and frequently involved the exploitation of kinship networks and contacts made through employment. One of the most peripatetic of the diarists analyzed was John James, the Cornish tin miner and (eventually) mine manager who lived in various locations in Britain and overseas (Chapter 9). In 1867 he moved alone from Cornwall to Newcastle-upon-Tyne, a distance of over 650 kms. Although he left Cornwall reluctantly, and returned at the earliest opportunity, the move was undertaken within a clearly defined structure. He was asked to go to

Newcastle by his London-based employers as a sampler of Norwegian tin ores (which were imported through Newcastle) and he intended his family to follow him to Newcastle. In practice the work in Newcastle only lasted five months and his family remained in Cornwall, but although he was many kilometres from his home area he clearly retained close contact with family and friends, whilst utilizing work contacts to provide him with some security in Newcastle. The fact that John James was a skilled worker with a reasonable if unpredictable income was typical of many long distance migrants who consisted mainly of professional and skilled workers who would be in the best position to use their expertise and relative affluence to ease the disruption of long distance migration.

The diaries of Rhona Little also provide detailed evidence of the ways in which long-distance moves were organized. She migrated alone from Londonderry (Northern Ireland) to London (England) in 1938 at the age of 18, but she did so within a well defined structure and moved to a constrained and regulated environment in London. Rhona was the eldest child in a middle-class Protestant family and before her migration to London she attended a small commercial college in Londonderry preparing for and taking Civil Service typing exams. She gained a post as a copy typist with the Inland Revenue in London, and preparations for her move were made late in 1937. She obviously knew that the exams she was taking would necessarily lead to movement to London and, although she expressed some unhappiness at leaving her family, she clearly looked forward to working in London, though she had never previously left Ireland. Her parents bought her ticket and found accommodation in London at a hostel run by the Young Womens' Christian Association and arranged through a friend of the family who had previously migrated to London. Rhona was accompanied on her journey to London by her father and a friend who was also migrating with her to work elsewhere in the Civil Service, and on the first weekend in London she and her father visited friends in the city and relatives who lived in Tonbridge. When left on her own in London Rhona had thus been put in touch with both family friends and relatives, and was established in a girls' hostel which provided a highly structured residential environment shared by other girls like herself. Thus a long-distance move undertaken alone was carefully planned with accommodation, employment and relevant contacts established before the move took place. It is clear from her diary that although the move caused changes to Rhona's life, these were mostly welcomed and in no way did she find the move traumatic or upsetting.

Although it was clearly harder to maintain close contact over long distances in the eighteenth century than in the twentieth, and inevitably such moves did create difficulties for some migrants, the overwhelming impression which emerges from both the aggregate data and individual case studies is that at all times most long distance moves were carefully planned, usually directed at specific work opportunities, often exploited existing contacts, and frequently involved family or friends moving together. Long distance moves were thus not necessarily much more traumatic than short distance moves, but they did present additional opportunities and challenges. In most cases a long distance move inevitably meant making new friends, settling in a strange environment and adapting to a new culture. Thus Rhona Little notes that she had difficulty communicating with the Cockney girl sat next to her in the typing pool as, though they both spoke English, their dialects were mutually incomprehensible. Rhona Little liked London and lived in the metropolis for most of her life, but many long distance migrants (like John James) returned to their community of origin when the opportunity arose. For such people a long distance move did not lead to the transference of loyalty to a new community.

The impact of migration on places

Aggregate studies of migration based on census and vital registration statistics, often emphasize the differential local and regional effects of migration. For instance, the net effects of migration flows led to sustained out-migration from many rural areas of Britain in the nineteenth century, and population loss from inner urban areas in the twentieth. Studies of rural depopulation have emphasized the impact of such loss on the age and social structures of rural communities, with population loss linked to economic decline in both agriculture and rural industry. Although there are some examples of communities which lost a large proportion of their population through out-migration, and declined dramatically in total population during the nineteenth century, the demographic impact of migration should not be over-stated (Saville, 1957; Lawton, 1968a, 1973; Champion and Fielding, 1992).

Data on life-time migration histories emphasizes the obvious but often neglected point that most settlements of all sizes both gained and lost population through migration. As shown in Chapter 4, at all times in the past people moved both down and up the urban hierarchy, with almost as much movement from large to small places as there was migration from small places to large. If the ratio of in-migration to out-migration is examined for places of different size, it was only settlements under 5,000 population which consistently experienced a net loss through out-migration in the eighteenth and nineteenth centuries (Table 10.1). In the twentieth century this pattern was reversed with small places experiencing a net gain in population through migration. Settlements in all other size categories gained more migrants than they lost in the eighteenth and nineteenth centuries, with London depending particularly heavily on in-migration in the late-eighteenth and early-nineteenth centuries, but London and other cities over 100,000 population experienced net loss through out-migration from the late-nineteenth or early-twentieth centuries.

306

Table 10.1 Ratio** of in-to out-migration by settlement size and year of migration

Settlement size	1750–1839	1840–79	1880–1919	1920–94	All moves
<5,000	63.3	73.1	91.3	118.4	89.8
5,000–9,999	109.6	108.9	100.9	124.6	110.2
10,000–19,999	134.5	105.1	111.1	114.4	113.3
20,000–39,999	128.7	108.0	120.2	102.6	112.5
40,000–59,999	115.0	146.4	120.7	102.8	115.1
60,000–79,999	155.6	141.8	94.6	105.6	115.0
80,000–99,999	139.0	132.1	110.4	100.3	110.8
100,000+	–	114.6	104.5	82.1	94.4
(excluding London*)					
London	189.9	135.2	78.6	58.6	92.7

* Including contiguous suburbs.
** In-migration/out migration × 100.
Data source: 16,091 life histories provided by family historians.

However, such net effects should not obscure the fact that, even in places of under 5,000 population in the late-eighteenth century, large numbers of people were moving into such settlements as well as moving out. Thus in the period 1750–1839, for every ten people who moved out of settlements of under 5,000 population, more than six moved in. The experience of living in most such settlements was not one of population haemorrhage, but rather of massive population turnover, with people moving both in and out, though with a net loss which gradually impacted upon population composition and structure. Similarly, in the same period, London not only sucked in migrants from elsewhere in Britain, but also lost population to other places. For every 19 migrants that entered the capital ten left for other destinations. The degree to which population structures were altered by such processes depended, of course, on the extent to which in- and out-flows were similar. None of this negates the fact that some small settlements did experience massive population losses over short periods of time – for instance some Highland areas affected by the clearances – but this was not the average experience of most places. The effects of net population loss through out-migration in small places were mostly quite gradual, and the experience of living in most settlements over a period of one or two decades would have been predominantly one of population turnover with in-movement amounting to at least two thirds of out migration.

The data collected from family historians is drawn from all parts of the country and, as shown in Chapter 2, is broadly representative of the distribution of population as a whole. However, although the total sample is large, there are not sufficient cases to examine in detail the migration histories of individual small settlements. Such work would have to be done through family reconstitution in specific localities. However, at the regional level, it is clear from analysis presented in Chapter 3 and subsequent chapters, that most parts of Britain experienced very similar migration trends. Although the effects of net population redistribution are apparent with, for instance, London and the South East gaining more migrants than they lost in the twentieth century and parts of the Highlands losing more through long distance out-migration in the nineteenth, the basic structure and characteristics of migration were remarkably similar in all parts of the country. Despite the fact that there were marked social and economic differences between various parts of Britain, the processes controlling migration were spatially very stable. This suggests that such processes were relatively independent of the economic and social structures of regions. This could be because migration was a process which was primarily affected by individual decisions, related to personal aspirations and life-course factors rather than local or regional economic and social conditions. Only in unusual circumstances – such as Irish famine migration, the clearance of Highland glens, or large-scale closure of local industry – was migration primarily affected by external structural forces.

It can thus be suggested that the impact of migration on particular settlements was rarely massive or catastrophic. Though many people left small settlements in the eighteenth and nineteenth centuries, some two thirds of this loss was replaced by in-migrants. Net loss appears dramatic when measured over several decades, but during the few years that many people remained in a settlement the effect would have been much less noticeable. Given that most migration in all places and time periods was over short distances, the majority of those moving in and out of small settlements would have been engaged in local circulatory moves. They were thus likely to have similar social, economic, cultural and demographic characteristics, and thus the high rate of population turnover did not necessarily cause rapid change in population structures. Again, this does not contradict the evidence of aggregate studies which have demonstrated the changing age structures of rural populations due both to differential out-migration and changing fertility. However, the evidence from family historians does place such changes into a broader perspective. Whilst a relatively small net difference in the characteristics of in- and out-migrants produced such demographic changes over a matter of decades, the everyday experience of living in such settlements would not have been one of massive change. Whilst many people moved, most came from nearby, and in many places in-migrants were similar to those who had left for other locations.

Temporal changes in the impact of population migration

There were massive shifts in the structure of British economy and society over the two centuries after 1750, including changes in demography, employment, transport and accessibility, housing, education and the role of women in society. In many respects the world of the 1950s would have been unrecognizable to a family growing up 200 years earlier. Given these well known changes it seems reasonable to suggest that there would also have been substantial alterations in the volume, structure and characteristics of residential migration over the same period. The ways in which different aspects of the pattern and process of population migration have changed over time have been investigated in Chapters 3 to 9. Whilst there have been some systematic shifts in the structure of migration over time (summarized below), what is most remarkable about the foregoing analysis is the degree of stability which occurred. When likely problems and biases in the evidence are taken into account, particularly the effects of limited evidence for the more distant past, the stability of migration characteristics is quite striking. In this section the nature and extent of changes in migration characteristics over time are examined in more detail by focusing on four cohorts of people: those born in the decades 1750–59, 1800–9, 1850–59, and 1900–9. First the aggregate characteristics of the life-time migration experiences of these four cohorts are examined and, second, the similarity of life-time experiences is emphasized through the comparison of selected individual migrants born into each cohort.

The characteristics of residential migration for the four selected cohorts (Table 10.2) effectively summarize data already presented in earlier chapters. In this section we simply highlight the main ways in which the experiences of the cohorts differ. There was a marked increase in the frequency with which people moved rising from an average of 2.5 moves during a lifetime for those born in the 1750s to 6.8 for those born 1900–9. However, it seems likely that some of this apparent increase in mobility is due to the lack of data on short distance moves in the late-eighteenth and early-nineteenth centuries, and the very full evidence that is available through oral sources for the twentieth century. The frequency with which people moved has probably increased, but not by as much as the figures suggest. There has also been a marked increase in the mean distances moved, with by far the greatest difference between those born in the 1900s and each of the other three cohorts. The main change is in fact in the proportion moving over long distances (over 100 kms). In each cohort the modal distance category was under 5 kms, and over half of all moves in each cohort were under 10 kms.

The proportion of moves between settlements of different sizes is also quite stable, with most of the changes reflecting alterations in the size hierarchy of urban settlements in Britain. For each cohort, most moves were either within the same settlement or to a place of a similar size. There have been some changes in the characteristics of migrants, with more people moving as dependent children in the cohorts born in the 1850s and 1900s, and consequently more migrants under the age of 20, but for each cohort movement in a family grouping was by far the most common experience.

Table 10.2 Selected migrant characteristics, reasons for movement, distance moved and settlement size of destination by birth cohort (%)

Migrant characteristics	1750–59	1800–09	1850–59	1900–09
Gender				
female	16.3	21.9	35.0	44.1
Age				
<20	14.5	11.6	23.4	25.5
20–39	56.3	48.3	48.8	46.2
40–59	18.7	26.1	17.7	16.2
60+	10.5	14.0	10.1	12.1
Marital status				
single	28.9	18.7	32.8	37.3
married	64.2	71.6	60.7	56.9
other	6.9	9.7	6.5	5.8
Companions				
alone	21.6	12.5	12.6	18.7
couple	24.4	19.9	17.1	22.2
nuclear family	47.9	58.0	60.3	49.7
extended family	3.5	5.4	5.7	5.0
other	2.6	4.2	4.3	4.4
Position in family				
child	7.8	7.7	19.9	22.7
male head	74.1	69.6	48.9	41.1
wife	11.4	14.5	20.3	21.6
female head	3.2	4.5	6.0	9.0
other	3.5	3.7	4.9	5.6
Reason				
work	45.7	48.7	36.8	30.4
marriage	24.3	19.1	18.7	11.7
housing	5.2	7.6	15.3	22.4
family	2.7	2.8	3.6	3.8
crisis	5.2	7.3	8.0	8.5
retirement	2.3	2.7	3.2	3.6
war service	1.9	1.6	2.0	4.8
other	12.7	10.2	12.4	14.8
Mean number of moves	2.5	3.5	4.7	6.8
Mean distance moved (km)	30.6	31.0	36.2	49.3
Distance band (km)				
<5	38.9	51.0	55.6	46.7
5–9.9	18.6	11.1	8.4	8.1
10–19.9	14.0	10.3	7.1	8.5
20–49.9	12.4	11.2	9.9	11.0
50–99.9	8.8	7.7	7.3	9.3
100–199.9	4.3	5.1	6.8	8.0
200+	3.0	3.6	4.9	8.4
Settlement size				
<5,000	82.4	58.6	38.2	30.3
5,000–9,999	3.7	5.9	5.7	4.9
10,000–19,999	3.8	4.8	5.9	7.4
20,000–39,999	2.0	6.0	7.6	8.3
40,000–59,999	0.2	2.6	4.2	7.0
60,000–79,999	1.5	3.5	2.4	4.4
80,000–99,999	0.2	1.2	2.4	3.3
100,000+ (excluding London*)	0.0	6.8	21.0	26.8
London	6.2	10.6	12.6	7.6
Total number	656	3,075	5,190	5,055

Data source: 16,091 life histories provided by family historians.

The reasons why people moved have also changed, with a decrease in the relative importance of movement for marriage and employment, and an increase in the proportion of moves for housing and family (life-course) reasons, as the result of a crisis, and on retirement. As outlined elsewhere, this may also partly reflect the changing quality of historical data, with less information on reasons for mobility in the late-eighteenth and early-nineteenth centuries. Increased movement for housing and life-course events also explains increases in the frequency with which young children were moving with their parents. When inevitable biases in the data are taken into account, differences in the characteristics of migration between the four cohorts are surprisingly small.

Such statistics are given more meaning when they are related to the lives of some individual migrants. In this sections three sets of four migrants are compared: three from each of the four cohorts selected for detailed examination. Each set of migrants has been chosen because of some similarity in their origins. Thus the first group (Table 10.3) were each born into the families of skilled workers living in London, the second group were born in rural areas into the families of agricultural workers (Table 10.4), and the third group were born in coastal settlements into families where the main wage earner had some connection with the sea (Table 10.5). By examining the lives of men and women with similar origins, but born at four different periods of time, the similarities and differences in migration experiences can be set into clearer perspective.

Obviously, the 12 individual case studies are a very small selection of those that could be used. For the most part they are chosen to represent typical and commonly occurring events in the data set. There are, of course, many examples of migrants where aspects of their lives were unusual. Thus we could have chosen Rebecca B., born in 1758, who worked as a domestic servant until in 1784 she was convicted of stealing from her employer, and in 1787 was transported to Botany Bay (Australia) with her baby daughter where she died the following year. Alternatively, we could have included Alfred C. (born 1850) who moved house 17 times during his life and fathered 22 children (from one mother) between 1874 and 1892, of whom 11 survived to adulthood. We could also have included Elizabeth G. (born 1852) who at the age of 22 was reputed to have walked barefoot from London to Bradford to get married. Such events are obviously atypical, but it should be emphasized that apart from specific unusual circumstances, other aspects of the lives of these individuals were also very similar to the experiences of the majority of people in the sample.

The lives of the four migrants born to skilled workers living in London contain many similarities despite the wide timespan separating their births. Each of the male children either followed their father's trade or entered a similarly skilled occupation, and the female married a man following the same trade as her father. In each case most moves were concentrated in and

311

Table 10.3 Migration life histories of four people born in London

Isaac T.	Elizabeth M.	Henry M.	Herbert S.
Born 1759, Holborn, London	Born 1806, Shoreditch, London	Born 1857, Westminster, London	Born 1904, Hackney, London
Father an engraver	Father a cabinet maker	Father a carpenter	Father a coach painter
1775+ employed as engraver, working from home 1796+ Independent Minister and engraver	No known paid employment – domestic duties Husband a cabinet maker	1869+ employed as french polisher in Bethnal Green	1918–25 employed as apprentice printer in Hackney 1925+ master printer in Bethnal Green
Married: 1781	Married: 1827	Married: 1878	Married: 1930, 1936
Moved: 1779 – Islington 1783 – Holborn 1786 – Lavenham (Suffolk) 1793 – Lavenham 1796 – Colchester (Essex) 1811 – Ongar (Essex) 1814 – Ongar 1822 – Ongar	Moved: 1827 – Hackney 1837 – Shoreditch 1870 – Hackney 1879 – Tottenham	Moved: 1878 – Bethnal Green 1879 – Bethnal Green 1885 – Bethnal Green 1893 – Poplar 1894 – Bethnal Green 1901 – Bethnal Green	Moved: 1907 – Homerton 1930 – Hackney 1933 – Homerton 1934 – Homerton 1936 Chingford (Essex)
Moves due to health (1779), housing (1786, 93, 1814, 22), employment (1783, 96, 1811)	Moves due to husband's employment (1827–70), widowhood (1879)	Moves due to marriage (1878) and housing/life-course reasons	Moves due to housing (1907, 33) and family/life-course reasons
Died 1829, Ongar	Died 1887, Tottenham	Died 1903, Bethnal Green	Died 1978, Chingford

All locations were in the built-up area of London at the relevant date (though not necessarily in the county of London) unless otherwise stated.
Data source: life histories provided by family historians.

around the same area of London, and it was the migrant born in the 1750s who in fact moved most frequently and over the furthest distance. These moves were directly related to his decision to become an Independent Minister later in life. All migrants moved for a variety of employment, life-course and housing reasons with each migrant recording between four and eight residential moves during their life-time. Skilled workers such as the cabinet maker, french polisher and printer in these examples would have earned a reasonable income – perhaps £2 per week in London in the 1850s (Mayhew, 1861/2; Stedman Jones, 1971) – but would also have invested quite heavily in their local community to establish a workshop and develop links with customers. Such employment ties would have constrained mobility

Table 10.4 Migration life histories of four people born in rural areas

Henry D.	David H.	James R.	Anne C.
Born 1753, Plumpton, Sussex	Born 1806, Kelmarsh, Northants	Born 1853, Millington, East Riding	Born 1901, Barton, Lincolnshire
Father a shepherd	Father an agricultural labourer	Father a farm labourer	Father a shepherd and agricultural labourer
1762+ employed as a shepherd's boy, then a shepherd	1820+ employed as a shepherd	1867–78 employed as an agricultural labourer 1878–88 innkeeper 1888–98 farm manager 1898+ tenant farmer	1915–19 employed as a tailor's apprentice No regular paid employment after 1919 but did some domestic service work and took in paying guests
Married: 1780	Married: 1830	Married: 1877	Married: 1920, 1952
Moved: 1762 – Falmer, Sussex	Moved: 1830 – Oxendon, Northants	Moved: 1857 – Great Givendale, East Riding	Moved: 1905 – Wootton, Lincolnshire
1769 – Hangleton, Sussex 1770 – Kingston by Lewes, Sussex 1773 – Lower Stoneham, Lewes, Sussex 1776 – Rottingdean, Sussex 1791 – Plumpton, Sussex ? – Chailey, Sussex	1851 – Kelmarsh, Northants	1871 – Great Givendale 1877 – Millington, East Riding 1878 – Pocklington, East Riding 1888 – Bramham, West Riding 1893 – Thorner, West Riding 1898 – Pudsey, West Riding	1908 – Broughton, Lincolnshire 1911 – Hull, East Riding 1912 – Cleethorpes, Lincolnshire 1915 – Broughton 1920 – Cleethorpes 1952 – Cleethorpes 1962 – Grimsby, Lincolnshire 1987 – Grimsby
Moves due to employment, probably also other moves between farms in Sussex	Moved on marriage and for employment	All adult moves for employment	Moves for father's work (1905–12), own work (1915), marriage (1920, 52), housing (1962), health (1987)
Died 1836, Chailey	Died 1888, Kelmarsh	Died 1919, Pudsey	Died 1993, Grimsby

Data source: life histories provided by family historians.

Table 10.5 Migration life histories of four migrants associated with coastal settlements and maritime occupations

Ann F.	Daniel F.	William W.	Annie F.
Born 1750, Bervie, Kincardineshire	Born 1803, Liverpool, Lancashire	Born 1858, Hull, East Riding	Born 1904, Plymouth, Devon
Father a fisherman	Father a ships pilot	Father a marine and fairstall proprietor	Father a trawlerman
1760–81 employed as a fisherwoman then household duties. Husband a fisherman	1823–35 employed as a boatman on Liverpool docks	1876–1914 employed as a fireman/engineer on trawler and fairstall proprietor when not at sea	1915–29 employed as a shop assistant
		1914–18 engineer on minesweeper (converted trawler) 1919–23 wagonette proprietor and fairstall proprietor	1943–64 shop product demonstrator
Married: 1773	Married c.1830	Married: 1902	Married: 1929
Moves: 1756 – St Vigeans, Angus 1767 – Arbroath, Angus c.1773 – St Vigeans	Moves: 1807 – Liverpool 1810 – Liverpool 1811 – Liverpool 1814 – Liverpool 1816 – Liverpool 1821 – Liverpool 1830 – Liverpool 1834 – Liverpool	Moves: 1860 – Hull 1876 – Hull 1878+ various places in Yorkshire 1902 – Hull 1904 – Hull 1914 – War service 1919 – Hull	Moves: 1908 – St Blazey, Cornwall 1910 – Plymouth 1929 – Plymouth 1930 – Plymouth 1937 – Gillingham, Kent 1942 – Plymouth 1943 – Plymouth 1945 – Plymouth 1951 – Plymouth 1961 – Plymouth 1966 – Plymouth 1968 – Plymouth
Moves for father's work (1750, 67) and marriage 1773	Moves with parents for housing and life-course moves (1807–21), marriage and employment (1830)	For much of his life lived in caravans and followed fair around Yorkshire when not at sea, renting a house in Hull as a base. In 1919 moved into terraced house	Most moves for father's and husband's employment until 1942 when husband was killed at sea. Later moves were for family and housing reasons
Died 1800, St Vigeans	Died 1835, Liverpool	Died 1923, Hull	Died 1968, Plymouth

Data source: Life histories provided by family historians.

options, and it is notable that apart from Isaac T. who combined his engraving with work as an Independent Minister, all migrants remained within easy commuting distance of the same part of London during their working lives. Herbert S. moved furthest from his workplace (in Bethnal Green) in 1936 when he moved to the popular and expanding suburb of Chingford (Essex). Here he lived next door to his brother (with whom he also worked) and family ties were clearly combined with the desire to suburbanize to an improved environment as motives for this move.

The life-time migration experiences of four migrants born into the families of agricultural workers in different parts of Britain also demonstrate many similarities. The three males each, initially, took unskilled agricultural work similar to that of their father, although James R. varied his occupation (becoming an innkeeper for ten years) and was eventually the most successful establishing himself as a farm manager and tenant farmer. Anne C. did not work on the land, and after her marriage took no regular paid employment, although she did contribute to the household's income by taking in paying guests in Cleethorpes in the summer and doing domestic work for neighbours. Each individual moved over short distances, mostly within the same locality, and slightly longer moves tended to be stimulated by external events. Thus in 1911 Anne C. moved with her parents from North Lincolnshire across the Humber to Hull due to her father's inability to find agricultural work. Within a year she had moved back to North Lincolnshire. Most moves were for employment reasons, with some agricultural workers moving frequently between farms wherever they could be hired. Thus Henry D. worked as a shepherd on at least four different Sussex farms between 1769 and 1776, and this frequent short distance mobility was typical of many agricultural workers. Because such moves rarely left records, the amount of mobility amongst agricultural workers is almost certainly understated in the data from family historians.

The migration profiles of four individuals born in coastal settlements to households where there was a strong link to the sea are both similar to each other and to other case studies cited above. Three of the individuals followed in their father's footsteps and, at least initially, earned their living from the sea. Only Annie F. (born in 1904) entered different employment (shop work), although she continued to live in the dockyard town of Plymouth and her life was affected by the sea in that her husband was killed at sea in 1942, and she was forced to move house on one occasion due to dockyard redevelopment. There would have been fewer opportunities for a woman to work with boats, although Ann F. was a fisherwoman assisting her father and husband in the eighteenth century. Once again all moves were within the same settlement or to nearby places, and longer distance moves were mostly stimulated by external events. Thus Ann F. moved in the 1750s with her father down the Scottish coast to a village outside Arbroath due to the collapse of the fishing industry in their home village, and the only time that William W. left the same locality of Yorkshire was when he was on active service in the navy during the First World War. The example of William W. is interesting as, although in one sense he was highly mobile – living for much of his life in a caravan and, when not working on a trawler, taking his fairground stall around Yorkshire – his movements clearly followed an established pattern within a clearly defined area, and he frequently rented land in Hull to provide him with a base from which to operate. In 1919, following his discharge from the navy, he moved into a rented house in Hull. Employment as a fisherman, mariner or boatman was often unpredictable with long periods of unemployment. William W.'s alternative occupation clearly helped him through such times, but for many, regular employment depended on being part of a local community where you were well known when work became available. Thus it is not surprising that Daniel F. remained in Liverpool all his short life. Although longer distance moves were often for employment reasons, most moves within a settlement were due to housing or life-course factors.

There are inevitably some unique features in every individual life history, and few personal life stories precisely mirror the aggregate characteristics of the whole data set. But, despite such differences, the overall structures of the life-time migration histories of these 12 people were remarkably similar and, in particular, showed a high degree of stability over time. Although, in the twentieth century, the effects of a widening labour market, increased affluence and improved communications were beginning to show (hence the suburbanization of Herbert S. to Chingford, and the temporary move of Annie F. from Plymouth to Gillingham (Kent) in connection with her husband's work), the similarities between the personal migration histories are rather greater than the differences. This reinforces the interpretation advanced earlier. The fact that the basic pattern and process of individual migration retained a high degree of stability over time and space despite large structural differences between localities and time periods, suggests that the individual and life-course factors which structured migration decisions within specific households were relatively unchanging. In both the eighteenth and twentieth centuries, in London and in Kincardineshire, most people moved over short distances stimulated by a combination of employment, housing and family life-course factors depending on their own particular circumstances. Longer distance moves were unusual, often stimulated in part by an external event, and those that did move away from a familiar area sometimes returned later in their lives.

However, the fact that the observable patterns and processes of migration remained similar over two centuries, does not necessarily imply that the meanings associated with migration were unchanging. One of the key features of the process of modernity is that attitudes, values and identities have altered (Giddens, 1990, 1991; Lash and Friedman, 1992; Therborn, 1995). It can therefore be suggested that patterns which appear to be similar in both the eighteenth and twentieth centuries, would have been regarded in different ways by the migrants themselves. Unfortunately such themes are difficult to discern, even using diary and life history evidence.

Varied attitudes to migration may have manifested themselves in a variety of ways. First, expectations about levels of migration and opportunities for movement have changed and knowledge of different places has increased dramatically. We live in an increasingly global society and thus it can be suggested that the expectation of movement is now much greater than it would have been in the past, even if the reality is surprisingly similar. Second, time–space compression has made daily and residential movement much easier. As noted above, the concept of what was a 'long distance' move must have altered in relation to changing transport technology and perceptions of distance. Thus, although the actual distances over which people move may have changed surprisingly little, the impact of such moves on people's lives may well have altered. Third, the changing relationship between home and workplace has influenced the frequency with which people need to move due to employment change. In the past most job changes necessitated residential change in order to remain near a workplace. In the late-twentieth century families are much more likely to choose a suitable residential location and commute over long distances. Thus the imperative to move has altered, and to some extent greater daily mobility has reduced the need for residential mobility. Many of these issues can remain little more than speculation because, for the most part, we do not have direct evidence on attitudes to, and the meaning of, migration in the past. At one level the data on migration suggest relatively little changed over a long period, but this apparent stability can be interpreted in a variety of different ways.

CHAPTER 11

Conclusion: a broader perspective on migration and mobility in Britain

Introduction

It is important to begin this concluding chapter by re-emphasizing the issues which this study does and does not address. Research based on a sample of individual longitudinal migration histories is not a substitute for aggregate studies of net and gross migration flows between administrative units derived from census and vital registration data. Such studies have been completed for those years when relevant data are available in Britain: they depict the broad outlines of regional demographic change and emphasize the net effects of inter-regional migration flows (Friedlander and Roshier, 1966; Lawton, 1968a; 1983). However, there are many issues which such studies cannot tackle, and it is these lacunae that data on life-time residential histories can begin to fill. In particular, data collected from family historians are ideally suited to investigation of the ways in which migration and mobility changed over the life-course; the links between life-time migration experiences and migrant characteristics; examination of the reasons why people moved; investigation of the relative significance of short distance and circulatory moves (which are not revealed in aggregate census-based studies); assessment of the significance of mobility between different places; and examination of the relative importance of movement within and between different parts of the country. Previous chapters have focused on such issues and, in doing so, have complemented the information available from most existing studies of migration.

The main reason why such issues have not previously been examined in Britain relates to the availability of suitable data. In some other European countries population registers have recorded complete migration histories of individuals for up to 200 years. Thus population registers are available in Sweden from the mid-eighteenth century, in Belgium from the 1840s, The Netherlands from the 1850s and Italy from the 1860s. Germany also has good quality migration statistics for urban areas from the early-nineteenth century (Gutmann and van de Walle, 1978; Kalvemark, 1979; Hochstadt and

Jackson, 1984; Kertzer and Hogan, 1985; Sperling, 1989; De Schaepdrijver, 1990; Hoppe and Langton, 1992; Hoerder and Moch, 1996). Such sources provide longitudinal life-time migration histories which allow population movement to be examined over the whole life-course. In countries without such registers, and for earlier time periods, the detailed reconstruction of population mobility has only been undertaken at the local level, usually through the painstaking reconstitution of small populations over specific periods of time. These studies have begun to yield some important information on local mobility in pre-industrial Britain, and elsewhere in Europe, but it has not previously been possible to examine issues such as the nature and extent of local circulation, the reasons for migration, and the links between mobility and the life-course for large areas or long periods of time (Terisse, 1974; Clark, 1979; Kussmaul, 1981; Moch, 1983; Souden, 1984; Whyte and Whyte, 1986; Schofield, 1987). The data provided by British family historians, in effect, reconstructs population registers for a sample of the population drawn from all parts of the country and spanning a 200-year time period. For the first time it has been possible to examine something approaching all residential moves undertaken by a large number of individuals during their life-courses, and examine the ways in which these varied over space and time.

Data provided by family historians are not perfect. As shown in Chapter 2 there are some biases, it is likely that some moves have been missed, and information on reasons for migration is partial. The data do, however, yield more detailed information about the process of migration than most previous British studies. There is also scope to extend the range of analysis undertaken. Material presented in this volume has deliberately focused on relatively simple quantitative analysis integrated with qualitative data used to explain and interpret trends. Future analysis could usefully extend the quantitative investigation through event history analysis and multivariate modelling (Davies, 1984; Allison, 1984; Hobcraft and Murphy, 1986; Courgeau and Lelièvre, 1992; Courgeau, 1995).

The remainder of this chapter attempts to place the material analyzed earlier in this volume in a broader perspective. This is achieved by, first, reassessing some of the main theories, hypotheses and questions formulated about migration in the past and examining the extent to which evidence from this sample of complete residential histories supports or challenges existing knowledge. Attention is focused particularly on the hypothesis of the mobility transition formulated by Zelinsky, on Ravenstein's 'laws' of migration and on a series of more recently-formulated research questions relevant to population migration. Second, some of the results from this study are placed in a broader perspective by comparing them with selected evidence from studies of migration trends in a series of European countries, to assess the extent to which British migration patterns were distinctive within the context of European demography. Third, the value of a longitudinal approach to migration, based on individual migration biographies, is assessed in the context of both contemporary and historical research on population movement in Britain. It is argued that, whilst there is clearly need to adopt a multi-method approach to research on population migration, the analysis of life-time residential histories has shed important new light on the migration process and provides an important alternative to more conventional census-based studies.

Was there a mobility transition?

As suggested in Chapter 1, the hypothesis of a mobility transition formulated by Zelinsky (1971) is superficially appealing, but it is also open to considerable criticism. Zelinsky (1979, 1983, 1993) has reappraised his views on several occasions, and his ideas have been critically assessed in a number of different contexts (Anderson, 1980; Skeldon, 1992; Cadwallader, 1993; Woods, 1993; Hoerder and Moch, 1996). Zelinsky originally put forward eight related statements which can be summarized as follows. First, that communities underwent a transition from limited physical and social mobility to higher rates during the process of modernization; second, that the course of the mobility transition paralleled and was related to other demographic transitions; third, that there were major and orderly changes in the form of spatial mobility at different stages of the transition; fourth, that there was an interaction between social and spatial mobility with the possibility of one being traded for the other; fifth, that changes in mobility patterns diffused through time and space; sixth, that the processes of mobility change accelerate over time; seventh, that due to this acceleration socio-spatial patterns of mobility in different countries will change depending on the relative timing of the mobility transition; and, eighth, that the process of mobility transition moves through an irreversible series of stages.

Evidence from Europe has clearly demonstrated that levels of pre-industrial mobility were much higher than Zelinsky assumed, and research in developing countries has shown the inappropriateness of modernization theory which assumes regular progression through a number of stages. Whilst recognizing these, and other, problems with his original conception, Zelinsky (1993, p. 218) has recently argued for 'the situating of the mobility transition within a considerably grander scheme' and the development of 'a general theory of the historical geography of modernization'. Neither the concept of the mobility transition nor its links to what Zelinsky terms 'a sizeable family of transitions, all serving as essential components of the modernization process' are completely dead. It is thus appropriate to use evidence from the study of life-time residential histories since 1750 to ascertain whether there is any evidence to support the notion of a mobility transition in Britain. Data collected in this study can be used to examine

321

four main aspects of a mobility transition. First, changes in the extent of mobility over time and alterations to the pace of any change that occurred; second, changes in the form and structure of mobility over time, including the distance, direction and duration of movement and the characteristics of migrants; third, spatial variations in the nature and pace of changes in mobility and the extent to which such changes diffused from one region to another; and, fourth (and to a more limited degree) the nature of any links between social and spatial mobility.

It can be suggested that changes in the extent of mobility can be measured in a variety of ways. These may include the number of times people moved in a life-time, the distances over which they moved and, related to this, the impact of mobility on people and places. Evidence presented in previous chapters suggests that each of these changed to only a limited extent over the 200 years from 1750. Thus, although the mean number of moves undertaken in a life-time increased from the eighteenth to the twentieth centuries, at least part of this change can be attributed to defects in the data for earlier time periods. The distances over which people moved changed little during the eighteenth and nineteenth centuries and, although the proportion of long distance moves increased in the twentieth century this was, arguably, much less than might have been expected given changes in transport technology and modes of communication. At all times most people moved frequently over short distances and within well defined geographical areas. Given the short distances over which most people moved it is also hard to argue that there were any significant changes in the impact of migration over time, especially as in terms of impact and meaning a move over (say 30 kms) in the 1760s may have been as disruptive as a move over 100 kms in the 1980s. Although it is possible to interpret any data set in a variety of ways, and there were undoubtedly some changes from the eighteenth to the twentieth centuries, the most compelling picture which emerges is one of relative stability rather than change.

There is also only limited evidence of the changes in form and characteristics of migration postulated by Zelinsky. The characteristics of migrants were particularly stable in the eighteenth and nineteenth centuries with migration dominated by young adults moving in family groupings. In general terms the same was true for the twentieth century, but in the period after 1920 there was more movement by the elderly and, related to this, an increase in the proportion of moves undertaken alone. There were also fewer moves by children in the twentieth century, suggesting increased mobility constraints for families with young children. The basic spatial pattern of migration was very stable over time, with most movement contained within well defined regional migration systems and only London consistently attracting substantial numbers of migrants over long distances. Likewise, the pattern and process of migration proved to be very similar between different parts of Britain. The most remote rural areas were more distinctive than most regions, and London was rather different from all other cities, but there was a high degree of stability over both time and space, and little evidence that particular mobility characteristics diffused from one region to another.

Although the data were not primarily designed to study social mobility, there is some evidence of a link between social and spatial mobility, and of some families operating a trade-off between the two. This is a topic which warrants much further investigation. Although most migration did not lead to a change in occupational status, long-distance moves and moves undertaken alone were more likely to lead to upward social mobility. It is also clear from evidence contained within the histories of individual families that the decision to move was often a family decision in which economic prospects for some (usually younger) members may have been improved, whilst older family members may have experienced downwards social mobility. As with most other aspects of the data set these trends are quite stable over time.

Thus although there were some changes in the volume and nature of migration within Britain in the past, it is hard to argue that these amounted to the mobility transition hypothesized by Zelinsky. Moreover, rather than these changes occurring in the nineteenth century in tune with other economic, technological and demographic shifts of the Victorian period, the most marked changes took place after 1920. These reflected well established trends in twentieth century society, as people lived longer and were more likely to move alone in old age, and as access to cheap personal transport enabled some people to move over rather longer distances. However, even in the mid-twentieth century the most common experience was of short distance movement in nuclear family groupings: a pattern that had been well established since the mid-eighteenth century. Interpretation of these data obviously depends on whether attention is focused on a core of apparent stability, or on more peripheral (but possibly significant) changes at the margins. However, evidence of fundamental change in the volume, direction, form and characteristics of migration and mobility is lacking. In the context of Britain since the mid-eighteenth century, it can be suggested that pursuit of a grand theory which 'binds the various transitions together into the immense tapestry of modernization' (Zelinsky, 1993, p. 218) may be misplaced.

Re-assessing Ravenstein's 'laws of migration'

Ravenstein's 'laws of migration', briefly stated in Chapter 1, have formed a starting point for much migration research. They were based largely on empirical analysis of the census birthplace tables in the British censuses of 1871 and 1881, with some use of data for North America and other European countries. Although some of Ravenstein's 'laws' were based directly on census evidence, many were really hypotheses for which, given the limitations of census birthplace tables, he could have had only limited evidence (Ravenstein, 1876, 1885/1889). The extent to which Ravenstein's ideas have been borne out by later research has been assessed in the context of British migration by Grigg (1977) who suggests that of Ravenstein's 11 'laws' at least six have been broadly substantiated by later researchers, whilst most of the others remain unsubstantiated due to lack of evidence. Grigg concludes that Ravenstein's 'work on migration in nineteenth-century Britain, although greatly elaborated by later writers, has not been superseded.' Other researchers, particularly those writing from a more global perspective, have been less kind. In developing a framework of global migration systems Hoerder (1996) accuses Ravenstein of producing work which lacks dynamism and sophistication, whilst Nugent (1996) includes Ravenstein's ideas in his critique of modernization theory. However, such critics may be extending Ravenstein's hypotheses beyond the level at which they can reasonably be used. They remain essentially a commentary on nineteenth-century British migration and, in many ways, they are an outdated perspective on the complexities of migration. But their prominence in textbooks and other work warrants some reappraisal of Ravenstein's work, and the data collected from family historians and analyzed in this volume allow the examination of the 'laws' in a way which has not previously been possible.

Ravenstein's assertion that the majority of migration was over short distances is clearly borne out by the evidence presented in previous chapters. Though, as discussed in Chapter 10, the definition of what was a long or short distance is clearly problematic (and varied over time and between groups of the population) Ravenstein's estimate of 25 per cent of migration taking place over long distances is close to the figure of one fifth derived from the family history data set. Given that Ravenstein relied on aggregate census tables (which mostly obscure short distance within-county flows) he probably underestimated the sheer importance of short distance circulatory moves over the life-course as a whole, but the predominance of short distance moves in all places and time periods is a consistent feature of many migration studies.

Linked to this theme is the hypothesis that those migrants who did travel over long distances mainly went to large industrial and commercial towns. This statement tends to over-simplify a complex pattern of migration and is only partly borne out by the data. It is true (Chapter 3) that London consistently attracted migrants over much longer distances than all other places, and that there was a substantial amount of inter-urban movement between most large cities. Because of the spacing of large places such movement had to be over relatively long distances. However, examination of movement up and down the urban hierarchy (Chapter 4) shows that distances moved to larger places were similar to those moved to smaller places. In comparison to circulatory moves between nearby settlements of the same size, movement both up and down the urban hierarchy tended to be over relatively long distances. Thus it is true that moves from small to large places were more likely to be over long distances than moves between small places, due to the high volume of short distance circulatory moves which (by definition) must have been between small places. However, long distance moves were fairly evenly distributed between settlements of varying size and could occur both up and down the urban hierarchy.

One of Ravenstein's best known but least substantiated 'laws' is the assertion that most migrants moved in a series of steps, creating a wave-like movement of migrants into urban centres. Validation of this depends, in part, on the interpretation of Ravenstein's use of the term stepwise migration. If all that was implied is that most people moved frequently over short distances, rather than infrequently over long distances, then the data from family historians largely support the hypothesis. However, if it is taken to mean (as is usually the case and Ravenstein implied) that there was a clear directionality to stepwise migration with migrants moving from small places to nearby larger settlements and so on up the urban hierarchy until they reached big cities, there is no evidence that this was the case. In fact the data on life-time residential histories paint precisely the opposite picture (Chapter 4). Although the balance changed over time, moves from large to small places were almost as common as those from small to large, and migrants moving to large cities were more likely to have moved directly from small places than to have moved up the urban hierarchy in a stepwise fashion. Given that it is impossible to determine stepwise moves from published census tables (indeed, only life-time residential histories can discover such patterns), it seems likely that this assertion was based upon misplaced observation of what Ravenstein thought was happening rather than on substantive evidence.

Ravenstein's hypothesis that every migratory current had a counter-current to some extent contradicts his own assertion of the attractive powers of large cities, but the longitudinal migration data clearly demonstrate the importance of flows both in and out of all settlements, and the fact that net differences between migration streams were very small in relation to gross flows (Chapter 4). Ravenstein was clearly correct in asserting that most migration currents had a substantial counter-current. However, his hypothesis that most movement was from rural to urban areas was, at best, an overstatement. As shown in Chapter 4, although there was a small net flow towards larger towns up to the 1880s, when gross flows are examined there was also substantial movement from large cities to smaller settlements, and this became a dominant trend in the twentieth century. Although there was differential movement to large towns this was a very small part of the total migration system. Ravenstein also suggested, without a great deal of evidence, that urban dwellers were less migratory than those who originated in the countryside. The implication is that country people moved to towns and stayed there, and those originating in towns did not move out. This is clearly not substantiated by the data presented in Chapters 3 and 4 which show high volumes of migration both within and between urban and rural areas. There is no evidence of any differences in the propensity to migrate by region or settlement size.

Ravenstein paid some attention to migration differentials, asserting that women were more migratory than men within Britain, but that men were more likely to move overseas. This has often been extended to suggest that men were also likely to move over longer distances in Britain (Grigg, 1977). He also argued that most migrants were adults and that families rarely migrated. These differentials are only partly supported by the longitudinal data. There is clear evidence that men were more likely to emigrate to the destinations that were attracting large numbers of overseas migrants during the nineteenth century (Chapter 9), though substantial numbers of women also emigrated; but there is no evidence that men moved over longer distances within Britain or that women had a greater propensity to move (Chapter 3). In fact, in all respects apart from the reasons for movement, the migration characteristics of men and women were very similar. Ravenstein was correct in stating that adults were (marginally) more likely to migrate than children, reflecting the constraints on migration imposed by young children, but this ignores the fact that at all times the majority of all types of moves were undertaken by people in family groupings. It is simply not true to say that families rarely moved home, but what is true (and what probably influenced Ravenstein's observations) is that lone adults were over-represented in longer distance (and more apparent and easily recorded) moves.

Ravenstein's assertion that towns grew more by migration than natural increase in the nineteenth century has been challenged by other writers (using census and vital registration data) and is clearly not substantiated by the longitudinal data (Cairncross, 1949; Lawton, 1968a). The small net differences between in- and out-migration in most settlements meant that natural increase must have been at least as important as migration for urban growth (Chapter 4). As demonstrated above, there is also only limited evidence to support Ravenstein's statement that the volume of migration increased over time in tune with the development of industry and new forms of transport. This is the aspect of Ravenstein's work which links most closely to the later (and much criticized) modernization theory and, as shown in Chapter 3, even the increases in migration distances which did occur were much less than might have been expected given changes in transport technologies. Finally, Ravenstein argued that the main reasons for migration were economic, with people being attracted to employment and higher wages in towns. Analysis of reasons for migration provided by family historians broadly supports this assertion, but with some interesting gender differences. However, the actual pattern of reasons for migration was rather more complex than Ravenstein assumed. For the period when Ravenstein was writing, although long-distance migration was mainly for employment reasons, the much more common short distance moves were just as likely to be for marriage or (increasingly) housing reasons. Even amongst long distance moves a substantial minority were due to other causes, including personal crises and family reasons.

Overall, although by no means all of Ravenstein's assertions have been substantiated by this analysis, his more important hypotheses have stood the test of time remarkably well, and his errors are understandable given the data with which he was working. Although he rightly emphasized the importance of short distance moves, most of his later observations about migration differentials, the role of towns in the migration system and the reasons for migration applied mainly to the much rarer phenomenon of long distance migration. This is not surprising, because the census data with which Ravenstein worked emphasizes longer distance inter-county flows, and such movement would also have been more obvious to contemporary observers. Urban dwellers were aware of, possibly distinctive, in-migrants but paid little attention to those who left the cities for other locations, or the high volume of local circulation that occurred in rural and urban areas. Such movement can only be revealed through life-time residential histories which help to place the observations of Ravenstein within a broader perspective.

Recent developments in migration theory and analysis

Recent work on both contemporary and historical migration studies has shifted attention away from the more mechanistic and quantitative approaches to migration, and has focused on migration as a process of social and cultural change affecting both individuals and communities. It has been argued that most migration studies (in common with much population geography) have been dominated by empirical studies which have lacked an adequate theoretical base, and that population geographers have failed to respond to changes taking place elsewhere in geography and in the social sciences in general (Findlay and Graham, 1991; White and Jackson, 1995). In a review of migration theory Woods (1985) has emphasized the need to relate the structural context of society to the behavioural responses of individuals, whilst a number of researchers have recently argued for the need to relate migration studies to processes of social and cultural change, and to embrace recent developments in cultural theory (Piore, 1980; Champion and Fielding, 1992; White and Jackson, 1995). Although mathematical models of migration behaviour still attract considerable attention, much more emphasis has been placed on the social and cultural meaning of the migration process set within the context of the development of modernity (Giddens, 1990, 1991). Whilst much of the analysis presented in this volume has been essentially empirical, the use of migration biographies linked to individual diaries and life histories has allowed some interpretation of the impact of migration and mobility on people and places (Chapter 10).

In a recent review of European migration in the past, Jackson and Moch (1996) have posed five questions which, together, they argue focus attention on key issues in 'defining the mechanisms and social meaning of migration'. These can be summarized as, first, examination of the selectivity of migration streams, the ways in which selectivity was related to and impacted upon local labour markets and communities, and the relationship of the propensity to migrate to individual migrant psychology and attitudes towards innovation and change. Second, they focus on the motivations for geographical mobility, the role of the family in decision-making, and the ways in which perceptions of opportunities were structured. Third, they suggest that greater attention should be paid to complete migration systems, focusing especially on return migration and multiple or circulatory moves within the context of complete migration systems. A fourth question relates to the impact of migration on origins and destinations, both in terms of the demographic and social impact of differential out migration and the impact of large scale migration on urban structures. Fifth, they focus attention on the impact of migration on individual movers, particularly the extent to which migration was associated with dislocation and alienation or took place within well established structures which minimized the impact of change. All these questions, and the links between them, have to some extent been addressed in this study.

The impact of migration selectivity, the disruptive influences of migration and its impacts on people and places depends very much on the perspective from which the migration process is viewed. In the context of overseas moves a degree of selectivity was clearly important, with men more likely to emigrate than women and with migration streams structured by labour market opportunities and (especially in the case of migration to Australia and New Zealand) the selection procedures of emigration companies. Thus the repeated overseas moves of nineteenth-century tin miner John James (to Norway, Newfoundland, Ireland and other locations) were prompted by lack of employment in his preferred location in Cornwall, by labour market opportunities elsewhere, and by his relatively privileged position as a skilled mining engineer and (latterly) mine manager (Chapter 9). Selection thus occurred on criteria of employment and skill, but there was also a strong push element, and throughout his life John James retained a long-term desire to live in Cornwall despite his own wide-ranging migrations and the fact that his family were widely scattered around the globe. He frequently returned to Cornwall between work overseas and moved back to his home village at the end of his working life. Although at times migrating alone, he was also keen to take his wife and family with him on overseas moves whenever possible. Thus although men did dominate emigration in the nineteenth century many women also moved overseas. It is suggested that the selective and disruptive nature of overseas migration should not be over-stated.

If the migration system is viewed in the context of all geographical mobility, ranging from short distance residential moves within one community to emigration overseas, then the impact of selectivity and dislocation for people and places takes on even more limited dimensions. One of the main advantages of longitudinal life-time migration histories is that they place mobility within the broader context of everyday events rather than emphasizing the relatively unusual but more easily recorded longer distance migrations. Evidence from the life-time residential histories collected from family historians emphasizes the relative lack of gender differences in the structure and process of migration though, to varying degrees, men and women did tend to move for different reasons, reflecting the gendered nature of particular responsibilities within society (women were most likely to move on marriage, as the result of personal crises and for family reasons, particularly to live near and care for relatives). The fact that most people moved as part of a family group, that this movement was over short distances within a well-known area, and that the pattern of migration over the life-course was essentially similar for all people, places and time periods, emphasizes the universality rather than selectivity of migration experiences as a whole. However, there is evidence that some longer distance and thus more noticeable and potentially disruptive moves were selective (particularly of lone adults, initially in younger age groups but in the twentieth century also including substantial proportions of older lone migrants, both male and female). It can also be suggested that it is the visible nature of such moves both to contemporaries and in surviving sources which has led to the somewhat misleading emphasis on the differential effects of migration.

The analysis of life-time residential histories, and the reasons why people moved, has also emphasized the stability of migration motivations over time and space. Analysis of detailed diaries and life histories has highlighted the complexity of most migration decisions, with movement often related to a combination of employment, family, housing and other factors, and with many moves generated by personal crises of one sort or another. Thus the Victorian Londoner, Henry Jaques, moved frequently within particular parts of the metropolis as a response to employment opportunities, life-course changes, family responsibilities and housing aspirations. Although the precise pattern of occurrences obviously varied from individual to individual, the ways in which different reasons for migration combined over the life-course appear very stable over space and time, and between different groups of the population. Migration decisions seem to have been structured in similar ways over much of the last two hundred years and changes in the relative importance of particular reasons can be explained partly by deficiencies in the data for the more distant past, and partly by large scale

329

structural changes in the nature of society in the twentieth century. In a society where most people are more affluent and have more opportunities it is not surprising that movement for reasons other than marriage and employment has become relatively more important.

It can thus be suggesed that too many studies in the past have, understandably, overstated the noticeable and atypical aspects of migration and mobility, and have neglected the everyday and commonplace dimensions of population movement. The analysis of life-time residential histories allows issues such as the selectivity, impact and motivations for migration to be placed within a broader perspective. With respect to the development of theory and further empirical research, we suggest that this study urges developments in two contrasting but related directions. First, there is strong evidence for the universal nature of the migration process with more evidence of stability than change over time and space, and between different groups of the population. It can thus be argued that migration should be viewed as a basic human event which almost everyone experienced at some time in their lives. Although obviously influenced by temporal and spatial specificities, by structural changes over time, and by the process of modernity, the main dimensions were remarkably stable and reflected deep-seated human responses to particular situations. Because the human life-course is essentially similar for all people (though the timing and nature of events may vary) the stabilizing influence of life-course events on the migration process has tended to outweigh the potentially dislocating effects of structural change within society. There is potential for much more work on the relationship between life-course events and broader changes in economy and society. Second, the longitudinal residential histories have emphasized the complex ways in which individuals and families made migration decisions. Although aggregate outcomes may have been similar, the decision-making processes through which individual migrants went were highly varied and personal. There is scope for much more work on the ways in which particular migration decisions were worked out in relation to personal circumstances, family structures and individual psychology. We suggest that more insights into such processes will be gained through the sensitive interpretation of qualitative evidence than through the construction of mathematical models of human behaviour.

Placing British migration in a broader context

The results of this study of mobility and migration in Britain are broadly consistent with recent research on migration in Europe, based mainly on population registers and other sources which provide longitudinal evidence (Chatelain, 1969; Todd, 1975; Poussou, 1983; Bardet, 1983; Moch, 1983, 1989; Hochstadt, 1983; Poitrineau, 1985; Kertzer and Hogan, 1985; Lucassen, 1987;

Bade, 1986; Reher, 1990; Hoerder and Moch, 1996). These studies emphasize the 'continuities and regional networks in migration history' (Jackson and Moch, 1996, p. 53) in a set of systems in which people moved frequently and predominantly over short distances or over well-established migration routes. Although short distance moves predominated, some well established seasonal migration systems moved workers over considerable distances, whilst large industrial and capital cities could also attract migrants over long distances. Studies also stress the temporary and circulatory nature of migration with many of those moving from rural to urban areas moving on to other destinations. The British data fit almost precisely into this picture, with most people moving over short distances, only London consistently attracting large numbers over long distances, and many people moving back to their areas of origin.

Data on the selectivity and gendered nature of migration in Europe are more varied, with several studies stressing the different migration experiences of men and women, related to regional labour demands, inheritance systems and societal structures (Brettell, 1987; Fuchs and Moch, 1990). Considerable emphasis is also placed on the migration of young adults, though the importance of family migration has also been recognized (Moch, 1986). This British study places greater emphasis on the similarities between male and female experiences (whilst recognizing important differences in the reasons why men and women moved and in gender specific labour markets such as the movement of domestic servants), and on family migration. It is difficult to assess whether there were any real differences in these respects between Britain and continental Europe, or whether the apparent contrasts reflect different emphases in the data and research.

Most commentators do agree that there were important differences between Britain and the rest of Europe with respect to the contribution of migration to urban growth. Urban mortality tended to be higher (and fertility lower) in continental Europe than in nineteenth-century Britain, and thus rural to urban migration played a more significant role in urban growth than in Britain (Moch, 1996). However, although continental Europe may have experienced a larger net transfer of migrants to towns than occurred in Britain, there is also evidence that (as in Britain) there was a substantial amount of movement out of cities as well as in, and it has been suggested in the context of nineteenth-century European cities that 'migration may have been much less important to the growth of some important administrative and commercial capitals than contemporary observers believed' (Jackson and Moch, 1996, p. 60). Most continental European studies also stress the significance of rural industrial development in generating labour migration both within and between regions (Lis and Soly, 1979; Hochstadt, 1981; de Vries, 1984; Hohenberg and Lees, 1985; Moch, 1996). Movements to sites of rural industry were also an important feature of the British migration system, but there were significant differences in timing with Britain achieving a high degree of urbanization and urban-based industry at an earlier date than most other European countries.

Although other studies of migration in Britain, especially those dealing with pre-industrial populations (Clark, 1979; Kussmaul, 1981; Souden, 1984; Whyte and Whyte, 1986; Schofield, 1987), have highlighted some of these themes, this analysis, based on a large number of longitudinal residential histories, allows more direct comparison with European evidence drawn from population registers than most other research. The similarities with much of continental Europe are striking, and the differences are consistent with variations in the timing and nature of urbanization and industrialization, and differences in economic and social structures including patterns of land inheritance. A similar picture of the nature and role of migration has also been painted for many developing nations (Skeldon, 1990; Chant, 1992), and it is significant that it is in such contexts that biographical or life history approaches to migration have been most widely used. Because of the relative lack of statistical evidence migration researchers in many developing countries have made greater use of ethnographic approaches which have involved the collection of detailed information on residential change together with other aspects of the respondents' lives. There has recently been increased interest in biographical approaches to research on contemporary migration in the developed world, recognizing that such data can yield important insights not available through official statistics or structured questionnaires (Halfacree and Boyle, 1993; Skeldon, 1995). Although in an historical context we must often rely on evidence reconstructed from a range of documents, rather than on the testimony of individual migrants (though both pose problems of bias and verification), the degree of individual detail which can be extracted from such sources, when combined with the aggregate analysis of national and regional trends, can shed important new light on migration in the past.

Although much of the analysis presented in this volume has been at the aggregate level, examining long-run trends in migration over space and time and between groups of the population, it is at the individual level that explanation and interpretation is most effective. This has been achieved through the selective use of individual case studies drawn from the 16,091 provided by family historians and the dozen detailed diaries and life histories. In this context it seems appropriate to conclude with one more life story. This is chosen not because it is unusual or startling, but because it reflects some of the universal themes that apply to almost all the life histories examined. Samuel E. was born in the small town of Chipping Sodbury, Gloucestershire (population 1,273 in 1841) in 1820. His father was a stonemason and his mother combined household duties with straw bonnet making. As the son of a skilled worker it seems likely that his childhood was unremarkable, living in a household which was neither particularly rich nor poor. Samuel remained with his parents in Chipping Sodbury (and as far as is known in the same house) until he was 14 years old, when he was apprenticed to a wheelwright in the much larger town of Bristol (population 122,296 in 1841) some 20 kms from his home. This was the nearest large labour market and it seems very likely that Samuel knew both the town and some people in it before he migrated.

Samuel E. remained at the same address in Bristol (living at his place of work) until he married in 1844. This life-course change forced him to find work and accommodation that was appropriate for a married man and his expected family. He moved back to his home town of Chipping Sodbury where he gained work as a wheelwright and suitable accommodation. It seems reasonable to assume that in doing this he exploited his network of contacts in his home town. In 1850, now with two daughters, he moved some 20 kms east to the small village of Brokenborough near the market town of Malmesbury, Wiltshire (population 2,443 in 1851). His reason for moving was the opportunity to establish his own business as a coachbuilder in the nearby village of Westbury St Mary. Again, it seems highly likely that this was an area known to Samuel and that he used local contacts to acquire his business opportunity. He remained at Brockenborough for 18 years (as far as is known in the same house), but in 1868 he moved with eight of his 12 children to the town of Worcester. This was a rather longer distance move taking him some 83 kms to a larger town (population 33,226 in 1871) and a completely new locality. The main reason for moving was that he changed his employment yet again, establishing himself as a small paint and colour manufacturer. Each change of employment represented a degree of specialization in Samuel's trade from wheelwright to coachbuilder to paint and colour manufacturer, and presumably Worcester was perceived as offering an attractive business opportunity. Samuel E. lived with his family in the same house in Worcester until 1884 when life-course events again encouraged him to move. His wife had died in 1863 and with a much reduced family (three daughters still lived at home) he moved within Worcester. At about the same time he also gave up his paint and colour business (he was now aged 64), but he took work as a commercial traveller in Worcester. He continued in the same house and employment for a further six years until his death in 1890.

This life history, which spans much of the nineteenth century, reflects many of the themes explored in this volume. Samuel was a skilled and successful man, but most of his movements were over relatively short distances and, having moved to a substantial city (Bristol), he returned to his home settlement. His longest move was made directly from a small village to a medium sized town, and all his moves were stimulated by either employment opportunities or life-course events (marriage, death of a spouse) which affected many people. In some respects every individual life history has elements that are unique, and we can never know everything about any life in the past. Even with the best documented life histories, interpretation of a residential history entails a degree of speculation and inference. Whilst it is suggested that the broad structure of migration patterns and processes in the past were relatively stable over time and space, and between groups of the population, the case studies also remind us of the importance of the individual migrant. At all times in the past and present changing home, be it a move to an adjacent street or emigration to the other side of the world, is essentially a personal event initiated and planned by individuals and their families and most directly affecting the individual migrants. Although broader societal and structural implications of migration may be important, it is the impact of migration on individuals like Samuel E., his family, and all the other women, men and children who moved in the past which should remain centre stage.

APPENDIX 1

Diaries and life histories used in the research

Diary of David Brindley 1856–91, provided by Mrs N. Chesters. See Lawton and Pooley (1975).

Diary of John James 1847–80, provided by Mr M.D. Smith.

Life history of Henry Jaques, 1842–1907, provided by Mrs A.M. Giller.

Life history of Amos Kniveton, 1835–1927, provided by Mr W. Melling.

Diary of Rhona Little 1932–59, provided by Mrs R.M.L. Ward.

Diary of Annie Madin, 1916–22, provided by Mr J.M. Smith.

Diary of George Osborne, 1841–42, provided by Mrs S.A. Moyle.

Life history of Mollie Potts, 1902–78, provided by Mrs D.M. Webster.

Diary of Cliff and Dora Powell, 1940–42, provided by Mrs W. Angove.

Life history of Thomas Roberts, 1814–65, provided by Mrs D.M. Webster.

Family History of Benjamin Shaw, 1772–1841. Preston Record Office, DDX/1154/1 and DDX/1154/2. See also Crosby (1991) and Pooley and D'Cruze (1994).

Life history of Thomas Thomas, 1788–1870, provided by Mrs M. Hann.

Diary of Albert Woolley, 1908–15, provided by Mrs B.C. Jenkins.

Diary of Joseph Yates, 1826–96, provided by Mrs J. Watson.

Family history and genealogy societies co-operating with the research project

Aberdeen & NE Scotland FHS
Birmingham & Midland FHS
Bradford FHS
Buckinghamshire FHS
Cambridgeshire FHS
Central Scotland FHS
Chesterfield & District FHS
Cleveland, North Yorks & S Durham FHS
Cornwall FHS
Cumbria FHS
Derbyshire FHS
Dorset FHS
Dumfries & Galloway FHS
Dyfed FHS
East of London FHS
East Surrey FHS
East Yorkshire FHS
Essex SFH
FHS of Cheshire
Glamorgan FHS
Glasgow & West of Scotland FHS
Gloucestershire FHS
Gwynned FHS
Hamilton & District FHS
Hampshire GS
Herefordshire FHS
Highland FHS
Huddersfield & District FHS
Huntingdonshire FHS
Kent FHS
Lancashire FHS

Leicestershire FHS
Lincolnshire FHS
Liverpool & South Lancashire FHS
London & North Middx FHS
Manchester & Lancs FHS
Norfolk FHS
North Cheshire FHS
North West Kent FHS
Northamptonshire FHS
Northumberland & Durham FHS
Nottinghamshire FHS
Oxfordshire FHS
Powys FHS
Quaker FHS
Ripon, Harrogate & District FHS
Royston & District FHS
Scottish Genealogical Society
Sheffield & District FHS
Shetland FHS
Shropshire FHS
Somerset & Dorset FHS
South Cheshire FHS
Suffolk FHS
Sussex FHS
Tay Valley FHS
The Borders FHS
The Society of Genealogists
Troon & District FHS
West Middlesex FHS
West Surrey FHS
Wiltshire FHS
Yorks Archaeology Society

In addition, the Federation of Family History Societies provided up-to-date lists of affiliated societies and publicized the project in its journal.

A. Personal Details

Surname		Firstname		Sex	

Date of Birth		Date of Death	

Place of Birth		Place of Death	
county		county	
settlement		settlement	
street		street	

Father's Occupation		Mother's Occupation	

B. Lifetime Occupational History

	Date entered workforce				Date left workforce	
	Date from	Date to	Job Description	Employer's Business	Employer's Business	Location of employer
1						
2						
3						
4						

C. Other Important Life Events

Marriage	Date of first marriage	
	Date(s) of subsequent marriage(s)	

Childbirth	
Date of birth of first child	
Date of birth of last child	
Number of children	

Major family events				
Type				
Date				

D. Residential History

Date	Age	Marital status	Origin	Companions	Reason	Destination	Assistance	Housing tenure	Occupation	Place of work	Sources used
1			county settlement street			county settlement street		B: A:	B: A:	B: A:	
2			county settlement street			county settlement street		B: A:	B: A:	B: A:	
3			county settlement street			county settlement street		B: A:	B: A:	B: A:	
4			county settlement street			county settlement street		B: A:	B: A:	B: A:	
5			county settlement street			county settlement street		B: A:	B: A:	B: A:	

Details of variables included in three data entry tables

Table 1 Name Table (data relating to an individual)

Field name	Field description
namenumber	The unique number allocated to the individual
famgroup	The unique number allocated to each respondent
fhsoc	Identifier for the fhsoc to which the respondent belongs
sex	Male or female (coded)
yearbirth	Year of birth
birthcounty	Three letter code for county of birth, or country if overseas
birthsettlement	Full name of settlement
birthstreet	Street and number of house if given
yeardeath	Year of death
deathcounty	County of death
deathsettlement	Full name of settlement
deathstreet	Street and number of house if given
fatheroccupation	Details of father's occupation at time of birth
codefathocc	As above (coded)
motheroccupation	Details of mother's occupation at time of birth
codemothocc	As above (coded)
datebeganwork	Year started work
dateleftwork	Year left work
numberoccs	Total number of occupations in life-time
datemarried	Year of first marriage
timesmarried	Number of marriages
yearfirstchild	Year first child was born
yearlastchild	Year last child was born
numberchild	Total number of live children born
crisis1type	First crisis (coded)
crisis1date	Year event took place
crisis2type	As crisis1type
crisis2date	As crisis1date
crisis3type	As crisis1type
crisis3date	As crisis1date
numbermoves	Total number of times the individual moved house in his/her life-time

Table 2 Move Table (data relating to each residential move)

Field name	Field description
namenumber	The unique number allocated to the individual
movenumber	The move number as entered on data collection form
yearmove	The year in which this move took place
agemove	The age at time of move
maritalstatus	Marital status
origcounty	Three letter county code
origsettlement	Full name of settlement
origstreet	Street and number of house (if given)
destcounty	Three letter county code
destsettlement	Full name of settlement
deststreet	Street and number of house (if given)
companions	Full details of all those people with whom the individual moved
codecompanions	Companions (coded)
relationhousehead	Relationship to household head (coded)
reason	Full details of the reason given for the move
codereason	Reason for moving (coded)
assistance	Full details of those who assisted in undertaking the move and the method used
codeassist	Assistance (coded)
codetenurebefore	Housing tenure before move (coded)
codetenureafter	Housing tenure after move (coded)
occbefore	Job title before the move
codeoccbefore	Occupation before move (coded)
indcodebefore	Industry before move (coded)
countywkbefore	County worked in before move
placeworkbefore	Settlement worked in before move
occafter	Job title after the move
codeoccafter	Occupation after move (coded)
indcodeafter	Industry after move (coded)
countywkafter	County worked in after move
placeworkafter	Settlement worked in after move
codeoccchange	Occupational change as result of move (coded)
indcodechange	Industrial change as result of move (coded)
codesources	Sources used (coded)

Table 3 Occupation Table (data relating to each change of occupation)

Field name	Field description
namenumber	The unique number allocated to the individual
occupnumber	Number of occupation as listed on form
date	The date at which the individual began this occupation
length	The number of years spent in this occupation
beforeocc	Job title of the occupation undertaken previously (if applicable)
beforeocccode	Previous occupation (coded)
beforeind	The industry in which they worked previously (if applicable)
beforeindcode	Previous industry (coded)
afterocc	Job title of new occupation
aftercodeocc	New occupation (coded)
afterind	New industry
afterindcode	New industry (coded)
changeocccode	Occupational change (coded)
changeindcode	Industrial change (coded)
statuschange	Change in socio-economic status (coded)
prevcounty	The county of previous occupation
prevlocation	The settlement of previous occupation
thiscounty	The county of this occupation
this location	The settlement of this occupation
residchange	Residential change (coded)

Maps of movement into, out of and within each region

Appendix 5.1 (a) Moves into London, 1750–1879; (b) moves out of London, 1750–1879; (c) Moves within London, 1750–1879

Appendix 5.2 (a) Moves into London, 1880–1994; (b) moves out of London, 1880–1994; (c) Moves within London, 1880–1994

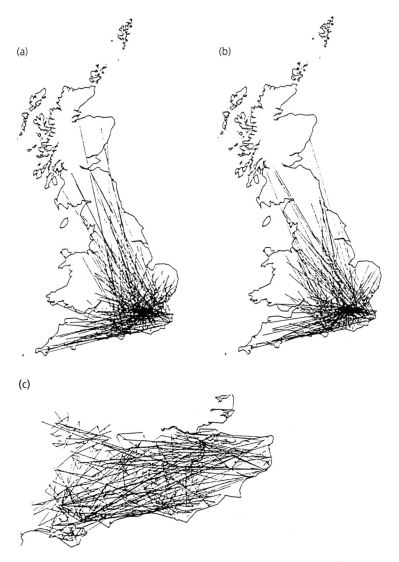

Appendix 5.3 (a) Moves into South-East England, 1750–1879; (b) moves out of South-East England, 1750–1879; (c) Moves within South-East England, 1750–1879

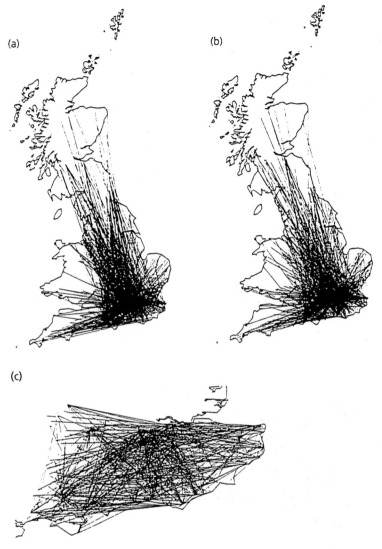

Appendix 5.4 (a) Moves into South-East England, 1880–1994; (b) moves out of South-East England, 1880–1994; (c) Moves within South-East England, 1880–1994

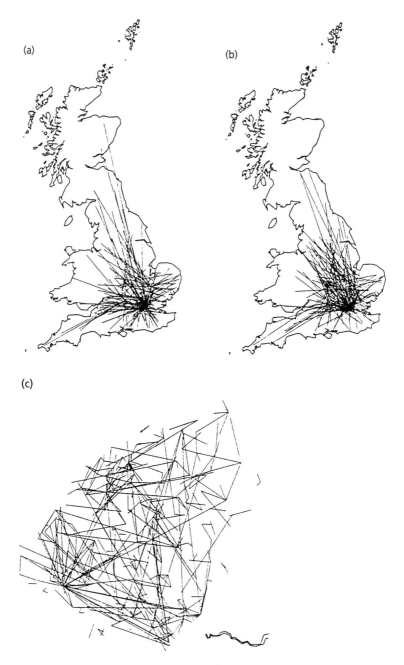

Appendix 5.5 (a) Moves into the South Midlands, 1750–1879; (b) moves out of the South Midlands, 1750–1879; (c) Moves within the South Midlands, 1750–1879

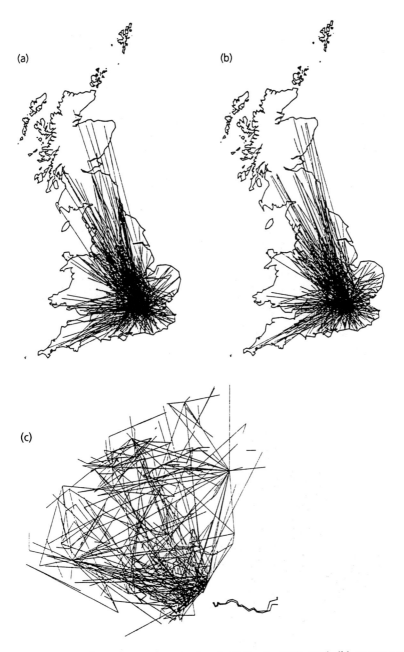

Appendix 5.6 (a) Moves into the South Midlands, 1880–1994; (b) moves out of the South Midlands, 1880–1994; (c) Moves within the South Midlands, 1880–1994

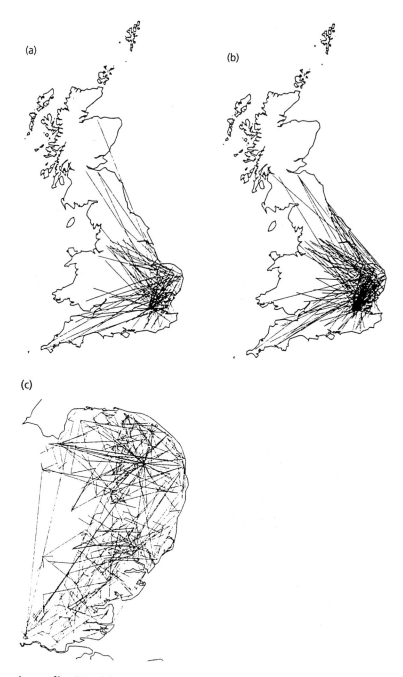

Appendix 5.7 (a) Moves into Eastern England, 1750–1879; (b) moves out of Eastern England, 1750–1879; (c) Moves within Eastern England, 1750–1879

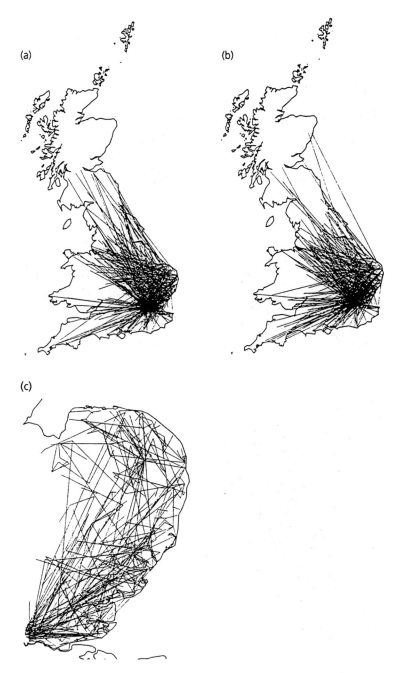

(a)

(b)

(c)

Appendix 5.8 (a) Moves into Eastern England, 1880–1994; (b) moves out of Eastern England, 1880–1994; (c) Moves within Eastern England, 1880–1994

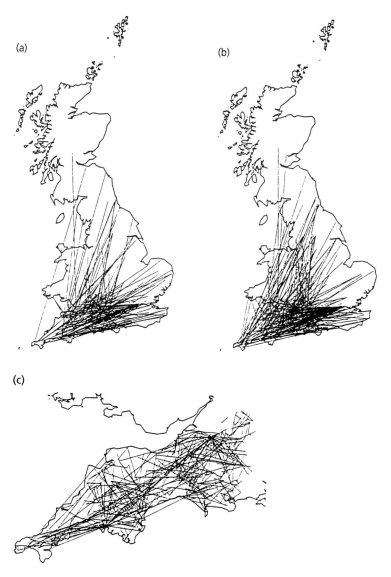

Appendix 5.9 (a) Moves into South-West England, 1750–1879; (b) moves out of South-West England, 1750–1879; (c) Moves within South-West England, 1750–1879

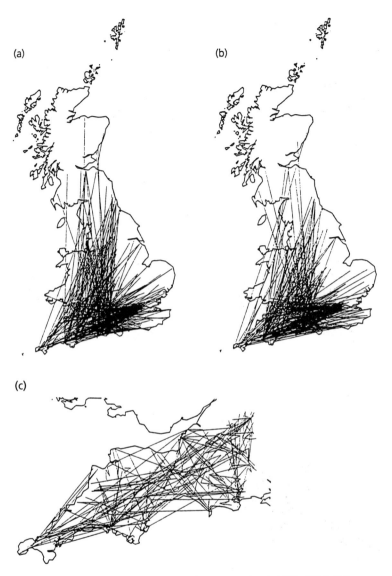

Appendix 5.10 (a) Moves into South-West England, 1880–1994; (b) moves out of South-West England, 1880–1994; (c) Moves within South-West England, 1880–1994

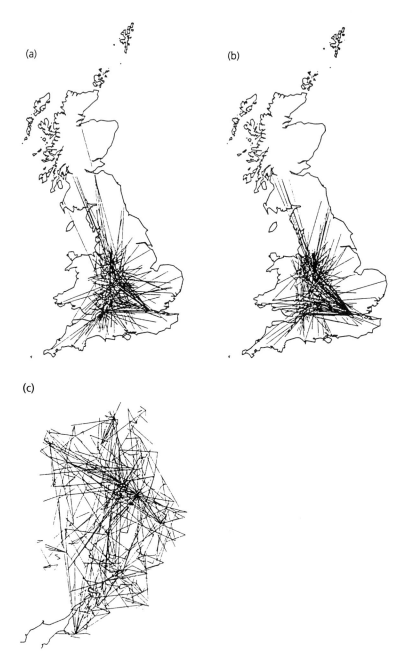

(a)

(b)

(c)

Appendix 5.11 (a) Moves into the West Midlands, 1750–1879; (b) moves out of the West Midlands, 1750–1879; (c) Moves within the West Midlands, 1750–1879

Appendix 5.12 (a) Moves into the West Midlands, 1880–1994; (b) moves out of the West Midlands, 1880–1994; (c) Moves within the West Midlands, 1880–1994

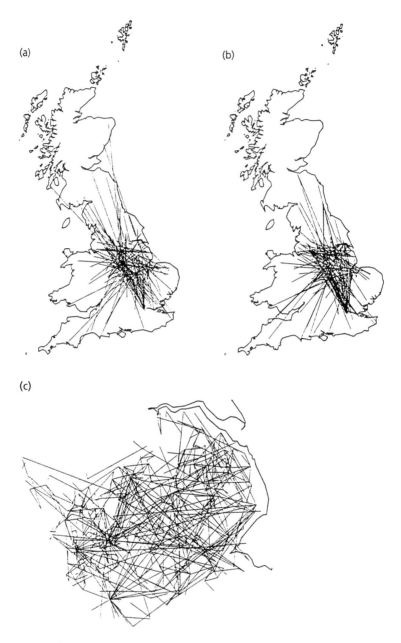

(a)

(b)

(c)

Appendix 5.13 (a) Moves into the North Midlands, 1750–1879; (b) moves out of the North Midlands, 1750–1879; (c) Moves within the North Midlands, 1750–1879

(a)

(b)

(c)

Appendix 5.14 (a) Moves into the North Midlands, 1880–1994; (b) moves out of the North Midlands, 1880–1994; (c) Moves within the North Midlands, 1880–1994

Appendix 5.15 (a) Moves into North-West England, 1750–1879; (b) moves out of North-West England, 1750–1879; (c) Moves within North-West England, 1750–1879

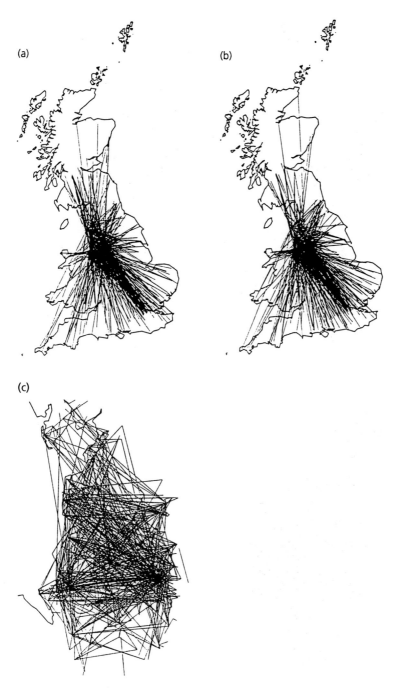

Appendix 5.16 (a) Moves into North-West England, 1880–1994; (b) moves out of North-West England, 1880–1994; (c) Moves within North-West England, 1880–1994

Appendix 5.17 (a) Moves into Yorkshire, 1750–1879; (b) moves out of Yorkshire, 1750–1879; (c) Moves within Yorkshire, 1750–1879

Appendix 5.18   (a) Moves into Yorkshire, 1880–1994; (b) moves out of Yorkshire, 1880–1994; (c) Moves within Yorkshire, 1880–1994

Appendix 5.19 (a) Moves into Northern England, 1750–1879; (b) moves out of Northern England, 1750–1879; (c) Moves within Northern England, 1750–1879

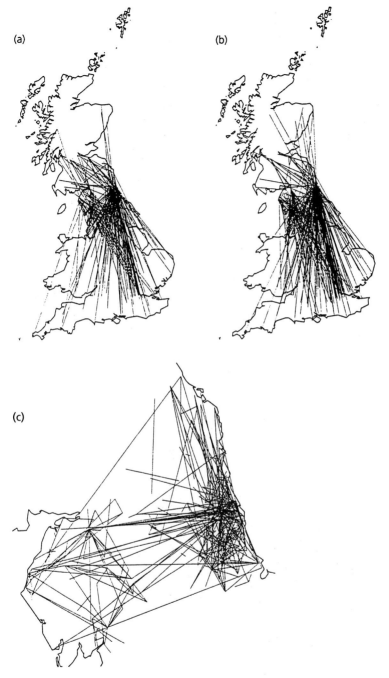

Appendix 5.20 (a) Moves into Northern England, 1880–1994; (b) moves out of Northern England, 1880–1994; (c) Moves within Northern England, 1880–1994

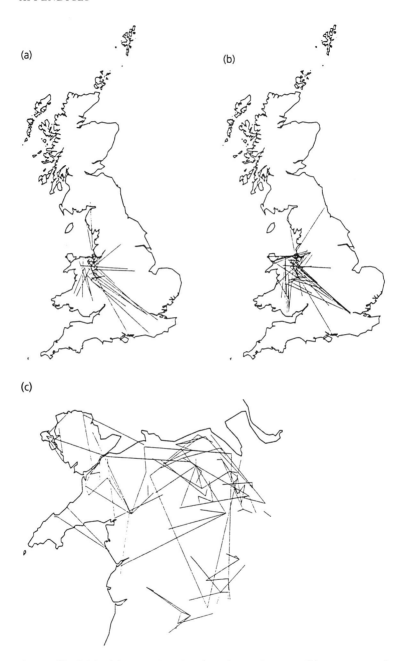

Appendix 5.21 (a) Moves into North Wales, 1750–1879; (b) moves out of North Wales, 1750–1879; (c) Moves within North Wales, 1750–1879

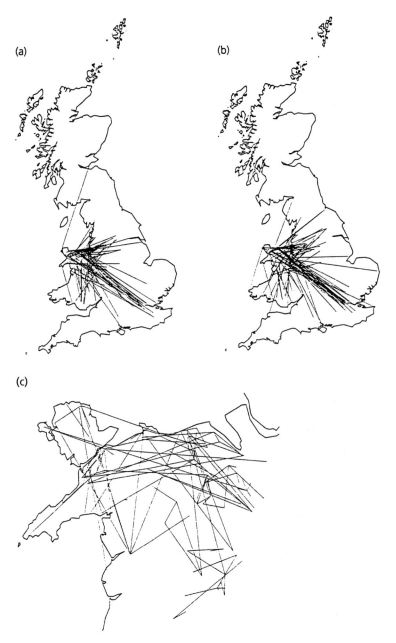

(a)

(b)

(c)

Appendix 5.22 (a) Moves into North Wales, 1880–1994; (b) moves out of North Wales, 1880–1994; (c) Moves within North Wales, 1880–1994

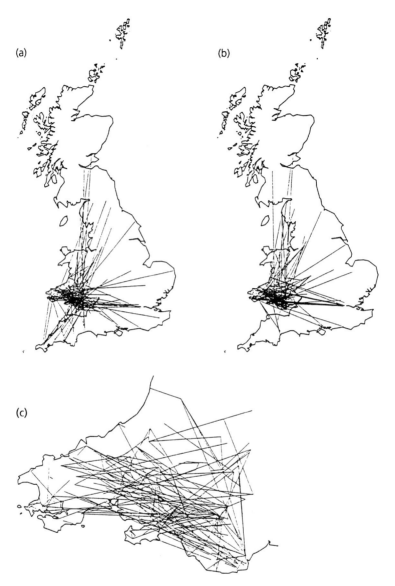

Appendix 5.23 (a) Moves into South Wales, 1750–1879; (b) moves out of South Wales, 1750–1879; (c) Moves within South Wales, 1750–1879

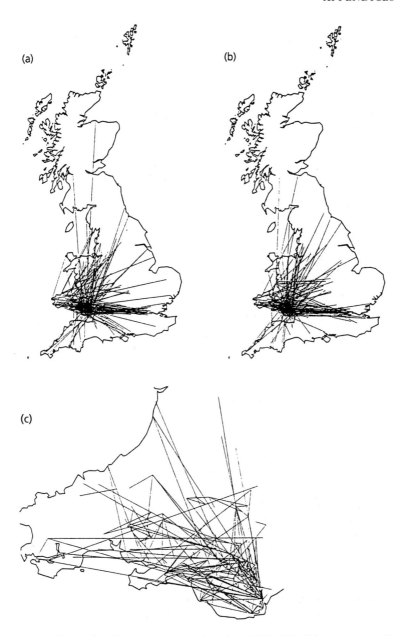

Appendix 5.24 (a) Moves into South Wales, 1880–1994; (b) moves out of South Wales, 1880–1994; (c) Moves within South Wales, 1880–1994

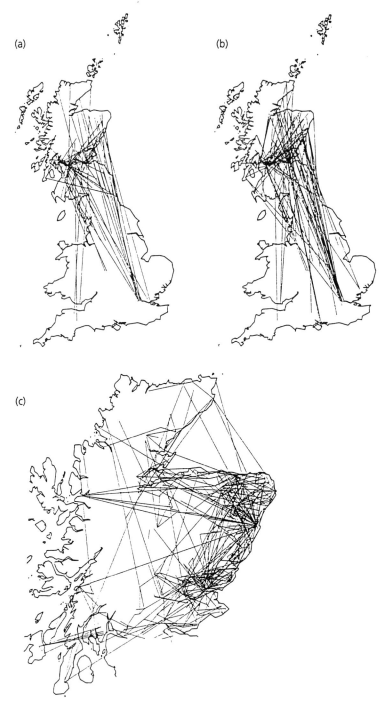

Appendix 5.25 (a) Moves into Northern Scotland, 1750–1879; (b) moves out of Northern Scotland, 1750–1879; (c) Moves within Northern Scotland, 1750–1879

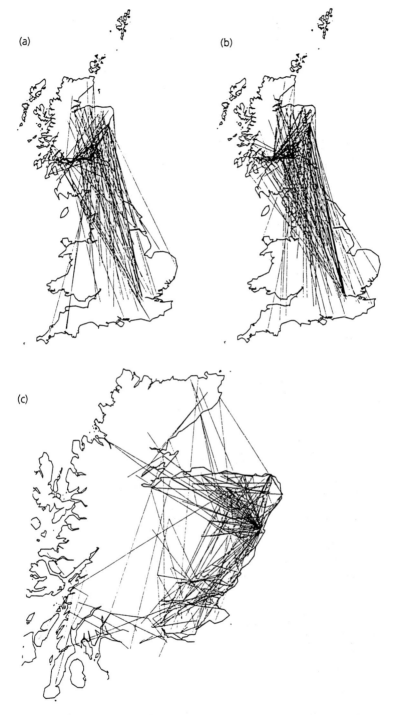

Appendix 5.26 (a) Moves into Northern Scotland, 1880–1994; (b) moves out of Northern Scotland, 1880–1994; (c) Moves within Northern Scotland, 1880–1994

371

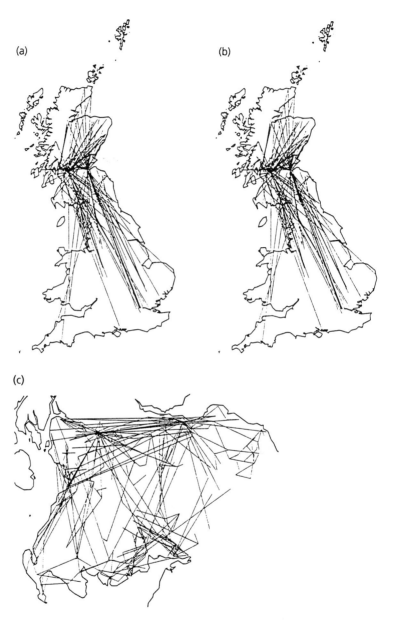

Appendix 5.27 (a) Moves into Southern Scotland, 1750–1879; (b) moves out of Southern Scotland, 1750–1879; (c) Moves within Southern Scotland, 1750–1879

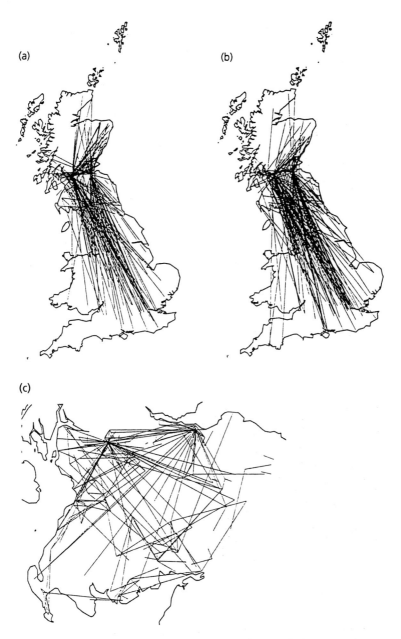

Appendix 5.28 (a) Moves into Southern Scotland, 1880–1994; (b) moves out of Southern Scotland, 1880–1994; (c) Moves within Southern Scotland, 1880–1994

Matrix of movement between settlements of different size, 1750–1839

						Size of origin settlement					
		<5,000	5,000–9,999	10,000–19,999	20,000–39,999	40,000–59,999	60,000–79,999	80,000–99,999	100,000+	London	Row Total
	Count	5,171	226	252	95	17	98	35		226	6,120
<5,000	Row %	84.5	3.7	4.1	1.6	0.3	1.6	0.6		3.7	74.7
	Column %	89.6	48.5	50.5	40.3	37.0	29.1	40.2		29.9	
	Total %	63.1	2.8	3.1	1.2	0.2	1.2	0.4		2.8	
		180	170	23	12	4	20	6		25	440
5,000–9,999		40.9	38.6	5.2	2.7	0.9	4.5	1.4		5.7	5.4
		3.1	36.5	4.6	5.1	8.7	5.9	6.9		3.3	
		2.2	2.1	0.3	0.1	0.0	0.2	0.1		0.3	
Size of		152	22	187	7		20	11		20	419
destination		36.3	5.3	44.6	1.7		4.8	2.6		4.8	5.1
settlement 10,000–19,999		2.6	4.7	37.5	3.0		5.9	12.6		2.6	
		1.9	0.3	2.3	0.1		0.2	0.1		0.2	

		(1)	(2)	(3)	(4)	(5)	(6)	(7)	(8)	Row Total
20,000–39,999		53	12	7	115	1	5		16	209
		25.4	5.7	3.3	55.0	0.5	2.4		7.7	2.5
		0.9	2.6	1.4	48.7	2.2	1.5		2.1	
		0.6	0.1	0.1	1.4	0.0	0.1		0.2	
40,000–59,999		18	1			23			1	43
		41.9	2.3			53.5			2.3	0.5
		0.3	0.2			50.0			0.1	
		0.2	0.0			0.3			0.0	
60,000–79,999		60	16	14	1		183		8	282
		21.3	5.7	5.0	0.4		64.9		2.8	3.4
		1.0	3.4	2.8	0.4		54.3		1.1	
		0.7	0.2	0.2	0.0		2.2		0.1	
80,000–99,999		24	1	5	2		3	30	6	71
		33.8	1.4	7.0	2.8		4.2	42.3	8.5	0.9
		0.4	0.2	1.0	0.8		0.9	34.5	0.8	
		0.3	0.0	0.1	0.0		0.0	0.4	0.1	
100,000+										
London		112	18	11	4	1	8	5	455	614
		18.2	2.9	1.8	0.7	0.2	1.3	0.8	74.1	7.5
		1.9	3.9	2.2	1.7	2.2	2.4	5.7	60.1	
		1.4	0.2	0.1	0.0	0.0	0.1	0.1	5.6	
Column Total		5,770	466	499	236	46	337	87	757	8,198
Row %		70.4	5.7	6.1	2.9	0.6	4.1	1.1	9.1	100

Data source: 16,091 life histories provided by family historians.
Table should be read as follows: 84.5% of migrants moving to a destination of under 5,000 population left a settlement of under 5,000 population.
89.6% of migrants leaving a settlement of under 5,000 population moved to a settlement of under 5,000 population.
63.1% of all moves were between settlements of under 5,000 population.

Matrix of movement between settlements of different size, 1840–79

		<5,000	5,000–9,999	10,000–19,999	20,000–39,999	Size of origin settlement 40,000–59,999	60,000–79,999	80,000–99,999	100,000+	London	Row Total
	<5,000										
Count		7,494	499	306	317	158	209	57	360	415	9815
Row %		76.4	5.1	3.1	3.2	1.6	2.1	0.6	3.7	4.2	49.9
Column %		81.5	32.9	27.3	23.6	23.7	22.1	15.8	38.0	16.4	
Total %		38.1	2.5	1.6	1.6	0.8	1.1	0.3	1.8	2.1	
	5,000–9,999	397	734	46	60	32	23	15	59	88	1,454
		27.3	50.5	3.2	4.1	2.2	1.6	1.0	4.1	6.1	7.4
		4.3	48.4	4.1	4.5	4.8	2.4	4.2	5.5	3.5	
		2.0	3.7	0.2	0.3	0.2	0.1	0.1	0.3	0.4	
Size of destination settlement	10,000–19,999	257	50	587	45	21	26	2	53	55	1,096
		23.4	4.6	53.6	4.1	1.9	2.4	0.2	4.8	5.0	5.6
		2.8	3.3	52.3	3.4	3.2	2.8	0.6	5.3	2.2	
		1.3	0.3	3.0	0.2	0.1	0.1	0.0	0.3	0.3	

	C1	C2	C3	C4	C5	C6	C7	C8	C9	Total
20,000–39,999	233 / 17.9 / 2.5 / 1.2	46 / 3.5 / 3.0 / 0.2	51 / 3.9 / 4.5 / 0.3	791 / 60.8 / 58.9 / 4.0	18 / 1.4 / 2.7 / 0.1	7 / 0.5 / 0.7 / 0.0	21 / 1.6 / 5.8 / 0.1	63 / 4.9 / 6.9 / 0.3	72 / 5.5 / 2.8 / 0.4	1,302 / 6.6
40,000–59,999	87 / 15.1 / 0.9 / 0.4	18 / 3.1 / 1.2 / 0.1	17 / 2.9 / 1.5 / 0.1	9 / 1.6 / 0.7 / 0.0	385 / 66.7 / 57.8 / 2.0	10 / 1.7 / 1.1 / 0.1	1 / 0.2 / 0.3 / 0.0	28 / 4.8 / 2.1 / 0.1	22 / 3.8 / 0.9 / 0.1	577 / 2.9
60,000–79,999	120 / 14.2 / 1.3 / 0.6	23 / 2.7 / 1.5 / 0.1	15 / 1.8 / 1.3 / 0.1	17 / 2.0 / 1.3 / 0.1	5 / 0.6 / 0.8 / 0.0	606 / 71.7 / 64.1 / 3.1	5 / 0.6 / 1.4 / 0.0	30 / 3.5 / 3.7 / 0.2	24 / 2.8 / 0.9 / 0.1	845 / 4.3
80,000–99,999	33 / 10.1 / 0.4 / 0.2	14 / 4.3 / 0.9 / 0.1	6 / 1.8 / 0.5 / 0.0	16 / 4.9 / 1.2 / 0.1	3 / 0.9 / 0.5 / 0.0	8 / 2.5 / 0.8 / 0.0	217 / 66.6 / 60.1 / 1.1	25 / 7.6 / 2.7 / 0.1	4 / 1.2 / 0.2 / 0.0	326 / 1.7
100,000+	294 / 15.5 / 3.2 / 1.5	56 / 7.1 / 3.7 / 0.3	40 / 4.9 / 3.5 / 0.2	47 / 4.4 / 3.5 / 0.0	26 / 1.9 / 4.0 / 0.1	29 / 3.4 / 3.1 / 0.1	41 / 5.3 / 11.3 / 0.2	1,327 / 52.7 / 31.5 / 6.8	34 / 5.0 / 1.4 / 0.2	1,894 / 9.7
London	276 / 11.7 / 3.0 / 1.4	78 / 3.3 / 5.1 / 0.4	54 / 2.3 / 4.8 / 0.3	41 / 1.7 / 3.1 / 0.2	18 / 0.8 / 2.7 / 0.1	27 / 1.1 / 2.9 / 0.1	2 / 0.1 / 0.6 / 0.0	32 / 1.3 / 4.3 / 0.2	1,821 / 77.5 / 71.8 / 9.3	2,349 / 11.9
Column Total	9,191	1,518	1,122	1,343	666	945	361	1,977	2,535	19,658
Row %	46.8	7.7	5.7	6.8	3.4	4.8	1.8	10.1	12.9	100

Data source: 16,091 life histories provided by family historians.

APPENDIX 6C

Matrix of movement between settlements of different size, 1880–1919

		<5,000	5,000–9,999	10,000–19,999	20,000–39,999	Size of origin settlement 40,000–59,999	60,000–79,999	80,000–99,999	100,000+	London	Row Total
	<5,000										
Count		4,879	308	399	380	180	99	86	580	271	7,182
Row %		67.9	4.3	5.6	5.3	2.5	1.4	1.2	8.1	3.8	34.3
Column %		69.9	23.8	27.0	19.7	16.8	17.1	15.1	28.2	11.4	
Total %		23.3	1.5	1.9	1.8	0.9	0.5	0.4	2.7	1.3	
	5,000–9,999	273	644	56	94	24	16	20	110	53	1,290
		21.2	49.9	4.3	7.3	1.9	1.2	1.6	8.5	4.1	6.2
		3.9	49.7	3.8	4.9	2.2	2.8	3.5	5.8	2.2	
		1.3	3.1	0.3	0.4	0.1	0.1	0.1	0.5	0.3	
Size of destination settlement	10,000–19,999	333	63	699	72	23	24	16	117	54	1,401
		23.8	4.5	49.9	5.1	1.6	1.7	1.1	8.3	3.9	6.7
		4.8	4.9	47.3	3.7	2.1	4.2	2.8	5.3	2.3	
		1.6	0.3	3.3	0.3	0.1	0.1	0.1	0.5	0.3	

	C1	C2	C3	C4	C5	C6	C7	C8	C9	Total
20,000–39,999	320	76	61	987	57	25	28	144	72	1,770
	18.1	4.3	3.4	55.8	3.2	1.4	1.6	8.1	4.1	8.5
	4.6	5.9	4.1	51.2	5.3	4.3	4.9	7.4	3.0	
	1.5	0.4	0.3	4.7	0.3	0.1	0.1	0.7	0.3	
40,000–59,999	131	16	30	48	639	11	6	76	39	996
	13.2	1.6	3.0	4.8	64.2	1.1	0.6	7.6	3.9	4.8
	1.9	1.2	2.0	2.5	59.7	1.9	1.1	3.5	1.6	
	0.6	0.1	0.1	0.2	3.1	0.1	0.0	0.4	0.2	
60,000–79,999	120	16	21	30	18	298	1	67	23	594
	20.2	2.7	3.5	5.1	3.0	50.2	0.2	11.2	3.9	2.8
	1.7	1.2	1.4	1.6	1.7	51.6	0.2	3.5	1.0	
	0.6	0.1	0.1	0.1	0.1	1.4	0.0	0.3	0.1	
80,000–99,999	71	22	14	31	8	4	336	50	12	548
	13.0	4.0	2.6	5.7	1.5	0.7	61.3	9.1	2.2	2.6
	1.0	1.7	0.9	1.6	0.7	0.7	58.9	2.8	0.5	
	0.3	0.1	0.1	0.1	0.0	0.0	1.6	0.3	0.1	
100,000+	546	104	122	134	89	56	54	2,922	101	4,605
	11.9	5.4	6.7	6.9	4.7	3.9	2.2	55.7	2.6	22.0
	7.8	8.0	8.2	6.9	8.3	9.7	9.5	37.2	4.3	
	2.6	0.5	0.6	0.7	0.4	0.3	0.2	14.0	0.5	
London	309	47	77	152	32	45	23	145	1,719	2,549
	12.1	1.8	3.0	6.0	1.3	1.8	0.9	5.7	67.4	12.1
	4.4	3.6	5.2	7.9	3.0	7.8	4.0	9.5	72.5	
	1.5	0.2	0.4	0.7	0.2	0.2	0.1	0.7	8.2	
Column Total	6,982	1,296	1,479	1,928	1,070	578	570	3,855	2,371	20,935
Row %	33.4	6.2	7.1	9.2	5.1	2.8	2.7	22.2	11.3	100

Data source: 16,091 life histories provided by family historians.

Matrix of movement between settlements of different size, 1920–94

						Size of origin settlement					
		<5,000	5,000–9,999	10,000–19,999	20,000–39,999	40,000–59,999	60,000–79,999	80,000–99,999	100,000+	London	Row Total
Size of destination settlement	<5,000										
	Count	3,144	270	279	281	236	152	110	629	148	5,249
	Row %	59.9	5.1	5.3	5.4	4.5	2.9	2.1	12.0	2.8	29.4
	Column %	55.8	30.4	20.5	18.7	19.5	16.5	17.9	30.9	14.0	
	Total %	17.6	1.5	1.6	1.6	1.3	0.9	0.6	3.5	0.8	
	5,000–9,999										
	Count	226	281	37	46	29	16	10	106	17	768
	Row %	29.4	36.6	4.8	6.0	3.8	2.1	1.3	13.8	2.2	4.3
	Column %	4.0	31.6	2.7	3.1	2.4	1.7	1.6	4.2	1.6	
	Total %	1.3	1.6	0.2	0.3	0.2	0.1	0.1	0.6	0.1	
	10,000–19,999										
	Count	266	51	596	59	49	39	39	146	19	1,264
	Row %	21.0	4.0	47.2	4.7	3.9	3.1	3.1	11.5	1.5	7.1
	Column %	4.7	5.7	43.8	3.9	4.0	4.2	6.4	6.5	1.8	
	Total %	1.5	0.3	3.3	0.3	0.3	0.2	0.2	0.8	0.1	

20,000–39,999	322	46	71	713	63	29	28	169	41	1,482
	21.7	3.1	4.8	48.1	4.3	2.0	1.9	11.4	2.8	8.3
	5.7	5.2	5.2	47.5	5.2	3.1	4.6	7.8	3.9	
	1.8	0.3	0.4	4.0	0.4	0.2	0.2	0.9	0.2	
40,000–59,999	244	34	73	73	545	36	14	146	27	1,192
	20.5	2.9	6.1	6.1	45.7	3.0	1.2	12.2	2.3	6.7
	4.3	3.8	5.4	4.9	45.0	3.9	2.3	7.5	2.6	
	1.4	0.4	0.4	0.4	3.0	0.2	0.1	0.8	0.2	
60,000–79,999	186	25	32	31	43	429	17	104	30	897
	20.7	2.8	3.6	3.5	4.8	47.8	1.9	11.6	3.3	5.0
	3.3	2.8	2.4	2.1	3.6	46.5	2.8	4.6	2.8	
	1.0	0.1	0.2	0.2	0.2	2.4	0.1	0.6	0.2	
80,000–99,999	96	15	37	32	22	25	289	73	23	612
	15.7	2.5	6.0	5.2	3.6	4.1	47.2	11.9	3.8	3.4
	1.7	1.7	2.7	2.1	1.8	2.7	47.1	3.1	2.2	
	0.5	0.1	0.2	0.2	0.1	0.1	1.6	0.4	0.1	
100,000+	901	143	183	209	169	134	73	3,054	112	5,037
	17.9	5.5	7.8	8.9	7.5	5.6	3.1	40.9	2.8	28.1
	16.0	16.1	13.4	13.9	13.9	14.6	11.9	23.5	10.6	
	5.0	0.8	1.1	1.2	1.0	0.8	0.4	17.1	0.6	
London	247	23	52	58	54	63	33	239	603	1,372
	18.0	1.7	3.8	4.2	3.9	4.6	2.4	17.4	44.0	7.7
	4.4	2.6	3.8	3.9	4.5	6.8	5.4	11.9	57.2	
	1.4	0.1	0.3	0.3	0.3	0.4	0.2	1.4	3.4	
Column Total	5,637	888	1,360	1,502	1,210	923	613	4,686	1,054	17,873
Row %	31.5	5.0	7.6	8.4	6.8	5.2	3.4	26.3	5.8	100

Data source: 16,091 life histories provided by family historians.

Bibliography

Abrams, P. and E.A. Wrigley (eds). *Towns in society* (Cambridge University Press, 1978).

Adams, I. *The making of urban Scotland* (London: Croom Helm, 1978).

Aldcroft, D. and M. Freeman (eds). *Transport in the industrial revolution* (Manchester University Press, 1983).

Allen, J. and C. Hamnett (eds). *A shrinking world? Global unevenness and inequality* (Oxford University Press, 1995).

Allison, P. *Event history analysis: regression for longitudinal event data* (London: Sage, 1984).

Anderson, B. *Internal migration during modernization in late-nineteenth century Russia* (Princeton University Press, 1980).

Anderson, G. *Victorian Clerks* (Manchester University Press, 1976).

Anderson, J., C. Brook and A. Cochrane (eds). *A global world? Reordering political space* (Oxford University Press, 1995).

Anderson, M. *Approaches to the history of the Western family* (London: Macmillan, 1980).

Anderson, M. The emergence of the modern life cycle in Britain. *Social History* 10, pp. 69–87, 1985.

Anderson, M. The social implications of demographic change. In *The Cambridge social history of Britain, 1750–1950: Volume 2. People and their environment*, F.M.L. Thompson (ed.) (Cambridge University Press, 1990) pp. 1–70.

Anderson, M. What is new about the modern family? In *Time, family and community*, M. Drake (ed.) (Oxford: Blackwell, 1994) pp. 67–90.

Armstrong, W.A. The uses of information about occupation. In *Nineteenth century Society*, E.A. Wrigley (ed.) (Cambridge University Press, 1972) pp. 191–310.

Armstrong, W.A. *Farmworkers: a social and economic history* (London: Batsford, 1988).

Bade, K. (ed.). *Population, labour and migration in nineteenth and twentieth century Germany* (Leamington Spa: Berg, 1986).

Baines, D. The use of published census data in migration studies. In *Nineteenth century society. Essays in the use of quantitative methods for the study of social data*, E.A. Wrigley (ed.) (Cambridge University Press, 1972) pp. 311–35.

Baines, D. Birthplace statistics and the analysis of internal migration. In *The census and social structure. An interpretative guide to nineteenth century censuses for England and Wales*, R. Lawton (ed.) (London: Frank Cass, 1978) pp. 146–64.

Baines, D. *Migration in a mature economy. Emigration and internal migration in England and Wales, 1861–1900* (Cambridge University Press, 1985).

Baines, D. *Emigration from Europe, 1815–1930* (London: Macmillan, 1991).

Baines, D. European emigration, 1815–1930: looking at the emigration decision again. *Economic History Review* 47, pp. 525–44, 1994.

Bardet, J.-P. *Rouen aux XVIIe et XVIIIe siècles: Les mutations d'un espace social* (Paris: Société d'Edition d'Enseignement Supèrieur, 1983).

Barke, M. The middle class journey to work in Newcastle upon Tyne, 1850–1913 *Journal of Transport History* 12, pp. 107–34, 1991.

Barker, T. and Robbins, M. *A history of London Transport: passenger travel and the development of the metropolis* (London: Allen and Unwin, 2 volumes 1963–74).

Bartholomew, K. Women migrants in mind: leaving Wales in the nineteenth and twentieth centuries. In *Migrants, emigrants and immigrants: a social history of migration*, C. Pooley and I. Whyte (eds) (London: Routledge, 1991) pp. 174–90.

Beckett, J. *The agricultural revolution* (Oxford: Blackwell, 1990).

Bellamy, J. Occupation statistics in the nineteenth century censuses. In *The census and social structure*, R. Lawton (ed.) (London: Frank Cass, 1978) pp. 165–78.

Berg, M. *The age of manufactures 1700–1820* (London: Fontana, 1985).

Berg, M., P. Hudson and M. Soneuscher (eds). *Manufacture in town and country before the factory* (Cambridge University Press, 1983).

Black, I. Geography, political economy and the circulation of finance capital in early industrial England *Journal of Historical Geography* 15, pp. 366–84, 1989.

Black, R. and V. Robinson (eds). *Geography and refugees: patterns and processes of change* (London: Belhaven, 1993).

Boddy, M. *The building societies* (London: Macmillan, 1980).

Bogue, D. Internal migration. In *The study of population: an inventory and appraisal*, P. Hauser and O. Duncan (eds) (University of Chicago Press, 1959) pp. 486–509.

Boyle, P. Modelling the relationship between tenure and migration in England and Wales. *Transactions of the Institute of British Geographers* 18, pp. 359–76, 1993.

Boyle, P. Public housing as a barrier to long distance migration. *International Journal of Population Geography* 1, pp. 147–64, 1995.

Braybon, G. and P. Summerfield. *Out of the cage: women's experiences in two world wars* (London: Pandora Press, 1987).

Brayshay, M. and Pointon, V. Migration and the social geography of mid-nineteenth century Plymouth. *The Devon Historian* 28, pp. 3–14, 1984.

Breiger, R. (ed.). *Social mobility and social structure* (Cambridge University Press, 1990).

Brettell, C. *Men who migrate, women who wait* (Princeton: Princeton University Press, 1987).

Briggs, A. *History of Birmingham. Volume 2: Borough and city, 1865–1938* (London: Oxford University Press, 1952).

Brown, L. and E. Moore. The intra-urban migration process: a perspective. *Geografiska Annaler*, 52B, pp. 1–13, 1970.

Brunskill, R. *Traditional buildings of England: an introduction to vernacular archi-tecture* (London: Gollancz, 1992).

Bryant, C. and Jary, D. (eds). *Giddens theory of structuration: a critical appreci-ation* (London: Routledge, 1991).

Buckatzsch, K. Places of origin of a group of immigrants into Sheffield, 1624–1799. *Economic History Review* 2, pp. 303–6, 1949–50.

Burnett, J *Useful toil* (London: Allen Lane, 1974).

Burnett, J. *Destiny obscure* (London: Allen Lane, 1982).

Burnett, J. *A social history of housing, 1815–1970* (London: Methuen, 2nd edition 1986).

Burnett, J., D. Vincent and D. Mayall (eds). *The autobiography of the working class: an annotated and critical bibliography* (Brighton: Harvester, 1984–7).

Butlin, R. Regions in England and Wales, c1600–1914. *In An historical geography of England and Wales*, R. Dodgshon and R. Butlin (eds) (London, Academic Press, 1990) pp. 223–54.

Cadwallader, M. A conceptual framework for analysing migration behaviour in the developed world. *Progress in Human Geography* 13, pp. 494–511, 1989.

Cadwallader, M. *Migration and residential mobility: macro and micro approaches* (Madison: University of Winsconsin Press, 1992).

Cadwallader, M. Classics in human geography revisited: commentary 2. *Progress in Human Geography* 17, pp. 215–17, 1993.

Cairncross, A.K. Internal migration in Victorian England. *The Manchester School* 17, pp. 67–87, 1949.

Carrier, N. and Jeffery, J. *External migration: A study of the available statistics, 1815–1950* (London: HMSO, 1953).

Carter, H. *The study of urban geography* (London: Arnold, 4th edition, 1995).

Chalklin, C. *The provincial towns of Georgian England* (London: Arnold, 1974).

Champion, A. (ed.). *Counterurbanization: the changing pace and nature of popu-lation deconcentration* (London: Arnold, 1989).

Champion, A. and A. Fielding (eds). *Migration processes and patterns. Vol. 1. Re-search progress and prospects* (London: Belhaven, 1992).

Chant, S. (ed.). *Gender and migration in developing countries.* (London: Belhaven, 1992).

Chapman, S.D. *The cotton industry in the industrial revolution* (London: Macmillan, 1972).

Chatelain, A. Migrations et domesticité féminine urbaine en France, XVIIIe siècle–XXe siècle *Revue d'histoire économique et sociale* 4, pp. 506–28, 1969.

Church, R.A. *The history of the British coal industry vol. 3: 1830–1913* (Oxford University Press, 1986).

Church, R.A. (ed.). *The industrial revolutions. Volume 10: the coal and iron indus-tries* (Oxford: Blackwell, 1994).

Clark, D. *Urban world/global city* (London: Routledge, 1996).

Clark, P. Migration in England during the late seventeenth and early eighteenth centuries. *Past and Present* 83, pp. 57–90, 1979.

Clark, P. and D. Souden (eds). *Migration and society in early-modern England* (London: Hutchinson, 1987).

Cloke, P. and G. Edwards. Rurality in England and Wales 1981: a replication of the 1971 index. *Regional Studies* 20, pp. 289–306, 1986.

Cloke, P. *et al. Writing the rural: five cultural geographies* (London: Chapman, 1994).

Coleman, A. *Utopia on trial* (London: Hilary Shipman, 1985).

Coleman, D. and J. Salt. The British population: patterns, trends and processes (Oxford University Press, 1992).

Constant, A. The geographical background of inter-village population movements in Northamptonshire and Huntingdonshire, 1754–1948. *Geography* 33, pp. 78–88, 1948.

Constantine, S. *Emigrants and Empire: British settlement in the dominions between the wars* (Manchester University Press, 1990).

Constantine, S. Empire migration and social reform 1880–1950. In *Migrants, emigrants and immigrants,* C. Pooley and I. Whyte (eds) (London: Routledge, 1991) pp. 62–86.

Cooke, P. and G. Rees. *The industrial restructuring of South Wales: the career of a state managed region.* (Cardiff: Department of Town Planning, UWIST, 1981).

Corfield, P. *The impact of English towns, 1700–1800* (Oxford University Press, 1982).

Courgeau, D. Migration theories and behavioural models. *International Journal of Population Geography* 1, pp. 19–28, 1995.

Courgeau, D. and E. Lelièvre. Analyse démographiques des biographies. *Population* 44, pp. 1233–8, 1989.

Courgeau, D. and E. Lelièvre. *Event history analysis in demography* (Oxford: Clarendon Press, 1992).

Crafts, N. *British economic growth during the industrial revolution* (Oxford: Clarendon, 1985).

Crafts, N., S. Leyburn and T. Mill. Trends and cycles in British industrial production, 1700–1913. *Journal of the Royal Statistical Society,* Series A, 152, pp. 43–60, 1989.

Craig, D. *On the crofters' trail: in search of the clearance Highlanders* (London: Cape, 1990).

Crosby, A. (ed.). *The family records of Benjamin Shaw, Mechanic of Dent, Dolphinholme and Preston, 1772–1841* (The Record Society of Lancashire and Cheshire, 130, 1991).

Crossick, G. (ed.). *The lower middle class in Britain, 1870–1914* (London: Croom Helm, 1977).

Cullingworth, J. *Town and country planning in Britain* (London: Routledge, 11th edition, 1994).

Danson, J.T. and T.W. Welton. On the population of Lancashire and Cheshire and its local distribution during the fifty years 1801–51. *Transactions of the Historic Society of Lancashire and Cheshire,* 9, pp. 195–212, 1857; 10, pp. 1–36, 1858; 11, pp. 31–70, 1859; 12, pp. 35–74, 1860.

Darby, H.C. The movement of population to and from Cambridgeshire between 1851 and 1861. *Geographical Journal,* 101, pp. 118–25, 1945.

Darroch, G. Migrants in the nineteenth century: fugitives or families in motion. *Journal of Family History* 6, pp. 257–77, 1981.

Daunton, M. *House and home in the Victorian city: working-class housing 1850–1914* (London: Arnold, 1983).

Daunton, M. (ed.). *Councillors and tenants* (Leicester University Press, 1984).

Daunton, M. *Royal Mail: the Post Office since 1840* (London: Athlone Press, 1985).

Daunton, M. *A property-owning democracy? Housing in Britain* (London: Faber, 1987).

Davidoff, L. The family in Britain. In *The Cambridge Social History of Britain, 1750–1950. Volume 2: People and their environment*, F.M.L. Thompson (ed.) (Cambridge University Press, 1990) pp. 71–130.

Davies, R. A generated beta-logistic model for longitudinal data with an application to residential mobility. *Environment and Planning A* 16, pp. 1375–86, 1984.

Davies, R. The analysis of housing and migration careers. In *Migration models*, J. Stillwell and P. Congdon (eds) (London: Belhaven, 1991) pp. 207–27.

Davies, R. and Flowerdew, R. Modelling migration careers using data from a British Survey *Geographical Analysis*, 24, pp. 35–57, 1992.

Davies, R. and Pickles, A. An analysis of housing careers in Cardiff. *Environment and Planning A*, 23, pp. 629–50, 1991.

Davis, G. *The Irish in Britain, 1815–1914* (Dublin: Gill and Macmillan, 1991).

Davis, G. The historiography of the Irish Famine. In *The Irish world wide. Volume six: the meaning of the Famine*, P. O'Sullivan (ed.) (Leicester University Press, 1997) pp. 15–39.

D'Cruze, S. Care, diligence and 'Usfull Pride' [sic]: gender, industrialization and the domestic economy c1770–1840. *Women's History Review* 3, pp. 315–46, 1994.

Deane, P. and W. Cole. *British economic growth, 1688–1959* (Cambridge University Press, 1967).

Dennis, R. Inter-censal mobility in a Victorian city. *Transactions of the Institute of British Geographers* NS2, pp. 349–63, 1977a.

Dennis, R. Distance and social interaction in a Victorian city. *Journal of Historical Geography* 3, pp. 237–50, 1977b.

Dennis, R. *English industrial cities of the nineteenth century* (Cambridge University Press, 1984a).

Dennis, R. The geography of Victorian values: philanthropic housing in London, 1840–1900. *Journal of Historical Geography* 15, pp. 40–54, 1984b.

Dennis, R. and Daniels, S. Community and the social geography of Victorian cities. *Urban History Yearbook* (Leicester University Press, 1981) pp. 7–23.

De Schaepdrijver, S. *Elites for the capital? Foreign migration to mid-nineteenth century Brussels* (Amsterdam: PDIS, 1990).

Devine, T. (ed.). *Scottish emigration and Scottish society* (Edinburgh: John Donald, 1992).

de Vries, J. *European Urbanization, 1500–1800* (London: Methuen, 1984).

Dickens, C. *Dombey and Son* (Oxford University Press, 1982 edition).

Doherty, J. *Short-distance migration in nineteenth-century Lancashire.* (PhD thesis, Department of Geography, Lancaster University, 1986).

Dotterer, R., D. Dash-Morre and S. Cohen (eds). *Jewish settlement and community in the modern world* (Susquehanna University Press, 1991).

Drake, M. (ed.). *Time, family and community: Perspectives on family and community history* (Oxford: Blackwell, 1994).

Dupaquier, J. and D. Kessler (eds). *La société Francaise au XIXe siècle* (Paris: Fayard, 1992).

Dyhouse, C. *Feminism and the family in England, 1880–1939* (Oxford: Blackwell, 1989).

Dyos, H.J. Workmen's fares in south London, 1860–1914. In *Exploring the urban past*, D. Cannadine and D. Reeder (eds) (Cambridge University Press, 1982 pp. 87–100).

Dyos, H.J. and Aldcroft, D. *British transport: an economic survey from the seventeenth century to the twentieth* (Leicester University Press, 1969).

Emsley, C. *Crime and society in England, 1750–1900* (London: Longman, 1987).

Erickson, C. *Invisible immigrants: the adaptation of English and Scottish immigrants in nineteenth-century America* (London: Weidenfeld and Nicolson, 1972).

Erickson, C. Emigration from the British Isles to the USA in 1841. Part I. Emigration from the British Isles. *Population Studies* 43, pp. 347–67, 1989.

Erickson, C. Emigration from the British Isles to the USA in 1841, Part II. Who were the English emigrants? *Population Studies* 44, pp. 21–40, 1990.

Erickson, C. *Leaving England: essays on British emigration in the nineteenth century* (Cornell U.P., 1994).

Escott, M. Residential mobility in a late-eighteenth century parish: Binfield, Berkshire, 1779–1801. *Local Population Studies* 40, pp. 20–35, 1988.

Evans, E. *The forging of the modern state: early industrial Britain, 1783–1870* (Harlow: Longman, 1983).

Everitt, A. Country, county and town: patterns of regional evolution in England. *Transactions of the Royal Historical Society* 29, pp. 79–108, 1979.

Farnie, D. *The English cotton industry and the world market, 1815–1896* (Oxford: Clarendon Press, 1979).

Featherstone, M, S. Lash and R. Robertson (eds). *Global modernities* (London: Sage, 1995).

Feinstein, C. and S. Pollard. *Studies in capital formation in the United Kingdom, 1750–1920* (Oxford: Clarendon, 1988).

Fielding, A. Counterurbanization in Western Europe. *Progress in Planning* 17, pp. 1–52, 1982.

Fielding, A. Migration and culture. In *Migration processes and patterns. Vol. 1. Research progress and prospects*, A. Champion and A. Fielding (eds) (London: Belhaven, 1992) pp. 201–14.

Finch, J. Do families support each other more or less than in the past? In *Time, family and community*, M. Drake (ed.) (Oxford: Blackwell, 1994) pp. 91–105.

Findlay, A. and Graham, E. The challenge facing population geography. *Progress in Human Geography* 15, pp. 149–62, 1991.

Finlay, R. *Population and metropolis: the demography of London 1580–1650.* (Cambridge University Press, 1981).

Fishman, W.J. *East End 1888* (London: Duckworth, 1988).

Flandrin, J.L. *Families in former times: kinship, household and sexuality* (Cambridge University Press, 1979).

Flinn, M.W. (ed.). *Scottish population history from the 17th century to the 1930s* (Cambridge University Press, 1977).

Flinn, M.W. *The history of the British coal industry, vol. 2: 1700–1830* (Oxford University Press, 1984).

Forrest, R. and Murie, A. *Selling the welfare state* (London: Routledge, 1988).

Fraser, D. *The evolution of the British Welfare State* (London: Macmillan, 2nd edition, 1984).

Freeman, M. The industrial revolution and the regional geography of England: a comment. *Transactions of the Institute of British Geographers* NS9, pp. 507–12, 1984.

Freeman, M. Transport. In *Atlas of industrializing Britain*, J. Langton and R. Morris (eds) (London: Methuen, 1986) pp. 80–93.

Friedlander, D. and R. Roshier. A study of internal migration in England and Wales. *Population Studies* 19, pp. 239–79, 1966.

Fuchs, R. and L.P. Moch Pregnant, single and far from home: migrant women in nineteenth-century Paris *American Historical review* 95, pp. 1007–35, 1990.

Gale, S. Explanations, theory and models of migration *Economic Geography* 49, pp. 257–74, 1973.

Gauldie, E. *Cruel habitations: a history of working-class housing 1780–1918* (London: Allen and Unwin, 1974).

Giddens, A. *The constitution of society: outline of the theory of structuration* (Cambridge University Press, 1984).

Giddens, A. *The consequences of modernity* (Cambridge: Polity Press, 1990).

Giddens, A. *Modernity and self identity: self and society in the late modern age* (Cambridge: Polity Press, 1991).

Gill, C. *History of Birmingham. Volume 1: Manor and Borough to 1865* (London: Oxford University Press, 1952).

Glenny, M. *The fall of Yugoslavia: the third Balkan war* (Harmondsworth: Penguin, 1993).

Gold, J. *An introduction to behavioural geography* (Oxford University Press, 1980).

Gold, K. We are family. *The Times Higher* pp. 17–18, July 1st, 1994.

Goldthorpe, J. *Social mobility and class structure in modern Britain* (Oxford: Clarendon Press, 2nd edition, 1987).

Golledge, R. A behavioural view of mobility and migration research. *Professional Geographer* 31, pp. 14–21, 1980.

Gordon, G. and B. Dicks (eds). *Scottish urban history* (Aberdeen University Press, 1983).

Gould, P. and R. White. *Mental maps* (London: Allen and Unwin, 2nd edition, 1986).

Gourvish, T. and R. Wilson. *The British brewing industry, 1830–1980* (Cambridge University Press, 1994).

Graham, B. and L. Proudfoot (eds). *An historical geography of Ireland* (London: Academic Press, 1993).

Green, A. *What contribution can labour migration make to reducing unemployment?* (Newcastle University: Centre for Urban and Regional Development Studies, Discussion Paper 73, 1985).

Green, D. Distance to work in Victorian London: a case study of Henry Poole, bespoke tailors. *Business History* 30, pp. 179–94, 1988.

Green, D. *From Artizans to Paupers* (Aldershot: Scolar Press, 1995).

Green, D. The nineteenth century metropolitan economy: a revisionist interpretation, *London Journal* 21, pp. 9–26; 1996.

Gregory, D. *Regional transformation and industrial revolution* (London: Macmillan, 1982).

Gregory, D. The friction of distance? Information circulation and the mails in early nineteenth-century England. *Journal of Historical Geography* 13, pp. 130–54, 1987.

Gregory, D. The production of regions in England's industrial revolution. *Journal of Historical Geography* 14, pp. 50–8, 1988a.

Gregory, D. The production of regions in England's industrial revolution. Reply to Langton. *Journal of Historical Geography* 14, pp. 174–6, 1988b.

Gregory, D. A new and differing face in many places: three geographies of industrialization. In *An historical geography of England and Wales*, R. Dodgshon and R. Butlin (eds) (London: Academic Press, 1990) pp. 351–99.

Gregson, N. Structuration threory: some thoughts on the possibilities of empirical research. *Environment and Planning D: Society and Space* 5, pp. 73–91, 1987.

Grigg, D.B. E.G. Ravenstein and the 'laws of migration'. *Journal of Historical Geography*, 3, pp. 41–54, 1977.

Grundy, E. Retirement migration and its consequences in England and Wales. *Ageing and society* 7, pp. 57–82, 1987.

Grundy, J. *The origins of the Liverpool cowkeepers* (M.Litt. thesis, Department of Geography, Lancaster University, 1982).

Gutmann M. and E. van de Walle New sources for social and demographic history: the Belgian population registers *Social Science History* 2, pp. 121–43, 1978.

Hagerstrand, T. Survival and arena: on the life history of individuals in relation to their geographic environment. In *Timing space and spacing time*. T. Carlstein, D. Parkes and N. Thrift (eds) (London: Arnold, 1978) pp. 122–45.

Haines, R. Indigent misfits or shrewd operators? Government assisted emigrants from the United Kingdom to Australia, 1831–1860. *Population Studies* 48, pp. 223–47, 1994.

Halfacree, K. and Boyle, P. The challenge facing migration research: the case for a biographical approach. *Progress in Human Geography* 17, pp. 333–48, 1993.

Hall, P. *The containment of urban England* (London: Allen and Unwin, 1973).

Hall, R. and White, P. *Europe's population: towards the next century* (London: UCL Press, 1995).

Hallas, C. Migration in nineteenth century Wensleydale and Swaledale. *Northern History* 27, pp. 139–61, 1991.

Hamnett, C., M. Harmer and P. Williams. *Safe as houses: housing inheritance in Britain* (London: Paul Chapman, 1991).

Handy, C. *Discrimination in housing* (London: Sweet and Maxwell, 1993).

Hareven, T. (ed.). *Transitions: the family and life course in historical perspective* (New York: Academic Press, 1978).

Harper, M. *Emigration from North-east Scotland* (Aberdeen University Press, 1988).

Harvey, D. *The urbanization of Capital. Studies in the history and theory of Capitalist urbanization* (Oxford: Blackwell, 1985).

Harvey, D. *The condition of postmodernity* (Oxford: Blackwell, 1989).

Higgs, E. *Making sense of the census: The manuscript returns of England and Wales, 1801–1901* (London: HMSO, 1989).

Hill, A.B. *Internal migration and its effects upon the death rates, with special reference to the county of Essex* (London: Medical Research Council, 1925).

Hobcraft, J. and Murphy, M. Demographic event history analysis: a selective review. *Population Index* 52, pp. 3–27, 1986.

Hobsbawm, E.J. The tramping artizan. *Economic History Review* 3, pp. 299–320, 1951.

390

Hochstadt, S. Migration and industrialization in Germany, 1815–1977. *Social Science History* 5, pp. 445–68, 1981.

Hochstadt, S. Migration in preindustrial Germany. *Central European History* 16, pp. 195–224, 1983.

Hochstadt, S. and J. Jackson. New sources for the study of migration in early nineteenth-century Germany *Historical Social Research* 31, pp. 85–92, 1984.

Hoerder, D. Migration in the Atlantic economies: Regional European origins and worldwide expansion. In *European migrants: global and local perspectives*, D. Hoerder and L.P. Moch (eds) (Boston: Northeastern University Press, 1996) pp. 21–51.

Hoerder, D. and L.P. Moch, (eds). *European migrants: global and local perspectives* (Boston: Northeastern University Press, 1996).

Hoggart, K. and Green, D. (eds). *London: a new metropolitan geography* (London: Arnold, 1992).

Hohenberg, P. and Lees, L. *The making of urban Europe, 1000–1950* (Cambridge, Mass.: Harvard University Press, 1985).

Holderness, B.A. Personal mobility in some rural parishes of Yorkshire, 1777–1822. *The Yorkshire Archaeological Journal* 42, pp. 444–54. 1970.

Holman, J. Apprenticeship as a factor in migration: Bristol 1675–1726. *Bristol and Gloucester Archaeological Journal*, 97, pp. 85–92, 1980.

Holmes, C. *John Bull's island: immigration and British society, 1871–1971* (London: Macmillan, 1988).

Hoppe, G. and J. Langton. *Flows of labour in the early phase of capitalist development: the time geography of longitudinal migration paths in nineteenth-century Sweden* (Historical Geography Research series, 29, 1992).

Hoppit, J. and E.A.Wrigley. *The industrial revolutions. Volumes 2 and 3: The industrial revolution in Britain* (Oxford: Blackwell, 1994).

Horn, P. *The rise and fall of the Victorian servant* (Dublin: Gill and Macmillan, 1975).

Horne, W.R. *The relationship between sense of place and new local government units* (PhD thesis, Department of Geography, Lancaster University, 1984).

Hudson, P. *The genesis of industrial capital: A study of the West Riding wool textile industry, 1750–1850* (Cambridge University Press, 1986).

Hudson, P. (ed.). *Regions and industries* (Cambridge University Press, 1989).

Hudson, P. *The industrial revolution* (London: Arnold, 1992).

Hudson, P. *English emigrants to New Zealand, 1839–1850: an analysis of the work of the New Zealand company* (PhD thesis; Departments of History and Geography, Lancaster University, 1996).

Hughes, G. and B. McCormick. Housing markets, unemployment and labour market flexibility in the U.K. *European Economic Review* 31, pp. 615–45, 1987.

Humphries, G. *South Wales.* (Newton Abbot: David and Charles, 1972).

Humphries, S. *The handbook of oral history: recording life stories* (London: interaction, 1984).

Hurley, B. and D. Steel (eds). *Learning from each other* (Devizes: Wiltshire Family History Society, 1994).

Jackson, A.A. *Semi-detached London: suburban development, life and transport, 1900–39* (London: Allen and Unwin, 1973).

Jackson, J.T. Long distance migrant workers in nineteenth-century Britain: a case of the St. Helens glassmakers. *Transactions of the Lancashire and Cheshire Historical Society* 131, pp. 113–37, 1982.

Jackson, J. and L.P. Moch. Migration and the social history of modern Europe. In *European Migrants: global and local perspectives.* D. Hoerder and L. Moch (eds) (Northeastern University Press, Boston, 1996) pp. 52–69.

Jefferys, J. *Retail trading in Britain, 1850–1950* (Cambridge University Press, 1954).

Jenkins, D.T. (ed.). *The industrial revolutions. Volume 8: the textile industries* (Oxford: Blackwell, 1994).

Johnson, J.H., J. Salt and P. Wood, *Housing and the migration of labour in England and Wales* (Farnborough: Saxon House, 1974).

Johnson, J.H. and J. Salt, (eds). *Labour migration: the internal geographical mobility of labour in the developed world* (London: David Fulton, 1990).

Johnston, W. The Welsh diaspora: emigrating around the world in the late-nineteenth century. *Journal of Welsh Labour History* 6, pp. 50–74, 1993.

Jones, H. *A population geography* (London: Harper & Row, 1981) .

Jones, M. The economic history of the regional problem in Britain. *Journal of Historical Geography* 10, pp. 385–95, 1984.

Jones, R.M. Welsh immigrants in the cities of north-west England, 1890–1930: some oral testimony. *Oral History* 9, pp. 33–41, 1981.

Kaelble, M. *Social mobility in the nineteenth and twentieth centuries: Europe and America in comparative perspective* (Leamington Spa: Berg, 1985).

Kalvemark, A.-S. The country that kept track of its population. In *Time, space and man: essays in microdemography* J. Sundin and E. Soderlund (eds) (Atlantic Highlands, N.J: Humanities Press, 1979).

Kemp, P. (ed.) *The private provision of rented housing* (London: Avebury, 1988).

Kertzer, D. and D. Hogan. On the move: migration in an Italian community, 1865–1921 *Social Science History* 9, pp. 1–24, 1985.

King, R. (ed.). *Mass migration in Europe: the legacy and the future* (London: Belhaven, 1993).

Kreidte, P., H. Medick and I. Schulmbohm. *Industrialization before industrialization* (Cambridge University Press, 1981).

Kussmaul, A. *Servants in husbandry in early-modern England* (Cambridge University Press, 1981).

Langton, J. The industrial revolution and the regional geography of England. *Transactions of the Institute of British Geographers* NS9, pp. 145–67, 1984.

Langton, J. The production of regions in England's industrial revolution: a response. *Journal of Historical Geography* 14, pp. 170–4, 1988.

Langton, J. People from the pits: the origins of colliers in eighteenth century S.W. Lancashire. In *Migration, mobility and modernization in Europe*, D. Siddle (ed.) (Liverpool University Press, 1998).

Langton, J. and P. Laxton. Parish registers and urban structure. *Urban History Yearbook*, pp. 74–84, 1978.

Langton, J. and R.J. Morris (eds). *Atlas of industrializing Britain, 1780–1914* (London: Methuen, 1986).

Lash, S. and J. Friedman (eds). *Modernity and identity* (Oxford: Blackwell, 1992).

Lash, S. and J. Urry. *The end of organized capitalism* (Cambridge, Polity Press, 1987).

392

Laslett, P. (ed.). *Household and family in past time* (Cambridge University Press, 1972).

Law, C. The growth of urban population in England and Wales, 1801–1911. *Transactions of the Institute of British Geographers* 41, pp. 125–43, 1967.

Lawton, R. The population of Liverpool in the mid-nineteenth century. *Transactions of the Historic Society of Lancashire and Cheshire* 107, pp. 89–120, 1955.

Lawton, R. Population movements in the West Midlands, 1841–1861. *Geography* 43, pp. 164–77, 1958.

Lawton, R. Irish migration to England and Wales in the mid-nineteenth century. *Irish Geography* 4, pp. 35–54, 1959.

Lawton, R. The journey to work in England and Wales: forty years of change. *Tidjschrift voor Economische en Sociale Geographie* 44, pp. 61–9, 1963.

Lawton, R. Population changes in England and Wales in the later nineteenth century. *Transactions of the Institute of British Geographers*, 44, pp. 55–75, 1968a.

Lawton, R. The journey to work in Britain: some trends and problems *Regional Studies* 2, pp. 27–40, 1968b.

Lawton, R. Rural depopulation in nineteenth-century England. In *English rural communities*, D. Mills (ed.) (London: Macmillan, 1973), pp. 195–219.

Lawton, R. Regional population trends in England and Wales, 1750–1971. In *Regional demographic development*, J. Hobcraft and P. Rees (eds) (London: Croom Helm, 1977), pp. 29–70.

Lawton, R. Urbanization and population change in nineteenth century England. In *The expanding city*, J. Patten (ed.) (London: Academic Press, 1983), pp. 179–24.

Lawton, R. and C. Pooley. David Brindley's Liverpool: an aspect of urban society in the 1880s. *Transactions of the Historic Society of Lancashire and Cheshire* 126, pp. 149–68, 1975.

Lawton, R. and C. Pooley. *Britain 1740–1950: an historical geography* (London: Arnold, 1992).

Laxton, P. Textiles. In *Atlas of industrializing Britain*, J. Langton and R. Morris (eds) (London: Methuen, 1986), pp. 106–13.

Lee, C.H. The service sector: regional specialization and economic growth in the Victorian economy. *Journal of Historical Geography* 10, pp. 139–56, 1984.

Lee, C.H. *The British economy since 1700: a macro-economic approach* (Cambridge University Press, 1986).

Lee, E. A theory of migration. *Demography* 3, pp. 47–57, 1966.

Levine, D. *Family formation in an age of nascent capitalism* (New York: Academic Press, 1977).

Lewis, G.J. *Human migration: a geographical perspective* (London: Croom Helm, 1982).

Lis, C. and H. Soly. *Poverty and capitalism in pre-industrial Europe* (Brighton: Harvester, 1979).

Loudon, J. *Medical care and the General Practitioner, 1750–1850* (Oxford: Clarendon, 1986).

Lovett, A., I.D. Whyte and K. Whyte. Poisson regression analysis and migration fields: the example of the apprenticeship records of Edinburgh in the seventeenth and eighteenth centuries. *Transactions of the Institute of British Geographers*, NS10, pp. 317–32, 1985.

Lucassen, J. *Migrant labour in Europe, 1600–1900: the drift to the North Sea* (London: Croom Helm, 1987).

Lummis, T. *Listening to history: the authenticity of oral evidence* (London, Hutchinson, 1987).

Macfarlane, A. Review of L. Stone 'The family, sex and marriage in England'. *History and Theory*, 18, pp. 103–26, 1979.

Marsden, T., P. Lowe and S. Whatmore (eds). *Rural restructuring: global processes and their responses* (London: David Fulton, 1990).

Marsden, T. *et al. Constructing the countryside* (London: UCL Press, 1993).

Marshall, J.D. *Furness and the industrial revolution* (Barrow-in-Furness Library and Museum Committee, 1958).

Marshall, J.D. *The Old Poor Law, 1795–1914* (London: Macmillan, 2nd edition, 1985).

Massey, D. and Jess, P. (eds). *A place in the world? Places, cultures and globalization* (Oxford University Press, 1995).

Mathias, P. *The first industrial nation: an economic history of Britain, 1700–1914* (London: Routledge, second edition, 1983).

Mathias, P. and J. Davis (eds). *The first industrial revolutions* (Oxford: Blackwell, 1989).

Mayhew, H. *London labour and the London poor* (London: Griffin, 1861/2).

McKeown, T. *The modern rise of population* (London: Arnold, 1976).

Medick, H. The proto-industrial family economy: the structural function of household and family during the transition from peasant to industrial capitalism. *Social History* 1, pp. 291–316, 1976.

Mendels, F. Protoindustrialization: the first phase of the process of industrialization *Journal of Economic History* 32, pp. 241–61, 1972.

Merrett, S. *State housing in Britain* (London: Routledge, 1979).

Michie, R. (ed.) *The industrial revolutions. Volume 11: commercial and financial services* (Oxford: Blackwell, 1994).

Mills, D. and C. Pearce, C. *People and places in the Victorian census* (Historical Geography Research Series, 23, 1989).

Mingay, G. The rural slum. In *Slums*, S.M. Gaskell (ed.) (Leicester University Press, 1990), pp. 92–143.

Mitchell, B. and P. Deane. *Abstract of British Historical Statistics* (Cambridge University Press, 1962).

Moch, L.P. *Paths to the city: regional migration in nineteenth-century France* (Beverly Hills: Sage, 1983).

Moch, L.P. The family and migration: News from the French *Journal of Family History* 11, pp. 193–203, 1986.

Moch, L.P. The importance of mundane movements: Small towns, nearby places and individual itineraries in the history of migration. In *Migration in modern France: Population mobility in the nineteenth and twentieth centuries*, P. Ogden and P. White (eds) (London: Unwin Hyman, 1989), pp. 97–117 .

Moch, L.P. *Moving Europeans: migration in Western Europe since 1650* (Bloomington: Indiana University Press, 1992).

Moch, L.P. The European perspective: changing conditions and multiple migrations, 1750–1914. In *European migrants: global and local perspectives*, D. Hoerder and L.P. Moch (eds) (Boston: Northeastern University Press, 1996), pp. 115–40.

Muchnick, D. *Urban renewal in Liverpool* (London: G. Bell, 1970).

Nicholas, S. and R. Shergold. Internal migration in England, 1818–39. *Journal of Historical Geography* 13, pp. 155–68, 1987.

Nugent, W. Demographic aspects of European migration worldwide. In *European migrants: global and local perspectives*, D. Hoerder and L.P. Moch (eds) (Boston: Northeastern University Press, 1996). pp. 70–89.

Nuttgens, P. Design for living. In *A new century of social housing*, S. Lowe and D. Hughes (eds) (Leicester University Press, 1991), pp. 100–20.

Oakley, A. *The sociology of housework* (London: Robertson, 1974).

O'Farrell, P. *Letters from Irish Australia, 1825–1929* (University of Sydney Press, 1984).

O'Grada, C. *Ireland before and after the Famine: explorations in economic history, 1800–1925* (Manchester University Press, 1988).

Olsson, G. Distance and human interaction: a migration study. *Geografiska Annaler* 47B, pp. 3–43; 1963.

O'Sullivan, P. (ed.) *The Irish world wide. Volume six: The meaning of the Famine* (Leicester University Press, 1997).

Overton, M. *Agricultural revolution in England: the transformation of the agricultural economy 1500–1850* (Cambridge University Press, 1996).

Panayi, P. *Immigration, ethnicity and racism in Britain, 1815–1945* (Manchester University Press: 1994).

Parton, A.G. The travels of Joseph Smith, well-sinker, 1877–1897: a study in personal migration for work. *North Staffordshire Journal of Field Studies* 20, pp. 33–40, 1980.

Parton A.G. Poor law settlement certificates and migration to and from Birmingham, 1726–57. *Local Population Studies* 38, pp. 23–9, 1987.

Patten, J. Patterns of migration and movement of labour to three pre-industrial East Anglian towns. *Journal of Historical Geography* 2, pp. 111–29, 1976.

Peach, C. (ed.). *Ethnicity in the 1991 Census. Vol. 2: The ethnic minority populations of Great Britain* (London: Office for National Statistics, 1996).

Perkin, H. *The origins of modern English Society* (London: Routledge, 1969).

Perks, R.B. A feeling of not belonging: interviewing European immigrants in Bradford. *Oral History* 12, pp. 64–7, 1984.

Perks, R. 'You're different: you're one of us.' The making of a British Asian. *Oral History*, 15, pp. 67–74, 1987.

Perks, R. *Oral history: talking about the past* (London: Historical Association, 1992).

Pickstone, J.V. *Medicine and industrial society: a history of hospital development in Manchester and its region 1752–1946* (Manchester University Press, 1985).

Pilgrim Trust. *Men without work* (Cambridge University Press, 1938).

Piore, M. *Birds of passage: migrant labour and industrial societies* (Cambridge University Press, 1980).

Poitrineau, A. *Les espagnols de l'Auvergne et du Limousin du XVIIIe au XXe siècle* (Aurillac, Malroux-Mazel, 1985).

Pollard, S. A new estimate of British coal production, 1750–1850 *Economic History Review* 33, pp. 212–35, 1980.

Pollard, S. (ed.) *The industrial revolutions, Volume 9: The metal fabrication and engineering industries* (Oxford: Blackwell, 1994).

Pooley, C. *Migration, mobility and residential areas in nineteenth-century Liverpool* (PhD thesis, Department of Geography, University of Liverpool, 1978) .

Pooley, C. Residential mobility in the Victorian city. *Transactions of the Institute of British Geographers* NS4, pp. 258–77, 1979.

Pooley, C. Welsh migration to England in the mid-nineteenth century. *Journal of Historical Geography* 9, pp. 287–305, 1983.

Pooley, C. Housing for the poorest poor: slum clearance and rehousing in Liverpool, 1890–1918. *Journal of Historical Geography* 11, pp. 70–88, 1985.

Pooley, C. England and Wales. In *Housing strategies in Europe, 1880–1930*, C. Pooley (ed.) (Leicester University Press, 1992) pp. 73–104.

Pooley, C. The mobility of criminals in North-West England, c.1880–1910. *Local Population Studies* 53, pp. 15–28, 1994.

Pooley, C. and S. D'Cruze. Migration and urbanization in North-West England, c.1760–1830. *Social History* 19, pp. 339–58, 1994.

Pooley, C. and J.C. Doherty. The longitudinal study of migration: Welsh migration to English towns in the nineteenth century. In *Migrants, emigrants and immigrants: a social history of migration*, C.G. Pooley and I.D. Whyte (eds) (London: Routledge, 1991) pp. 143–73.

Pooley, C. and S. Irish. *The development of corporation housing in Liverpool, 1869–1945* (Lancaster University: Centre for N.W. Regional Studies, 1984).

Pooley, C. and J. Turnbull. Counterurbanization: the nineteenth century origins of a late-twentieth century phenomenon. *Area* 28, pp. 514–24, 1996.

Pooley, C. and J. Turnbull. Changing home and workplace in Victorian London: the life of Henry Jaques, shirtmaker. *Urban History* 24, pp. 148–78, 1997a.

Pooley, C. and J. Turnbull. Leaving home: the experience of migration from the parental home in Britain since c.1770. *Journal of Family History* 22, pp. 390–424, 1997b.

Pope, R. *War and society in Britain, 1894–1948* (London: Longman, 1991).

Poussou, J.-P. *Bordeaux et le sud-ouest au XVIIIe siècle: Croissance économique et attraction urbaine* (Paris: Editions de l'Ecole des Hautes Etudes en Sciences Sociales, 1983).

Prandy, K. *The Family, occupation and social stratification: 1840 to the present* (Final report to the ESRC, 1997).

Pred, A. *Making histories and constructing human geographies: the local transformation of practice, power relations and consciousness* (Boulder, Co.: Westview Press, 1990).

Price, S. *Building societies. Their origin and history* (London: Franey, 1958).

Pritchard, R.M. *Housing and the spatial structure of the city* (Cambridge University Press, 1976).

Pryor, R. *Residence history analysis* (Canberra: Australian National University, Department of Demography, 1979).

Raistrick, A. and B. Jennings. *A history of lead mining in the Pennines* (London: Longman, 1965).

Ravenstein, E.G. Census of the British Isles 1871: Birthplaces and migration. *Geographical Magazine* 3, pp. 173–7, 1876.

Ravenstein, E.G. The laws of migration. *Journal of the Royal Statistical Society*, 48 pp. 167–227, 1885; 52 pp. 214–301, 1889.

Ravetz, A. *Remaking cities: contradictions of the recent urban environment* (London: Croom Helm, 1980).

Redford, A. *Labour migration in England, 1800–1850* (Manchester University Press, 1926).

Reher, D. Mobility and migration in pre-industrial urban areas. The case of nineteenth-century Cuenca. In *Urbanization in History*, A. van der Woude, J. de Vries and A. Hayami (eds) (Oxford: Clarendon Press, 1990), pp. 164–85.

Rendall, J. *Women in an industrialising society: England 1750–1880* (Oxford: Blackwell, 1990).

Richards, E. *A history of the Highland clearances. 2 volumes* (London: Croom Helm, 1982–5).

Richards, E. Voices of British and Irish migrants in nineteenth-century Australia. In *Migrants, emigrants and immigrants: a social history of migration*, C. Pooley and I. Whyte (eds) (London: Routledge, 1991), pp. 19–41.

Richards, E. How did poor people emigrate from the British Isles to Australia in the nineteenth century? *Journal of British Studies* 32, pp. 250–79, 1993.

Robbins, K. *The eclipse of a great power: modern Britain, 1870–1975* (Harlow: Longman, 1983).

Roberts, E. Working class housing in Barrow and Lancaster, 1880–1930. *Transactions of the Historic Society of Lancashire and Cheshire* 127, pp. 109–32, 1978.

Roberts, E. *Women's work 1840–1940* (London: Macmillan, 1988).

Roberts, E. Women and the domestic economy, 1890–1970: the oral evidence. In *Time, family and community*, M. Drake (ed.) (Oxford: Blackwell, 1994), pp. 129–46.

Robinson, V. *Transients, settlers, and refugees : Asians in Britain* (Oxford: Clarendon, 1986).

Robson B.T. *Urban Growth: An approach* (London: Methuen, 1973).

Rodger, R. (ed.). *Scottish housing in the twentieth century* (Leicester University Press, 1989a).

Rodger, R. *Housing in urban Britain, 1780–1914* (London: Macmillan, 1989b).

Rose, M. (ed.) *The Lancashire cotton industry: a history since 1700* (Preston: Lancashire County Books, 1996).

Rose, M.E. *The relief of poverty, 1834–1914* (London: Macmillan, 2nd edition, 1986).

Rossi, P. *Why families move: a study in the social psychology of residential mobility* (Glencoe, Ill.: The Free Press, 1955).

Rostow, W.W. *The stages of economic growth: a non-communist manifesto* (Cambridge University Press, 1960).

Royle, E. *Modern Britain: a social history, 1750–1985* (London: Arnold, 1987) .

Russell, D. and G. Walker. *Trafford Park 1896–1939* (Manchester Studies, Manchester Polytechnic, 1979).

Salt, J. The geography of unemployment in the U.K. in the 1980s. *Espaces, populations, sociétés* 7, pp. 349–56, 1985.

Saunders, M. The influence of job vacancy advertising upon migration: some empirical evidence. *Environment and Planning A* 17, pp. 1581–9, 1985.

Saunders, M. *Job vacancy information and employment related migration in contemporary Britain*. (PhD thesis, Department of Geography, Lancaster University, 1986).

Saville, J. *Rural depopulation in England and Wales, 1851–1951* (London: Routledge, 1957).

Schofield, R. Age-specific mobility in an eighteenth-century English rural parish. In *Migration and society in early-modern England*, Clark and Souden (eds) (London: Hutchinson, 1987), pp. 261–74.

Schmiechen, J. *Sweated industries and sweated labour: the London clothing trades, 1860–1914* (London: Croom Helm, 1984).

Schumpeter, J.A. *Business cycles: a theoretical, historical and statistical analysis of the capitalist process* (New York: McGraw Hill, 1939).

Schurer, K. The role of the family in the process of migration. In *Migrants, emigrants and immigrants: a social history of migration*, C. Pooley and I. Whyte (eds) (London: Routledge, 1991), pp. 106–142.

Shannon, H.A. Migration and the growth of London, 1841–1891. *Economic History Review* 5, pp. 78–86, 1935.

Shaw, G. and A. Tipper. *British directories* (Leicester University Press, 1988).

Shephard, J.A. East Yorkshire's agricultural labour force in the mid-nineteenth century. *Agrarian History Review* 9, pp. 43–54, 1961.

Shorter, E. *The making of the modern family* (London: Collins, 1976).

Shumsky, N. Return migration in American novels of the 1920s and 1930s. In *Writing across worlds: literature and migration*, R. King, J. Connell and P. White (eds) (London: Routledge, 1995), pp. 198–215.

Simmons, J. *The railways of Britain* (London: Sheldrake Press, 3rd edition, 1990).

Singleton, J. *Lancashire on the scrapheap: the cotton industry 1945–1970* (Oxford University Press, 1991).

Skeldon, R. *Population mobility in developing countries: a reinterpretation* (London: Belhaven, 1990).

Skeldon, R. On mobility and fertility transitions in east and south-east Asia. *Asian and Pacific Migration Journal* 1, pp. 220–49, 1992.

Skeldon, R. The challenge facing migration research: a case for greater awareness. *Progress in Human Geography* 19, pp. 91–6, 1995.

Skelley, A. *The Victorian army at home* (London: Croom Helm, 1977).

Smith, C.T. The movement of population in England and Wales in 1851 and 1861. *Geographical Journal*, 107, pp. 200–10, 1951.

Smith, D. The British hosiery industry at the middle of the nineteenth century: An historical study in economic geography. *Transactions of the Institute of British Geographers* 32, pp. 125–42, 1963.

Souden, D. Mover and stayers in family reconstitution populations, 1660–1780. *Local Population Studies* 33, pp. 11–28, 1984.

Southall, H. The tramping artizan revisits: labour mobility and economic distress in early Victorian England. *Economic History Review* 44, pp. 272–91, 1991a.

Southall, H. Mobility, the artizan community and popular politics in early nineteenth-century England. In *Urbanising Britain*, G. Kearns and C. Withers (eds) (Cambridge University Press, 1991b) pp. 103–30.

Sperling, S. (ed.) *Studier och handlingar rörande Stockholms historia, 4* (Stockholm: Stockholms stadsarkiv, 1989).

Stedman Jones, G. *Outcast London* (Oxford University Press, 1971).

Stillwell, J.P. Rees and P. Boden (eds). *Migration processes and patterns: population redistribution in the United Kingdom* (London: Belhaven, 1992).

Stouffer, S. Intervening opportunities: a theory relating mobility and distance. *American Sociological Review* 5, pp. 845–67, 1940.

Stouffer, S. Intervening opportunities and competing migrants. *Journal of Regional Science*, 2, pp. 1–26, 1960.

Summerfield, P. *Women workers in the second world war* (London: Routledge, 1989).

Sutcliffe, A. (ed.) *Multi-storey living* (London: Croom Helm, 1974).

Sutcliffe, A. and R. Smith. *History of Birmingham. Volume 3: Birmingham 1939–1970* (London: Oxford University Press, 1974).

Swenarton, M. *Homes fit for heroes* (London: Heinemann, 1981).

Swenarton, M. and S. Taylor. The scale and nature of the growth of owner occupation in Britain between the wars *Economic History Review* 38, pp. 373–92, 1985.

Swift, R. and S. Gilley (eds). *The Irish in the Victorian city* (Beckenham: Croom Helm, 1985).

Swift, R. and S. Gilley (eds). The Irish in Britain, 1815–1939 (London: Pinter, 1989).

Tarn, J.N. *Working-class housing in nineteenth-century Britain* (London: Lund Humphries, 1971).

Taylor, J.S. *Poverty, migration and settlement in the industrial revolution: sojourners narratives*, (Palo Alto, Ca.: Society for the promotion of science and scholarship, 1989).

Térisse, M. Méthode de recherches démographiques en milieu urban ancien (XVIIe–XVIIIe) *Annales de démographie historique* pp. 249–62, 1974.

Thane, P. *Foundations of the Welfare State* (London: Longman, 1980).

Therborn, G. *European modernity and beyond: the trajectory of European societies, 1945–2000* (London: Sage, 1995).

Thomas, B. The migration of labour into the Glamorganshire coalfield. *Economica* 10, pp. 275–94, 1930.

Thompson, A. *The dynamics of the industrial revolution* (London: Arnold, 1973).

Thompson, E.P. *The making of the English Working Class* (Harmondsworth: Penguin, 1968).

Thompson, F.M.L. (ed.). *The rise of suburbia* (Leicester University Press, 1982).

Thompson, F.M.L. (ed.) *The Cambridge Social History of Britain* (Cambridge University Press, 1990).

Thompson, P. *The voice of the past: oral history* (Oxford University Press, 2nd edition, 1988).

Thrift, N. Transport and communications, 1730–1914. In *An historical geography of England and Wales*, R. Dodgshon and R. Butlin (eds) (London: Academic Press, 1990) pp. 453–86.

Tillot, P. Sources of inaccuracy in the 1851 and 1861 censuses. In *Nineteenth century society. Essays in the use of quantitative methods for the study of social data*, E.A. Wrigley (ed.) (Cambridge University Press, 1972) pp. 82–133.

Todd, E. Mobilité géographique et cycle de vie en Artois et en Toscane au XVIIIe siècle *Annales: Economies, Sociétés, Civilisations* 30, pp. 726–44, 1975.

Tranter, N. *British population in the twentieth century* (London: Macmillan, 1996).

Trench, W.S. *Report on the health of Liverpool* (Liverpool, 1886).

Turnbull, L. *A history of lead mining in the North East of England* (Newcastle: H. Hill, 1975).

United Nations. Methodology and evaluation of population registers and similar systems. *Studies in methods, series F, no. 15* (1959).

Vincent, D. *Bread, Knowledge and Freedom: A study of nineteenth-century working class autobiography* (London: Europa, 1981).

Von Tunzelmann, G.N. *Steam power and British industrialization to 1860* (Oxford: Clarendon Press, 1978).

Wall, R. *Family forms in historic Europe.* (Cambridge University Press, 1983).

Wall, R. and Winter, J. (eds). *The upheaval of war: family, work and welfare in Europe, 1914–1918* (Cambridge University Press, 1988).

Waller, P. *Town, city and nation. England 1850–1914* (Oxford University Press, 1983).

Wallerstein, E. *The capitalist world economy* (Cambridge University Press, 1979).

Walton, J. *Lancashire: A social history.* (Manchester University Press, 1987).

Walvin, J. *Passage to Britain: immigration in British history and politics* (Harmondsworth: Penguin, 1984).

Waring, J. Changes in the geographical distribution of the recruitment of apprentices to the London companies, 1486–1750. *Journal of Historical Geography* 6, pp. 241–50, 1980.

Warnes, A. Early separation of homes from workplaces and the urban structure of Chorley, 1780–1850. *Transactions of the Historic Society of Lancashire and Cheshire* 122, pp. 105–35, 1970.

Warnes, A. Estimates of journey to work distances from census statistics. *Regional Studies* 6, pp. 315–26, 1972.

Warnes, A. Migration and the lifecourse. In *Migration processes and patterns*, A. Champion and A. Fielding (eds) (London: Belhaven, 1992), pp. 175–87.

Weber, A.F. *The growth of cities in the nineteenth century: a study in statistics* (London: Macmillan, 1899).

Welton, T. *England's recent progress: an investigation of the statistics of migrations, mortality, etc. in the twenty years from 1881–1901 as indicating tendencies towards the growth or decay of particular communities.* (London: Chapman & Hall, 1911).

White, M.B. Family migration in Victorian Britain: the case of Grantham and Scunthorpe. *Local Population Studies* 41, pp. 41–50, 1988.

White, P. and P. Jackson. Research review 1: (Re)theorising population geography. *International Journal of Population Geography*, 1, pp. 111–24, 1995.

White, P. and R. Woods (eds). *The geographical impact of migration* (London: Longman, 1980).

Whitelegg, J. *Critical mass: transport, environment and equity in the twenty-first century* (London: Pluto, 1997).

Whyte, I. Migration in early-modern Scotland and England: a comparative perspective. In *Migrants, emigrants and immigrants: a social history of migration*, C. Pooley and I. Whyte (eds) (London, Routledge, 1991).

Whyte, I. and K. Whyte. Patterns of migration of apprentices to Aberdeen and Inverness during the seventeenth and eighteenth centuries. *Scottish Geographical Magazine* 102, pp. 81–91, 1986.

Williamson, J. *Coping with city growth during the British industrial revolution* (Cambridge University Press, 1989).

Willis, K. *Problems in migration analysis* (Farnborough: Saxon House, 1974).

Willis, K. Regression models of migration: an econometric reappraisal of some techniques, *Geografiska Annaler* 57B, pp. 42–54, 1975.

Winstanley, M. *The shopkeepers' world, 1830–1914* (Manchester University Press, 1983).

Winter, J. *The Great War and the British people* (London: Macmillan, 1986).

Withers, C. Destitution and migration: labour mobility and relief from famine in Highland Scotland, 1836–1850. *Journal of Historical Geography* 14, pp. 128–50, 1988.

Withers, C. and A. Watson. Stepwise migration and Highland migration to Glasgow, 1852–1898. *Journal of Historical Geography* 17, pp. 35–55, 1991.

Wojciechowska, B. Brenchley: a study of migratory movements in a mid-nineteenth century rural parish. *Local Population Studies* 41, pp. 28–30, 1988.

Wolpert, J. Behavioural aspects of the decision to migrate. *Papers of the Regional Science Association* 15, pp. 159–69, 1965.

Woods, R. Towards a general theory of migration. In *Contemporary studies of migration*, P. White and P. van der Knap (eds) (Norwich: Geobooks, 1985) pp. 1–5.

Woods, R. Approaches to the fertility transition in Victorian England. *Population Studies* 41, pp. 283–311, 1987.

Woods, R. *The population of Britain in the nineteenth century* (London: Macmillan, 1992).

Woods, R. Classics in human geography revisited: commentary 1. *Progress in Human Geography* 17, pp. 213–15, 1993.

Woods, R. and J. Woodward (eds) *Urban disease and mortality in nineteenth-century England* (London, Batsford, 1984).

Woodward, J. (ed.). *Health care and popular medicine in nineteenth century England: essays in the social history of medicine* (London: Croom Helm, 1977).

Wrigley, E.A. A simple model of London's importance in changing England's society and economy, 1650–1750. *Past and Present* 37, pp. 44–70, 1967.

Wrigley, E.A. The process of modernization and the industrial revolution in England *Journal of Interdisciplinary History* 3, pp. 225–59, 1972.

Wrigley, E.A. (ed.). *Identifying people in the past* (London: Arnold, 1973).

Wrigley, E.A. *People, cities and wealth: the transformation of traditional society* (Oxford: Blackwell, 1987).

Wrigley, E.A. *Continuity, chance and change: the character of the industrial revolution in England* (Cambridge University Press, 1988).

Wrigley, E.A. and Schofield, R.S. *The population history of England, 1541–1871: A reconstruction* (London: Arnold, 1981).

Yarborough, A. Geographical and social origins of Bristol apprentices, 1542–85. *Bristol and Gloucester Archaeological Journal* 98, pp. 113–30, 1980.

Zelinsky, W. The hypothesis of the mobility transition. *Geographical Review* 61, pp. 219–49, 1971.

Zelinsky, W. The demographic transition: changing patterns of migration. In *Population science in the service of mankind* P. Morrison (ed.) (Liege: International Union for the Scientific Study of Population, 1979). pp. 165–88.

Zelinsky, W. The impasse in migration theory: a sketch map for potential escapees. In *Population movements: their forms and functions in urbanization and development* P. Morrison (ed.) (Brussells: Ordina, 1983), pp. 21–49.

Zelinsky, W. Classics in Human Geography revisited: author's response. *Progress in Human Geography* 17, pp. 217–19, 1993.

Zipf, G. The P1P2/D hypothesis: on the intercity movement of persons. *American Sociological Review* 11, pp. 677–86, 1946.

Index